T0254700

# Multiplicative Differential Geometry

# Multiplicative
# Differential Geometry

Svetlin G. Georgiev

**CRC Press**
Taylor & Francis Group
Boca Raton London New York

CRC Press is an imprint of the
Taylor & Francis Group, an **informa** business

A CHAPMAN & HALL BOOK

First edition published 2022
by CRC Press
6000 Broken Sound Parkway NW, Suite 300, Boca Raton, FL 33487-2742

and by CRC Press
2 Park Square, Milton Park, Abingdon, Oxon, OX14 4RN

© 2022 Svetlin G. Georgiev

*CRC Press is an imprint of Taylor & Francis Group, LLC*

*Library of Congress Cataloging-in-Publication Data*
A catalog record has been requested for this book

ISBN: 978-1-032-29060-7 (hbk)
ISBN: 978-1-032-29041-6 (pbk)
ISBN: 978-1-003-29984-4 (ebk)

DOI: 10.1201/9781003299844

Typeset in Palatino
by MPS Limited, Dehradun

# Contents

# *Preface*

The non-Newtonian calculi provide a wide variety of mathematical tools for use in science, engineering and mathematics. They appear to have considerable potential for use as alternatives to the classical calculus of Newton and Leibniz. It may well be that these calculi can be used to define new concepts, to yield new or simpler laws, or to formulate or solve problems.

In this book, we explain the differential geometry in a modern language, using the Grossmann-Katz calculi. The book is intended for senior undergraduate students and beginning graduate students of engineering and science courses. The book contains ten chapters.

In Chapter 1 we introduce and investigate the multiplicative vector space $\mathbb{R}^2_\star$, multiplicative inner product space $\mathbb{R}^2_\star$ and the multiplicative Euclidean space $E^2_\star$. We define multiplicative lines and deduct their equations. They are given conditions for perpendicular, parallel and intersecting multiplicative lines. In the chapter are defined the space $E^3_\star$, the multiplicative orthonormal bases and multiplicative planes.

Chapter 2 is devoted to multiplicative Frenet curves. We define multiplicative tangent, multiplicative normal and multiplicative normal plane for a multiplicative Frenet curve. In the chapter are introduced the multiplicative Frenet frames and they are deducted the multiplicative Frenet formulae. As applications of the multiplicative Frenet formulae, they are defined multiplicative curvature, multiplicative curvature vector and multiplicative torsion. They are investigated the local behaviours of a multiplicative parameterized curve around multiplicative biregular points. In the chapter are defined multiplicative Bertrand curves and they are investigated some of their properties. It is introduced a multiplicative rigid motion and they are formulated and proved existence and uniqueness theorems.

Chapter 3 deals with the multiplicative plane curves. We define multiplicative envelopes of families of multiplicative plane curves depending on one parameter. In the chapter it is introduced multiplicative evolute of a multiplicative plane curve and they are deducted the equations of the multiplicative evolute. We define a multiplicative complex structure and deduct some of its properties. We give definitions for multiplicative curvature and multiplicative rotation angle of multiplicative plane curves.

In Chapter 4 we introduce multiplicative surfaces and define multiplicative regular surfaces, multiplicative curvilinear coordinate system, multiplicative chart. They are defined and deducted the equations of multiplicative tangent planes on multiplicative surfaces, multiplicative normals on surfaces. We introduce the multiplicative spherical map, the multiplicative shape operator, the first and second multiplicative fundamental forms. We define the multiplicative normal curvature, multiplicative principal directions and multiplicative principal curvatures.

Chapter 5 introduces the multiplicative Christoffel and Weingarten coefficients and we deduct the multiplicative Gauss and Godazzi-Mainardi equations. In the chapter are defined geodesic lines of multiplicative surfaces. They are given the equations of the multiplicative geodesics of the multiplicative planes, multiplicative unit sphere and the multiplicative Liouville surfaces.

In Chapter 6, we define multiplicative ruled surfaces, and we deduct the equations of their tangent planes, the equation of their first multiplicative fundamental form and we deduct a representation of their multiplicative Gauss curvatures. In the chapter are defined multiplicative minimal surfaces and it is given a criterion for multiplicative minimality of a multiplicative surface.

Chapter 7 deals with multiplicative differential forms. They are defined multiplicative exact differential forms and multiplicative closed differential forms.

In Chapter 8 it is introduced the multiplicative covariant differentiation and they are deducted some of its properties.

In Chapter 9, we define multiplicative manifold, multiplicative open sets and multiplicative differentiable maps on multiplicative manifolds. In the chapter, they are introduced multiplicative tangent vector, multiplicative vector field, multiplicative Lie brackets and multiplicative Riemannian connection. It is deducted the multiplicative Koszul formula and then they are given expressions of the multiplicative Christoffel coefficients.

Chapter 10 introduces the multiplicative tensors and they are given some of their properties. In the chapter it is defined the multiplicative curvature tensor. As applications, they are introduced the multiplicative Ricci and Einstein tensors.

The aim of this book is to present a clear and well-organized treatment of the concept behind the development of mathematics and solution techniques. The text material of this book is presented in highly readable, mathematically solid format. Many practical problems are illustrated displaying a wide variety of solution techniques.

**Svetlin G. Georgiev**
Paris, December 2021

# Author Bio

**Svetlin G. Georgiev** is a mathematician who has worked in various areas of the study. He currently focuses on harmonic analysis, functional analysis, partial differential equations, ordinary differential equations, Clifford and quaternion analysis, integral equations and dynamic calculus on time scales. He is also the author of *Dynamic Geometry of Time Scales*, CRC Press. He is a co-author of *Conformable Dynamic Equations on Time Scales*, with Douglas R. Anderson, and: *Multiple Fixed-Point Theorems and Applications in the Theory of ODEs, FDEs and PDE; Boundary Value Problems on Time Scales*, Volume 1 and Volume 2, all with Khalid Zennir and published by CRC Press.

# 1

## Elements of the Multiplicative Euclidean Geometry

### 1.1 The Multiplicative Vector Space $\mathbb{R}_\star^2$

**Definition 1.1:** Each ordered pair $(p_1, p_2)$ of elements of $\mathbb{R}_\star^2$ determines exactly one point $P$ of the multiplicative plane. The point determined by the point $0_\star = (0_\star, 0_\star) = (1, 1)$ is called the multiplicative origin. The ordered pair $(p_1, p_2)$ is also referred to as the multiplicative coordinate vector of $P$.

We will regard the words "point" and "vector" as interchangeable. The set of all vectors is denoted by $\mathbb{R}_\star^2$.

**Definition 1.2:** Let $c \in \mathbb{R}_\star$, $x = (x_1, x_2) \in \mathbb{R}_\star^2$, $y = (y_1, y_2) \in \mathbb{R}_\star^2$. Then define

$$x +_\star y = (x_1 +_\star y_1, x_2 +_\star y_2)$$
$$= (x_1 y_1, x_2 y_2),$$
$$c \cdot_\star x = (c \cdot_\star x_1, c \cdot_\star x_2)$$
$$= (e^{\log c \log x_1}, e^{\log c \log x_2}),$$
$$x -_\star y = (x_1 -_\star y_1, x_2 -_\star y_2)$$
$$= \left(\frac{x_1}{y_1}, \frac{x_2}{y_2}\right),$$
$$-_\star x = (-_\star x_1, -_\star x_2)$$
$$= \left(\frac{1}{x_1}, \frac{1}{x_2}\right).$$

**Example 1.1:** Let $x = (1, 2)$, $y = (1, 4) \in \mathbb{R}_\star^2$, $c = 2$. Then

DOI: 10.1201/9781003299844-1

$$x +_\star y = (1 +_\star 1, 2 +_\star 4)$$
$$= (1{\cdot}1, 2{\cdot}4)$$
$$= (1, 8),$$
$$x -_\star y = (1 -_\star 1, 2 -_\star 4)$$
$$= \left(\frac{1}{1}, \frac{2}{4}\right)$$
$$= \left(1, \frac{1}{2}\right)$$

and

$$2{\cdot}_\star x = (2{\cdot}_\star 1, 2{\cdot}_\star 2)$$
$$= (e^{\log 2 \log 1}, e^{(\log 2)^2})$$
$$= (1, e^{(\log 2)^2}).$$

**Example 1.2:** Let $x = (2, 4) \in \mathbb{R}_\star^2$. Then $-_\star x = \left(\frac{1}{2}, \frac{1}{4}\right)$.

**Exercise 1.1:** Let $x = (3, 5)$, $y = (1, 4) \in \mathbb{R}_\star^2$. Find

1. $x +_\star y$.
2. $3{\cdot}_\star (x -_\star y)$.

**Answer 1.1:**

1. $(3, 20)$.
2. $(e^{(\log 3)^2}, e^{\log 3 \log \frac{5}{4}})$.

Below, we will deduct some of the properties of the defined operations. Let

$$x = (x_1, x_2), \quad y = (y_1, y_2), \quad z = (z_1, z_2) \in \mathbb{R}_\star^2, \quad c, d \in \mathbb{R}_\star.$$

1. $(x +_\star y) +_\star z = x +_\star (y +_\star z)''$.

*Proof.* We have

$$(x +_\star y) +_\star z = (x_1 +_\star y_1, x_2 +_\star y_2) +_\star (z_1, z_2)$$
$$= (x_1 y_1, x_2 y_2) +_\star (z_1, z_2)$$
$$= ((x_1 y_1) +_\star z_1, (x_2 y_2) +_\star z_2)$$
$$= ((x_1 y_1) z_1, (x_2 y_2) z_2)$$
$$= (x_1 (y_1 z_1), x_2 (y_2 z_2))$$
$$= (x_1 +_\star (y_1 z_1), x_2 +_\star (y_2 z_2))$$
$$= (x_1, x_2) +_\star (y_1 z_1, y_2 z_2)$$
$$= x +_\star (y_1 +_\star z_1, y_2 +_\star z_2)$$
$$= x +_\star ((y_1, y_2) +_\star (z_1, z_2))$$
$$= x +_\star (y +_\star z).$$

This completes the proof.

2. $x +_\star y = y +_\star x$.

*Proof.* We have

$$x +_\star y = (x_1, x_2) +_\star (y_1, y_2)$$
$$= (x_1 +_\star y_1, x_2 +_\star y_2)$$
$$= (x_1 y_1, x_2 y_2)$$
$$= (y_1 x_1, y_2 x_2)$$
$$= (y_1 +_\star x_1, y_2 +_\star x_2)$$
$$= (y_1, y_2) +_\star (x_1, x_2)$$
$$= y +_\star x.$$

This completes the proof.

3. $x +_\star 0_\star = x$.

*Proof.* We have

$$x +_\star 0_\star = (x_1, x_2) +_\star (1, 1)$$
$$= (x_1 +_\star 1, x_2 +_\star 1)$$
$$= (x_1, x_2)$$
$$= x.$$

This completes the proof.

4. $x +_\star (-_\star x) = 0_\star$.

*Proof.* We have

$$x +_\star (-_\star x) = (x_1, x_2) +_\star \left(\frac{1}{x_1}, \frac{1}{x_2}\right)$$
$$= \left(x_1 +_\star \frac{1}{x_1}, x_2 +_\star \frac{1}{x_2}\right)$$
$$= (1, 1)$$
$$= (0_\star, 0_\star)$$
$$= 0_\star.$$

This completes the proof.

5. $e \cdot_\star x = x''.$

*Proof.* We have

$$e \cdot_\star x = e \cdot_\star (x_1, x_2)$$
$$= (e \cdot_\star x_1, e \cdot_\star x_2)$$
$$= (e^{\log e \log x_1}, e^{\log e \log x_2})$$
$$= (x_1, x_2)$$
$$= x.$$

This completes the proof.

6. $c \cdot_\star (x +_\star y) = c \cdot_\star x +_\star c \cdot_\star y.$

*Proof.* We have

$$c \cdot_\star (x +_\star y) = c \cdot_\star ((x_1, x_2) +_\star (y_1, y_2))$$
$$= c \cdot_\star (x_1 +_\star y_1, x_2 +_\star y_2)$$
$$= c \cdot_\star (x_1 y_1, x_2 y_2)$$
$$= (c \cdot_\star (x_1 y_1), c \cdot_\star (x_2 y_2))$$
$$= (e^{\log c \log(x_1 y_1)}, e^{\log c \log(x_2 y_2)})$$
$$= (e^{\log c (\log x_1 + \log y_1)}, e^{\log c (\log x_2 + \log y_2)})$$
$$= (e^{\log c \log x_1} e^{\log c \log y_1}, e^{\log c \log x_2} e^{\log c \log y_2})$$
$$= ((c \cdot_\star x_1) +_\star (c \cdot_\star y_1), (c \cdot_\star x_2) +_\star (c \cdot_\star y_2))$$
$$= (c \cdot_\star x_1, c \cdot_\star x_2) +_\star (c \cdot_\star y_1, c \cdot_\star y_2)$$
$$= c \cdot_\star (x_1, x_2) +_\star c \cdot_\star (y_1, y_2)$$
$$= c \cdot_\star x + c \cdot_\star y.$$

This completes the proof.

7. $(c +_\star d) \cdot_\star x = c \cdot_\star x +_\star d \cdot_\star x.$

*Proof.* We have

$$(c +_\star d)\cdot_\star x = (cd)\cdot_\star x$$
$$= (cd)\cdot_\star (x_1, x_2)$$
$$= ((cd)\cdot_\star x_1, (cd)\cdot_\star x_2)$$
$$= (e^{\log(cd)\log x_1}, e^{\log(cd)\log x_2})$$
$$= (e^{(\log c + \log d)\log x_1}, e^{(\log c + \log d)\log x_2})$$
$$= (e^{\log c \log x_1} e^{\log d \log x_1}, e^{\log c \log x_2} e^{\log d \log x_2})$$
$$= (c\cdot_\star x_1 +_\star d\cdot_\star x_1, c\cdot_\star x_2 +_\star d\cdot_\star x_2)$$
$$= (c\cdot_\star x_1, c\cdot_\star x_2) +_\star (d\cdot_\star x_1, d\cdot_\star x_2)$$
$$= c\cdot_\star (x_1, x_2) +_\star d\cdot_\star (x_1, x_2)$$
$$= c\cdot_\star x +_\star d\cdot_\star x.$$

This completes the proof.

8. $c\cdot_\star (d\cdot_\star x) = (c\cdot_\star d)\cdot_\star x.$

*Proof.* We have

$$c\cdot_\star (d\cdot_\star x) = c\cdot_\star (d\cdot_\star x_1, d\cdot_\star x_2)$$
$$= (c\cdot_\star (d\cdot_\star x_1), c\cdot_\star (d\cdot_\star x_2))$$
$$= ((c\cdot_\star d)\cdot_\star x_1, (c\cdot_\star d)\cdot_\star x_2)$$
$$= (c\cdot_\star d)\cdot_\star x.$$

This completes the proof.

## 1.2 The Multiplicative Inner Product Space $\mathbb{R}_\star^2$

**Definition 1.3:** Let $x = (x_1, x_2)$, $y = (y_1, y_2) \in \mathbb{R}_\star^2$. Define the multiplicative inner product of $x$ and $y$ as follows:

$$\langle x, y \rangle_\star = x_1\cdot_\star y_1 +_\star x_2\cdot_\star y_2.$$

We have

$$\langle x, y \rangle_\star = x_1\cdot_\star y_1 +_\star x_2\cdot_\star y_2$$
$$= e^{\log x_1 \log y_1} +_\star e^{\log x_2 \log y_2}$$
$$= e^{\log x_1 \log y_1 + \log x_2 \log y_2}.$$

**Example 1.3:** Let

$$x = (2, 3), \quad y = (3, 4) \in \mathbb{R}_\star^2.$$

Then

$$\langle x, y \rangle_\star = e^{\log 2 \log 3 + \log 3 \log 4}$$
$$= e^{\log 2 \log 3 + 2 \log 2 \log 3}$$
$$= e^{3 \log 2 \log 3}.$$

**Exercise 1.2:** Let

$$x = (3, 8), \quad y = (2, 4).$$

Find

1. $x +_\star y$.
2. $x -_\star y$.
3. $\langle x, y \rangle_\star$.

**Answer 1.2:**

1. $(6, 32)$.
2. $\left( \frac{3}{2}, 2 \right)$.
3. $e^{\log 3 \log 2 + 6(\log 2)^2}$.

Below, we will deduct some of the properties of the multiplicative inner product.

**Theorem 1.3:** *Let*

$$x = (x_1, x_2), \quad y = (y_1, y_2), \quad z = (z_1, z_2) \in \mathbb{R}_\star^2.$$

*Then*

$$\langle x, y +_\star z \rangle_\star = \langle x, y \rangle_\star +_\star \langle x, z \rangle_\star.$$

*Proof.* We have

$$y +_\star z = (y_1, y_2) +_\star (z_1, z_2)$$
$$= (y_1 +_\star z_1, y_2 +_\star z_2)$$
$$= (y_1 z_1, y_2 z_2),$$
$$\langle x, y \rangle_\star = e^{\log x_1 \log y_1 + \log x_2 \log y_2},$$
$$\langle x, z \rangle_\star = e^{\log x_1 \log z_1 + \log x_2 \log z_2}.$$

Hence,

$$
\begin{aligned}
\langle x, y \rangle_\star +_\star \langle x, z \rangle_\star &= e^{\log x_1 \log y_1 + \log x_2 \log y_2} e^{\log x_1 \log z_1 + \log x_2 \log z_2} \\
&= e^{\log x_1 \log y_1 + \log x_2 \log y_2 + \log x_1 \log z_1 + \log x_2 \log z_2} \\
&= e^{\log x_1 (\log y_1 + \log z_1) + \log x_2 (\log y_2 + \log z_2)} \\
&= e^{\log x_1 \log (y_1 z_1) + \log x_2 \log (y_2 z_2)} \\
&= \langle x, y +_\star z \rangle_\star.
\end{aligned}
$$

This completes the proof.

**Theorem 1.4:** *Let*

$$
x = (x_1, x_2), \; y = (y_1, y_2) \in \mathbb{R}^2_\star, \; c \in \mathbb{R}_\star.
$$

*Then*

$$
c \cdot_\star \langle x, y \rangle_\star = \langle c \cdot_\star x, y \rangle_\star = \langle x, c \cdot_\star y \rangle_\star. \tag{1.1}
$$

*Proof.* We have

$$
\begin{aligned}
c \cdot_\star x &= c \cdot_\star (x_1, x_2) \\
&= (c \cdot_\star x_1, c \cdot_\star x_2) \\
&= (e^{\log c \log x_1}, e^{\log c \log x_2})
\end{aligned}
$$

and as above,

$$
c \cdot_\star y = \left( e^{\log c \log y_1}, e^{\log c \log y_2} \right).
$$

Next,

$$
\langle x, y \rangle = e^{\log x_1 \log y_1 + \log x_2 \log y_2}
$$

and

$$
\begin{aligned}
c \cdot_\star \langle x, y \rangle_\star &= c \cdot_\star e^{\log x_1 \log y_1 + \log x_2 \log y_2} \\
&= e^{\log c \log e^{\log x_1 \log y_1 + \log x_2 \log y_2}} \\
&= e^{\log c (\log x_1 \log y_1 + \log x_2 \log y_2)}
\end{aligned} \tag{1.2}
$$

and

$$\begin{aligned}
\langle c \cdot_\star x, y \rangle_\star &= e^{(\log e^{\log c \log x_1}) \log y_1 + (\log e^{\log c \log x_2}) \log y_2} \\
&= e^{\log c \log x_1 \log y_1 + \log c \log x_2 \log y_2} \\
&= e^{\log c (\log x_1 \log y_1 + \log x_2 \log y_2)},
\end{aligned}$$
(1.3)

and

$$\begin{aligned}
\langle x, c \cdot_\star y \rangle_\star &= e^{\log x_1 \log e^{\log c \log y_1} + \log x_2 \log e^{\log c \log y_2}} \\
&= e^{\log x_1 \log c \log y_1 + \log x_2 \log c \log y_2} \\
&= e^{\log c (\log x_1 \log y_1 + \log x_2 \log y_2)}.
\end{aligned}$$
(1.4)

• By (1.2), (1.3) and (1.4), we get (1.1). This completes the proof.

**Corollary 1.1:** *Suppose that all conditions of Theorem 1.4 hold. Then*

$$\langle x, y \rangle_\star = \langle y, x \rangle_\star.$$

*Proof.* We have

$$\langle x, y \rangle_\star = e^{\log x_1 \log y_1 + \log x_2 \log y_2}$$

and

$$\begin{aligned}
\langle y, x \rangle_\star &= e^{\log y_1 \log x_1 + \log y_2 \log x_2} \\
&= e^{\log x_1 \log y_1 + \log x_2 \log y_2} \\
&= \langle x, y \rangle_\star.
\end{aligned}$$

This completes the proof.

**Theorem 1.5:** *If* $\langle x, y \rangle_\star = 0_\star$ *for any* $x \in \mathbb{R}_\star^2$, *then* $y = 0_\star$.

*Proof.* We have

$$\begin{aligned}
0_\star &= 1 \\
&= \langle x, y \rangle_\star \\
&= e^{\log x_1 \log y_1 + \log x_2 \log y_2} \\
&= e^0
\end{aligned}$$

for any $x_1, x_2 \in \mathbb{R}_\star$. For $x_1 = x_2 = e$, we get

$$\begin{aligned}
e^0 &= e^{\log y_1 + \log y_2} \\
&= e^{\log(y_1 y_2)},
\end{aligned}$$

whereupon

$$y_1 y_2 = 1. \tag{1.5}$$

For $x_1 = e$, $x_2 = e^{-1}$, we find

$$e^{\log y_1 - \log y_2} = e^0$$

or

$$e^{\log \frac{y_1}{y_2}} = e^0.$$

Therefore

$$\frac{y_1}{y_2} = 1.$$

By the last equation and (1.5), we arrive at the system

$$\frac{y_1}{y_2} = 1$$
$$y_1 y_2 = 1.$$

Hence,

$$y_1 = y_2 = 1$$

and

$$y = (y_1, y_2)$$
$$= (1, 1)$$
$$= 0_\star.$$

This completes the proof.

**Definition 1.4:** For a vector $x \in \mathbb{R}_{\star}^2$, define the multiplicative length of $x$ as follows:

$$|x|_\star = e^{((\log x_1)^2 + (\log x_2)^2)^{\frac{1}{2}}}.$$

Note that

$$\langle x, x \rangle_\star = e^{(\log x_1)^2 + (\log x_2)^2}$$

and

$$(\langle x, x \rangle_\star)^{\frac{1}{2}_\star} = e^{\left(\log e^{(\log x_1)^2 + (\log x_2)^2}\right)^{\frac{1}{2}}}$$
$$= e^{((\log x_1)^2 + (\log x_2)^2)^{\frac{1}{2}}}$$
$$= |x|_\star .$$

**Example 1.4:** Let $x = (2, 4)$. Then

$$|x|_\star = e^{((\log 2)^2 + (\log 4)^2)^{\frac{1}{2}}}$$
$$= e^{((\log 2)^2 + 4(\log 2)^2)^{\frac{1}{2}}}$$
$$= e^{5^{\frac{1}{2}} \log 2}.$$

**Theorem 1.6:** *The multiplicative length has the following properties.*

1. $|x|_\star \geq_\star 0_\star.$
2. If $|x|_\star = 0_\star$, then $x = 0_\star.$
3. $|c \cdot_\star x|_\star = |c|_\star \cdot_\star |x|_\star.$

*Proof.*

1. We have

$$|x|_\star = e^{((\log x_1)^2 + (\log x_2)^2)^{\frac{1}{2}}}$$
$$\geq e^0$$
$$= 1$$
$$= 0_\star.$$

2. Let $|x|_\star = 0_\star$. Then

$$e^{((\log x_1)^2 + (\log x_2)^2)^{\frac{1}{2}}} = 0_\star$$
$$= 1$$
$$= e^0.$$

Hence,

$$(\log x_1)^2 + (\log x_2)^2 = 0$$

and

$$\log x_1 = 0$$
$$\log x_2 = 0.$$

Therefore

$$x_1 = x_2 = 1$$

and $x = 0_\star$.

3. Let $c \geq 1$. Then $|c|_\star = c$ and

$$c \cdot_\star x = \left(e^{\log c \, \log x_1}, \, e^{\log c \, \log x_2}\right),$$

and

$$|c \cdot_\star x|_\star = e^{\left(\left(\log e^{\log c \, \log x_1}\right)^2 + \left(\log e^{\log c_1 \, \log x_2}\right)^2\right)^{\frac{1}{2}}}$$
$$= e^{\left((\log c \, \log x_1)^2 + (\log c \, \log x_2)^2\right)^{\frac{1}{2}}}$$
$$= e^{\log c \, ((\log x_1)^2 + (\log x_2)^2)^{\frac{1}{2}}},$$

and

$$c \cdot_\star |x|_\star = c \cdot_\star e^{((\log x_1)^2 + (\log x_2)^2)^{\frac{1}{2}}}$$
$$= e^{\log c \, \log e^{\left((\log x_1)^2 + (\log x_2)^2\right)^{\frac{1}{2}}}}$$
$$= e^{\log c \, ((\log x_1)^2 + (\log x_2)^2)^{\frac{1}{2}}}.$$

Therefore

$$|c \cdot_\star x|_\star = c \cdot_\star |x|_\star.$$

The case $c \in (0, 1]$ we leave to the reader as an exercise. This completes the proof.

**Theorem 1.7: (The Multiplicative Cauchy-Schwartz Inequality).** *For any* $x, y \in \mathbb{R}^2_\star$, *we have*

$$|\langle x, y \rangle_\star|_\star \leq |x|_\star \cdot_\star |y|_\star .$$

*The equality holds if and only if $x$ and $y$ are multiplicative proportional.*

*Proof.* Let $x = (x_1, x_2)$ and $y = (y_1, y_2)$. Then

$$\langle x, y \rangle_\star = e^{\log x_1 \log y_1 + \log x_2 \log y_2}$$

and

$$|\langle x, y \rangle_\star|_\star = \begin{cases} e^{\log x_1 \log y_1 + \log x_2 \log y_2} \\ \text{if } \log x_1 \log y_1 + \log x_2 \log y_2 \geq 0 \\ e^{-(\log x_1 \log y_1 + \log x_2 \log y_2)} \\ \text{if } \log x_1 \log y_1 + \log x_2 \log y_2 < 0. \end{cases}$$

$$\leq e^{((\log x_1)^2 + (\log x_2)^2)^{\frac{1}{2}} \left((\log y_1)^2 + (\log y_2)^2\right)^{\frac{1}{2}}}$$

$$= |x|_\star \cdot_\star |y|_\star .$$

Next,

$$|\langle x, y \rangle_\star|_\star = |x|_\star \cdot_\star |y|_\star$$

if and only if

$$e^{|\log x_1 \log y_1 + \log x_2 \log y_2|} = e^{((\log x_1)^2 + (\log x_2)^2)^{\frac{1}{2}} \left((\log y_1)^2 + (\log y_2)^2\right)^{\frac{1}{2}}},$$

if and only if

$$(\log x_1 \log y_1 + \log x_2 \log y_2)^2 = ((\log x_1)^2 + (\log x_2)^2)\left((\log y_1)^2 + (\log y_2)^2\right)$$

if and only if

$$(\log x_1)^2 (\log y_1)^2 + (\log x_2)^2 (\log y_2)^2 + 2 \log x_1 \log x_2 \log y_1 \log y_2$$
$$= (\log x_1)^2 (\log y_1)^2 + (\log x_1)^2 (\log y_2)^2 + (\log x_2)^2 (\log y_1)^2 + (\log x_2)^2 (\log y_2)^2 ,$$

if and only if

$$\log x_1 \log x_2 \log y_1 \log y_2 - (\log x_2)^2 (\log y_1)^2$$
$$= - \log x_1 \log x_2 \log y_1 \log y_2 + (\log x_1)^2 (\log y_2)^2 ,$$

if and only if

$$\log x_2 \log y_1 (\log x_1 \log y_2 - \log x_2 \log y_1)$$
$$= \log x_1 \log y_2 (\log x_1 \log y_2 - \log x_2 \log y_1),$$

if and only if

$$\log x_2 \log y_1 = \log x_1 \log y_2,$$

if and only if

$$e^{\log x_2 \log y_1} = e^{\log x_1 \log y_2},$$

if and only if

$$x_2 \cdot_\star y_1 = x_1 \cdot_\star y_2,$$

if and only if

$$x_1 /_\star y_1 = x_2 /_\star y_2.$$

This completes the proof.

**Corollary 1.2:** *For any* $x, y \in \mathbb{R}_\star^2$, *we have*

$$|x +_\star y|_\star \leq |x|_\star \cdot_\star |y|_\star .$$

*Proof.* We have, using the multiplicative Cauchy-Schwartz inequality,

$$|x +_\star y|_\star^{2_\star} = \langle x +_\star y, x +_\star y \rangle_\star$$
$$= \langle x, x \rangle_\star +_\star e^{2\cdot}{}_\star \langle x, y \rangle_\star +_\star \langle y, y \rangle_\star$$
$$= |x|_\star^{2_\star} +_\star e^{2\cdot}{}_\star \langle x, y \rangle_\star +_\star |y|_\star^{2_\star}$$
$$\leq |x|_\star^{2_\star} +_\star e^{2\cdot}{}_\star |x|_\star \cdot_\star |y|_\star +_\star |y|_\star^{2_\star}$$
$$= (|x|_\star +_\star |y|_\star)^{2_\star},$$

whereupon

$$|x +_\star y|_\star \leq |x|_\star +_\star |y|_\star \, .$$

This completes the proof.

## 1.3 The Multiplicative Euclidean Plane $E_\star^2$

Suppose that $P(x_1, x_2), Q(y_1, y_2) \in \mathbb{R}_\star^2$.

**Definition 1.5:** Define the multiplicative distance between the points $P$ and $Q$ as follows:

$$d_\star(P, Q) = |P -_\star Q|_\star \, .$$

The symbol $E_\star^2$ will be used to denote the set $\mathbb{R}_\star^2$ equipped with the multiplicative distance $d_\star$. In fact, we have

$$P -_\star Q = (x_1, x_2) -_\star (y_1, y_2)$$
$$= (x_1 -_\star y_1, x_2 -_\star y_2)$$
$$= \left( \frac{x_1}{y_1}, \frac{x_2}{y_2} \right)$$

and

$$d_\star(P, Q) = |P -_\star Q|_\star$$
$$= \left| \left( \frac{x_1}{y_1}, \frac{x_2}{y_2} \right) \right|_\star$$
$$= e^{\left( \left( \log \frac{x_1}{y_1} \right)^2 + \left( \log \frac{x_2}{y_2} \right)^2 \right)^{\frac{1}{2}}} \, .$$

**Example 1.5:** Let $P(4, 16)$, $Q(2, 4)$. Then

$$d_\star(P, Q) = e^{\left(\left(\log\frac{4}{2}\right)^2 + \left(\log\frac{16}{4}\right)^2\right)^{\frac{1}{2}}}$$

$$= e^{\left((\log 2)^2 + 4(\log 2)^2\right)^{\frac{1}{2}}}$$

$$= e^{5^{\frac{1}{2}}\log 2}.$$

**Exercise 1.3:** Let $P(2, 3)$, $Q(4, 2)$ and $R(8, 2)$. Find

1. $d_\star(P, Q)$.
2. $d_\star(P, R)$.
3. $d_\star(Q, R)$.

**Answer 1.8:**

1. $e^{\left((\log 2)^2 + \left(\log\frac{2}{3}\right)^2\right)^{\frac{1}{2}}}$.
2. $e^{\left(4(\log 2)^2 + \left(\log\frac{2}{3}\right)^2\right)^{\frac{1}{2}}}$.
3. $2$.

By the properties of the multiplicative length of a vector, they follow the following properties of the multiplicative distance.

1. $d_\star(P, Q) \geq_\star 0_\star$ for any $P, Q \in E_\star^2$.
2. $d_\star(P, Q) = 0_\star$ if and only if $P = Q$.
3. $d_\star(P, Q) = d_\star(Q, P)$ for any $P, Q \in E_\star^2$.
4. $d_\star(P, Q) \leq_\star d_\star(P, R) +_\star d_\star(R, Q)$ for any $P, Q, R \in E_\star^2$.

## 1.4 Multiplicative Lines

For a given vector $v = (v_1, v_2)$, let

$$[v]_\star = \{t \cdot_\star v : t \in \mathbb{R}_\star\}.$$

**Definition 1.6:** If $P$ is any point and $v$ is a vector, different than the vector $0_\star$, then

$$l = \{X : X -_\star P \in [v]_\star\}$$

is called a multiplicative line through $P$ with multiplicative direction $[v]_\star$. We will also write $l = P +_\star [v]_\star$.

In Fig. 1.1, it is shown a multiplicative line through the point $P(4, 9)$ and multiplicative direction $(2, 3)$. If $l$ is a multiplicative line and $X$ is a point, there are some phrases used to express the relationship $X \in l$.

1. $X \in l$.
2. $l$ contains $X$.
3. $X$ lies on $l$.
4. $l$ passes through $X$.
5. $X$ and $l$ are incident.
6. $X$ is incident with $l$.
7. $l$ is incident with $X$.

A fundamental property of a multiplicative line is that it is uniquely determined by two points that lie on it.

**Theorem 1.9:** *Let $P$ and $Q$ be distinct points of $E_\star^2$. Then there is a unique multiplicative line containing $P$ and $Q$.*

*Proof.* Let $v$ be a nonzero vector in $E_\star^2$. The multiplicative line $P +_\star [v]_\star$ passes through $Q$ if and only if $Q -_\star P \in [v]_\star$. Therefore

$$[Q -_\star P]_\star = [v]_\star.$$

Hence, the multiplicative line

$$P +_\star [Q -_\star P]_\star$$

is the unique multiplicative line. This completes the proof.

Suppose that $l$ is a multiplicative line through the points $P$ and $Q$. By Theorem 1.9, it follows that a typical point $X$ on the line $l$ is written as

$$\begin{aligned}
\alpha(t) &= P +_\star t \cdot_\star (Q -_\star P) \\
&= (1_\star -_\star t) \cdot_\star P +_\star t \cdot_\star Q.
\end{aligned}$$

Note that

$$d_\star(\alpha(t_1), \alpha(t_2)) = |t_1 -_\star t_2|_\star \cdot_\star d_\star(P, Q).$$

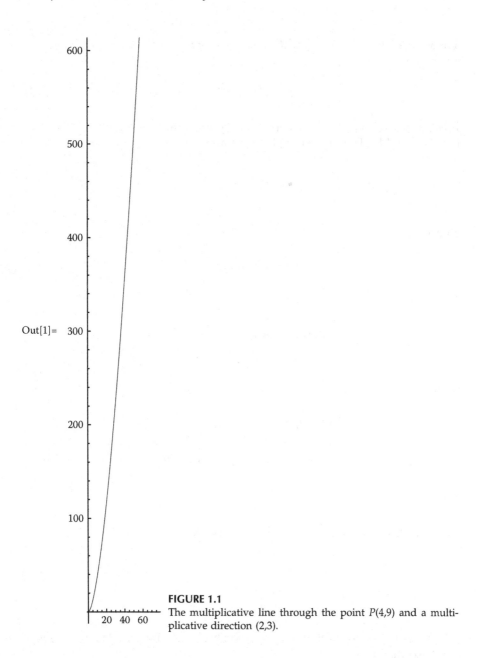

**FIGURE 1.1**
The multiplicative line through the point $P(4,9)$ and a multiplicative direction $(2,3)$.

**Definition 1.7:** If

$$X = (1_\star -_\star t) \cdot_\star P +_\star t \cdot_\star Q \tag{1.6}$$

and $t \in (0_\star, 1_\star)$ (or $t \in (1, e)$), we say that $X$ is between $P$ and $Q$.

**Theorem 1.10:** *If P, X and Q be distinct points of $E_\star^2$, then X is between P and Q if and only if*

$$d_\star(P, X) +_\star d_\star(X, Q) = d_\star(P, Q). \tag{1.7}$$

*Proof.* Suppose that $X$ is between $P$ and $Q$. Then there is some $t \in (0_\star, 1_\star)$ so that (1.6) holds. Then $t \in (1, e)$, $|t|_\star = t$ and

$$1_\star -_\star t = \frac{e}{t}$$
$$> 1.$$

Hence,

$$|1_\star -_\star t|_\star = 1_\star -_\star t.$$

Then

$$d_\star(P, X) = |t|_\star \cdot_\star |P -_\star Q|_\star$$
$$= t \cdot_\star |P -_\star Q|_\star .$$

Next,

$$d_\star(X, Q) = |Q -_\star X|_\star$$
$$= |1_\star -_\star t|_\star \cdot_\star |Q -_\star P|_\star$$
$$= (1_\star -_\star t) \cdot_\star |Q -_\star P|_\star .$$

Hence,

$$d_\star(P, X) +_\star d_\star(X, Q) = t \cdot_\star |P -_\star Q|_\star +_\star (1_\star -_\star t) \cdot_\star |P -_\star Q|_\star$$
$$= 1_\star \cdot_\star |P -_\star Q|_\star$$
$$= |P -_\star Q|_\star$$
$$= d_\star(P, Q).$$

Conversely, suppose that $X \in E_\star^2$ is such that (1.7) holds. Then there is a $\lambda > 0_\star$ so that

$$X -_\star P = \lambda \cdot_\star (Q -_\star X).$$

Hence,

$$(1_\star +_\star \lambda) \cdot_\star X = P +_\star \lambda \cdot_\star Q$$

and

$$X = (1_\star /_\star (1_\star +_\star \lambda)) \cdot_\star P +_\star (\lambda /_\star (1_\star +_\star \lambda)) \cdot_\star Q$$

Let

$$t = \lambda /_\star (1_\star +_\star \lambda).$$

We have

$$
\begin{aligned}
t &= \lambda /_\star (e +_\star \lambda) \\
&= \lambda /_\star (e\lambda) \\
&= e^{\frac{\log \lambda}{\log(e\lambda)}} \\
&= e^{\frac{\log \lambda}{1 + \log \lambda}} \\
&\leq e \\
&= 1_\star
\end{aligned}
$$

and $t \geq 0_\star$. Moreover,

$$
\begin{aligned}
1_\star -_\star t &= e -_\star e^{\frac{\log \lambda}{1 + \log \lambda}} \\
&= \frac{e}{e^{\frac{\log \lambda}{1 + \log \lambda}}} \\
&= e^{\frac{1}{1 + \log \lambda}}
\end{aligned}
$$

and

$$
\begin{aligned}
1_\star /_\star (1_\star +_\star \lambda) &= e /_\star (e +_\star \lambda) \\
&= e /_\star (e\lambda) \\
&= e^{\frac{\log e}{\log(e\lambda)}} \\
&= e^{\frac{1}{1 + \log \lambda}} \\
&= 1_\star -_\star t.
\end{aligned}
$$

Thus,

$$X = (1_\star -_\star t) \cdot_\star P +_\star t \cdot_\star Q$$

and $X$ is between $P$ and $Q$. This completes the proof.

**Example 1.6:** Let $P(3, 4)$, $Q(2, 7)$ and

$$R\left(e^{\log 2(1-\log 2+\log 3)},\ e^{2(\log 2)^2+(1-\log 2)\log 7}\right).$$

Then

$$2\cdot_* P +_* (1_* -_* 2)\cdot_* Q = 2\cdot_*(3, 4) +_* \tfrac{e}{2}\cdot_*(2, 7)$$

$$= \left(e^{\log 2\log 3},\ e^{2(\log 2)^2}\right) +_* \left(e^{(1-\log 2)\log 2},\ e^{(1-\log 2)\log 7}\right)$$

$$= \left(e^{\log 2(1-\log 2+\log 3)},\ e^{2(\log 2)^2+(1-\log 2)\log 7}\right)$$

$$= R.$$

Thus, $R$ is between $P$ and $Q$.

**Exercise 1.4:** Let $P(7, 9)$, $Q(3, 11)$. Check if the point $R(2, 3)$ is between the points $P$ and $Q$.

**Answer 1.11:** *No.*

**Definition 1.8:** Let $P$ and $Q$ be two distinct points. The set consisting of $P, Q$ and all points that are between $P$ and $Q$ is said to be multiplicative segment. It will be denoted by $PQ$.

**Definition 1.9:** Let $P$ and $Q$ be two distinct points. The point

$$M = 1_* /_* e^2\cdot_* (P +_* Q)$$

will be said the multiplicative midpoint of the multiplicative segment $PQ$.
    Let $P(x_1, x_2)$ and $Q(y_1, y_2)$. Then

$$P +_* Q = (x_1, x_2) +_* (y_1, y_2)$$

$$= (x_1 y_1,\ x_2 y_2)$$

and

$$M = 1_* /_* e^2 \cdot_* (P +_* Q)$$

$$= e /_* e^2 \cdot_* (x_1 y_1, x_2 y_2)$$

$$= e^{\frac{\log e}{\log e^2}} \cdot_* (x_1 y_1, x_2 y_2)$$

$$= e^{\frac{1}{2}} \cdot (x_1 y_1, x_2 y_2)$$

$$= (e^{\log e^{\frac{1}{2}} \log(x_1 y_1)}, e^{\log e^{\frac{1}{2}} \log(x_2 y_2)})$$

$$= (e^{\frac{\log(x_1 y_1)}{2}}, e^{\frac{\log(x_2 y_2)}{2}}).$$

We will show that $M$ is between $P$ and $Q$. We have

$$1_* /_* e^2 \cdot_* P +_* (1_* -_* 1_* /_* e^2) \cdot_* Q$$

$$= e^{\frac{1}{2}} \cdot_* (x_1, x_2) +_* \left( e -_* e^{\frac{1}{2}} \right) \cdot_* (y_1, y_2)$$

$$= (e^{\log e^{\frac{1}{2}} \log x_1}, e^{\log e^{\frac{1}{2}} \log x_2}) +_* e^{\frac{1}{2}} \cdot_* (y_1, y_2)$$

$$= \left( e^{\frac{\log x_1}{2}}, e^{\frac{\log x_2}{2}} \right) +_* \left( e^{\log e^{\frac{1}{2}} \log y_1}, e^{\log e^{\frac{1}{2}} \log y_2} \right)$$

$$= \left( e^{\frac{\log x_1}{2}}, e^{\frac{\log x_2}{2}} \right) +_* \left( e^{\frac{\log y_1}{2}}, e^{\frac{\log y_2}{2}} \right)$$

$$= \left( e^{\frac{\log x_1 + \log y_1}{2}}, e^{\frac{\log x_2 + \log y_2}{2}} \right)$$

$$= \left( e^{\frac{\log(x_1 y_1)}{2}}, e^{\frac{\log(x_2 y_2)}{2}} \right)$$

$$= M.$$

Moreover, $1_* /_* e^2 \in [0_*, 1_*]$.

**Definition 1.10:** If two multiplicative lines pass through a point $P$, we say that they intersect at $P$ and $P$ is their point of intersection.

**Definition 1.11:** If three or more multiplicative lines pass through a point $P$, we say that the multiplicative lines are multiplicative concurrent.

**Definition 1.12:** If three or more points lie on some multiplicative line, the points are said to be multiplicative collinear.

---

## 1.5 Multiplicative Orthonormal Pairs

**Definition 1.13:** Let $x, y \in E_*^2$. We say that $x$ and $y$ are multiplicative orthogonal and we will write $x \perp_* y$ if

$$\langle x, y \rangle_* = 0_*.$$

Let $x, y \in E_*^2$, $x = (x_1, x_2)$, $y = (y_1, y_2)$ and $x \perp_* y$. Then

$$
\begin{aligned}
1 &= 0_* \\
&= \langle x, y \rangle_* \\
&= e^{\log x_1 \log y_1 + \log x_2 \log y_2},
\end{aligned}
$$

whereupon

$$\log x_1 \log y_1 + \log x_2 \log y_2 = 0. \qquad (1.8)$$

If (1.8) holds, then $x \perp_* y$. For $x \in E_*^2$, $x = (x_1, x_2)$, with $x^{\perp_*}$ denote

$$x^{\perp_*} = \left( \frac{1}{x_2}, x_1 \right).$$

Then

$$
\begin{aligned}
\langle x, x^{\perp_*} \rangle_* &= e^{\log x_1 \log \frac{1}{x_2} + \log x_2 \log x_1} \\
&= e^{-\log x_1 \log x_2 + \log x_1 \log x_2} \\
&= e^0 \\
&= 1 \\
&= 0_*.
\end{aligned}
$$

Thus, $x \perp_* x^{\perp_*}$.

**Definition 1.14:** A vector of multiplicative length $1_*$ is said to be a multiplicative unit vector.

**Example 1.7:** Let $x = \left( e^{\frac{1}{\sqrt{2}}}, e^{\frac{1}{\sqrt{2}}} \right)$. Then

$$|x|_\star = e^{\left(\left(\log e^{\frac{1}{\sqrt{2}}}\right)^2 + \left(\log e^{\frac{1}{\sqrt{2}}}\right)^2\right)^{\frac{1}{2}}}$$
$$= e^{\left(\frac{1}{2}+\frac{1}{2}\right)^{\frac{1}{2}}}$$
$$= e$$
$$= 1_\star.$$

Thus, $x$ is a multiplicative unit vector.

**Definition 1.15:** A pair $\{x, y\}$ of multiplicative unit vectors in $E_\star^2$ will be called a multiplicative orthonormal pair if $x \perp_\star y$.

**Definition 1.16:** A pair $\{x, y\}$ of vectors in $E_\star^2$ is said to be multiplicative linearly independent if

$$\lambda_1 \cdot_\star x +_\star \lambda_2 \cdot_\star y = 0_\star \tag{1.9}$$

holds for $\lambda_1 = \lambda_2 = 0_\star$. Otherwise, if there are $\lambda_1, \lambda_2 \in \mathbb{R}_\star$, $(\lambda_1, \lambda_2) \ne (0_\star, 0_\star)$, then $\{x, y\}$ are said to be multiplicative linearly dependent.

**Theorem 1.12:** *Let $\{x, y\}$ be an orthonormal pair. Then they are multiplicative linearly independent.*

*Proof.* Firstly, note that $x \ne 0_\star$, $y \ne 0_\star$. Let $\lambda_1, \lambda_2 \in \mathbb{R}_\star$ be such that

$$\lambda_1 \cdot_\star x +_\star \lambda_2 \cdot_\star y = 0_\star$$

or

$$(\lambda_1 \cdot_\star x_1, \lambda_1 \cdot_\star x_2) +_\star (\lambda_2 \cdot_\star y_1, \lambda_2 \cdot_\star y_2) = (1, 1),$$

or

$$\left(e^{\log \lambda_1 \log x_1}, e^{\log \lambda_1 \log x_2}\right) +_\star \left(e^{\log \lambda_2 \log y_1}, e^{\log \lambda_2 \log y_2}\right) = (1, 1),$$

or

$$\left(e^{\log \lambda_1 \log x_1 + \log \lambda_2 \log y_1}, e^{\log \lambda_1 \log x_1 + \log \lambda_2 \log y_2}\right) = (1, 1),$$

or

$$\log\lambda_1\log x_1 + \log\lambda_2\log y_1 = 0$$
$$\log\lambda_1\log x_2 + \log\lambda_2\log y_2 = 0. \qquad (1.10)$$

Let

$$\Delta = \begin{vmatrix} \log x_1 & \log y_1 \\ \log x_2 & \log y_1 \end{vmatrix}$$
$$= \log x_1\log y_2 - \log x_2\log y_1.$$

Since $x\perp_\star y$, we have that

$$\langle x, y\rangle_\star = 0_\star$$

or

$$e^{\log x_1\log y_1 + \log x_2\log y_2} = e^0,$$

or

$$\log x_1\log y_1 + \log x_2\log y_2 = 0.$$

Without loss of generality, suppose that $\log y_1 \neq 0$. Then

$$\log x_1 = -\frac{\log x_2\log y_2}{\log y_1}.$$

Hence,

$$\Delta = -\frac{\log x_2(\log y_2)^2}{\log y_1} - \log x_2\log y_1$$
$$= -\frac{\log x_2((\log y_1)^2 + (\log y_2)^2)}{\log y_1}.$$

If $\log x_2 = 0$, then $\log x_1 = 0$ and $x = 0_\star$. This is impossible. Therefore $\log x_2 \neq 0$. Next, if

$$(\log y_1)^2 + (\log y_2)^2 = 0,$$

then

$$\log y_1 = 0,$$
$$\log y_2 = 0$$

and $y = 0_\star$, which is a contradiction. Consequently $\Delta \neq 0$ and the system (1.10) has unique solution

$$\log \lambda_1 = \log \lambda_2 = 0,$$

i.e.,

$$\lambda_1 = \lambda_2 = 0_\star.$$

Thus, $x$ and $y$ are multiplicative linearly independent. This completes the proof.

**Theorem 1.13:** *Let $\{x, y\}$ be a multiplicative orthonormal pair of vectors in $E_\star^2$. Then, any $z \in E_\star^2$ can be represented in the form*

$$z = \langle z, x \rangle_\star \cdot_\star x +_\star \langle z, y \rangle_\star \cdot_\star y.$$

*Proof.* We will search a representation of the vector $z$ in the form

$$z = a_1 \cdot_\star x +_\star a_2 \cdot_\star y.$$

Then

$$\langle z, x \rangle_\star = a_1 \cdot_\star \langle x, x \rangle_\star +_\star a_2 \cdot_\star \langle y, x \rangle_\star$$
$$= a_1$$

and

$$\langle z, y \rangle_\star = a_1 \cdot_\star \langle x, y \rangle_\star +_\star a_2 \cdot_\star \langle y, y \rangle_\star$$
$$= a_2.$$

This completes the proof.

## 1.6 Equations of a Multiplicative Line

**Definition 1.17:** Let $l$ be a multiplicative line with multiplicative direction $v$. The vector $v^{\perp_\star}$ will be called a multiplicative normal vector to $l$.

Note that any two multiplicative normal vectors to the same multiplicative line are multiplicative proportional.

**Theorem 1.14:** *Let $l$ be a multiplicative line and $P \in l$. Let also, $\{v, N\}$ be a multiplicative orthonormal pair. Then*

$$P +_\star [v]_\star = \{X: \langle X -_\star P, N\rangle_\star = 0_\star\}.$$

*Proof.* By Theorem 1.13, we have the identity

$$X -_\star P = \langle X -_\star P, v\rangle_{\star\cdot\star} v +_\star \langle X -_\star P, N\rangle_{\star\cdot\star} N \qquad (1.11)$$

for any point $X \in E_\star^2$. Suppose that

$$X = P +_\star t \cdot_\star v.$$

Then

$$\begin{aligned}
\langle X -_\star P, N\rangle_\star &= \langle t \cdot_\star v, N\rangle_\star \\
&= t \cdot_\star \langle v, N\rangle_\star \\
&= 0_\star.
\end{aligned}$$

Let now,

$$\langle X -_\star P, N\rangle_\star = 0_\star.$$

By (1.11), we find

$$X -_\star P = \langle X -_\star P, v\rangle_{\star\cdot\star} v$$

or

$$X = P +_\star \langle X -_\star P, v\rangle_{\star\cdot\star} v.$$

This completes the proof.

**Remark 1.1:** By Theorem 1.14, it follows that

$${X: \langle X -_\star P, N \rangle_\star = 0_\star}$$ (1.12)

is the multiplicative line through $P$ with multiplicative normal vector $N$ and multiplicative direction $N^{\perp_\star}$.

Let $l$ be a multiplicative line with multiplicative normal vector $N = (N_1, N_2)$. Let also, $P = (P_1, P_2) \in l$ and $X = (X_1, X_2)$ be an arbitrary point on $l$. Then

$$X -_\star P = \left( \frac{X_1}{P_1}, \frac{X_2}{P_2} \right)$$

and

$$1 = 0_\star$$
$$= \langle X -_\star P, N \rangle_\star$$
$$= e^{\log \frac{X_1}{P_1} \log N_1 + \log \frac{X_2}{P_2} \log N_2},$$

whereupon

$$\log \frac{X_1}{P_1} \log N_1 + \log \frac{X_2}{P_2} \log N_2 = 0,$$

or

$$\log \left( \frac{X_1}{P_1} \right)^{\log N_1} + \log \left( \frac{X_2}{P_2} \right)^{\log N_2} = 0,$$

or

$$\log \left( \left( \frac{X_1}{P_1} \right)^{\log N_1} \left( \frac{X_2}{P_2} \right)^{\log N_2} \right) = 0,$$

or

$$\left( \frac{X_1}{P_1} \right)^{\log N_1} \left( \frac{X_2}{P_2} \right)^{\log N_2} = 1.$$ (1.13)

If $l$ is a multiplicative line with a multiplicative direction $v = (v_1, v_2)$, then $v^{\perp_\star} = \left( \frac{1}{v_2}, v_1 \right)$ is a multiplicative normal vector to $l$ and by (1.13), we get

$$\left(\frac{X_1}{P_1}\right)^{\log\frac{1}{v_2}}\left(\frac{X_2}{P_2}\right)^{\log v_1} = 1$$

or

$$\left(\frac{X_2}{P_2}\right)^{\log v_1} = \left(\frac{X_1}{P_1}\right)^{\log v_2}. \tag{1.14}$$

**Example 1.8:** Let $l$ be a multiplicative line through the point $P(2, 3)$ and multiplicative direction $v = (4, 8)$. We will find its equation. By (1.14), we get

$$\left(\frac{X_2}{3}\right)^{\log 4} = \left(\frac{X_1}{2}\right)^{\log 8}$$

or

$$\left(\frac{X_2}{3}\right)^{2\log 2} = \left(\frac{X_1}{2}\right)^{3\log 2}.$$

In Fig. 1.2 it is shown the multiplicative $l$.

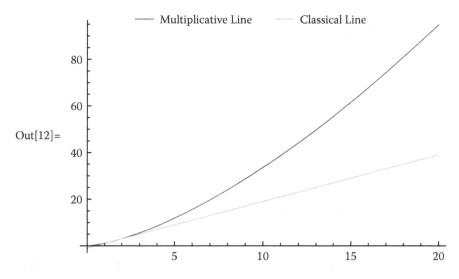

**FIGURE 1.2**
The multiplicative line l.

**Exercise 1.5:** Find the equation of the multiplicative line through $P(4, 5)$ and multiplicative direction $v = (10, 15)$.

**Answer 1.15:**

$$\left(\frac{X_2}{5}\right)^{\log 10} = \left(\frac{X_1}{4}\right)^{\log 15}.$$

**Example 1.9:** Let $l$ be a multiplicative line through $P(1, 5)$ with multiplicative normal vector $N = (e, 4)$. Its equation is

$$X_1^{\log e}\left(\frac{X_2}{5}\right)^{\log 4} = 1$$

or

$$X_1\left(\frac{X_2}{5}\right)^{2\log 2} = 1.$$

**Exercise 1.6:** Find the equation of the multiplicative line $l$ through $P(1, 4)$ with multiplicative normal vector $N = (7, 2)$.

**Answer 1.16:**

$$X_1^{\log 7}\left(\frac{X_2}{4}\right)^{\log 2} = 1.$$

**Definition 1.18:** For two points $P(p_1, p_2)$, $Q(q_1, q_2) \in E_*^2$, define a multiplicative vector **PQ** with initial point $P$ and terminal point $Q$ as follows:

$$\begin{aligned}\mathbf{PQ} &= (q_1 -_* p_1, q_2 -_* p_2) \\ &= \left(\frac{q_1}{p_1}, \frac{q_2}{p_2}\right).\end{aligned}$$

**Theorem 1.17:** *Let* $P(p_1, p_2)$, $Q(q_1, q_2)$ *and* $R(r_1, r_2) \in E_*^2$ *be given points. Then*

$$\begin{aligned}\mathbf{PQ} +_* \mathbf{QR} &= \mathbf{PR}, \\ \mathbf{PQ} -_* \mathbf{PR} &= \mathbf{RQ}.\end{aligned}$$

*Proof.* We have

$$PQ = \left(\frac{q_1}{p_1}, \frac{q_2}{p_2}\right),$$

$$QR = \left(\frac{r_1}{q_1}, \frac{r_2}{q_2}\right),$$

$$PR = \left(\frac{r_1}{p_1}, \frac{r_2}{p_2}\right),$$

$$RQ = \left(\frac{q_1}{r_1}, \frac{q_2}{r_2}\right).$$

Then

$$PQ +_\star QR = \left(\frac{q_1}{p_1}, \frac{q_2}{p_2}\right) +_\star \left(\frac{r_1}{q_1}, \frac{r_2}{q_2}\right)$$

$$= \left(\frac{r_1}{p_1}, \frac{r_2}{p_2}\right)$$

$$= PR$$

and

$$PQ -_\star PR = \left(\frac{q_1}{p_1}, \frac{q_2}{p_2}\right) -_\star \left(\frac{r_1}{p_1}, \frac{r_2}{p_2}\right)$$

$$= \left(\frac{q_1}{r_1}, \frac{q_2}{r_2}\right)$$

$$= RQ.$$

This completes the proof.

**Example 1.10:** Let $P(3, 4)$ and $PQ = (7, 4)$. We will find the coordinates of the point $Q$. Let $Q(q_1, q_2)$, where $q_1, q_2 \in \mathbb{R}_\star$. Then

$$PQ = \left(\frac{q_1}{3}, \frac{q_2}{4}\right).$$

Hence,

$$\left(\frac{q_1}{3}, \frac{q_2}{4}\right) = (7, 4)$$

and

$$q_1 = 21, \quad q_2 = 16.$$

Thus, $Q(21, 16)$.

**Exercise 1.7:** Let $P(1, 2)$ and $\mathbf{PQ} = (3, 2)$. Find the coordinates of the point $Q$.

**Answer 1.18:**

$$Q(3, 4).$$

**Example 1.11:** *Let $P(e, 4)$ and $v = (1, e)$. We will find the coordinates of the point $Q$ so that $|\mathbf{PQ}|_\star = e^2$ and $\mathbf{PQ} \perp_\star v$. Let $Q(q_1, q_2)$, where $q_1, q_2 \in \mathbb{R}_\star$. Then*

$$\mathbf{PQ} = \left( \frac{q_1}{p_1}, \frac{q_2}{p_2} \right)$$

*and*

$$|\mathbf{PQ}|_\star = e^{\left( \left( \log \frac{q_1}{p_1} \right)^2 + \left( \log \frac{q_2}{p_2} \right)^2 \right)^{\frac{1}{2}}}$$
$$= e^2.$$

*Next,*

$$0_\star = e^0$$
$$= e^{\log \frac{q_1}{e} \log 1 + \log \frac{q_2}{4} \log e}$$
$$= e^{\log \frac{q_2}{4}}.$$

*Thus, we get the system*

$$\left( \left( \log \frac{q_1}{p_1} \right)^2 + \left( \log \frac{q_2}{p_2} \right)^2 \right)^{\frac{1}{2}} = 2$$
$$\log \frac{q_2}{4} = 0,$$

*where upon*

$$\left( \log \frac{q_1}{p_1} \right)^2 + \left( \log \frac{q_2}{p_2} \right)^2 = 4$$
$$q_2 = 4$$

*and*

$$\log \frac{q_1}{e} = \pm 2,$$

*and*

$$\frac{q_1}{e} = e^{\pm 2},$$

*and*

$$q_1 = e^3 \text{ or } q_1 = e^{-1}.$$

*Thus, $Q(e^3, 4)$ or $Q(e^{-1}, 4)$.*

**Exercise 1.8:** *Let $P(e^2, e^3)$, $v = (e, e^2)$. Find the coordinates of the point $Q$ so that* **$PQ \perp_\star v$** *and* $|\mathbf{PQ}|_\star = e^9$.

*Solution. Let $Q(q_1, q_2)$, where $q_1, q_2 \in \mathbb{R}_\star$. Then*

$$\mathbf{PQ} = \left( \frac{q_1}{p_1}, \frac{q_2}{p_2} \right)$$

*and*

$$
\begin{aligned}
0_\star &= e^0 \\
&= \langle \mathbf{PQ}, v \rangle_\star \\
&= e^{\log \frac{q_1}{e^2} \log e + \log \frac{q_2}{e^3} \log e^2} \\
&= e^{\log \frac{q_1}{e^2} + 2 \log \frac{q_2}{e^3}} \\
&= e^{\log q_1 - 2 + 2 \log q_2 - 6} \\
&= e^{\log q_1 + 2 \log q_2 - 8},
\end{aligned}
$$

*i.e.,*

$$\log q_1 + 2 \log q_2 = 8.$$

*Next,*

$$
\begin{aligned}
|\mathbf{PQ}|_\star &= e^{\left( \left( \log \frac{q_1}{e^2} \right)^2 + \left( \log \frac{q_2}{e^3} \right)^2 \right)^{\frac{1}{2}}} \\
&= e^9,
\end{aligned}
$$

*whereupon*

$$(\log q_1 - 2)^2 + (\log q_2 - 3)^2 = 81.$$

*Therefore we get the system*

$$
\begin{aligned}
\log q_1 + \log q_2 &= 8 \\
(\log q_1 - 2)^2 + (\log q_2 - 3)^2 &= 81.
\end{aligned}
$$

*We have*

$$\log q_1 = 8 - 2\log q_2$$
$$81 = (\log q_1 - 2)^2 + (\log q_2 - 3)^2.$$

*Thus, we get the equation*

$$81 = (6 - 2\log q_2)^2 + (\log q_2 - 3)^2$$
$$= 36 - 24\log q_2 + 4(\log q_2)^2 + (\log q_2)^2 - 6\log q_2 + 9$$
$$= 5(\log q_2)^2 - 30\log q_2 + 45$$

*or*

$$5(\log q_2)^2 - 30\log q_2 - 36 = 0,$$

*and*

$$\log q_2 = \frac{15 \pm \sqrt{225 + 180}}{5}$$
$$= \frac{15 \pm \sqrt{405}}{5}$$
$$= \frac{15 \pm 9\sqrt{5}}{5},$$

*and*

$$\log q_1 = 8 - \frac{30 \pm 18\sqrt{5}}{5}$$
$$= \frac{10 \mp 18\sqrt{5}}{5}.$$

*Therefore*

$$q_1 = e^{\frac{10 \mp 18\sqrt{5}}{5}}$$
$$q_2 = e^{\frac{15 \pm 9\sqrt{5}}{5}}.$$

*Consequently*

$$Q_1\left(e^{\frac{10-18\sqrt{5}}{5}}, e^{\frac{15+9\sqrt{5}}{5}}\right), \quad Q_2\left(e^{\frac{10+18\sqrt{5}}{5}}, e^{\frac{15-9\sqrt{5}}{5}}\right).$$

*This completes the solution.*

We can find the equation of a multiplicative line if we know two points that lie on it. Let $P(p_1, p_2)$ and $Q(q_1, q_2)$ be two points on the line $l$. Then the vector

$$\mathbf{PQ} = \left(\frac{q_1}{p_1}, \frac{q_2}{p_2}\right)$$

is a multiplicative direction of $l$. Hence, applying (1.14), we find

$$\left(\frac{X_2}{P_2}\right)^{\log \frac{q_1}{p_1}} = \left(\frac{X_1}{P_1}\right)^{\log \frac{q_2}{p_2}}$$

or

$$\left(\frac{X_2}{q_2}\right)^{\log \frac{q_1}{p_1}} = \left(\frac{X_1}{q_1}\right)^{\log \frac{q_2}{p_2}}.$$

**Example 1.12:** Let $P(2, 3), Q(4, 11), R\left(\frac{1}{8}, \frac{1}{2}\right), S\left(\frac{1}{8}, 5\right)$ are given points and $l_1, l_2, l_3, l_4$ are multiplicative lines so that

$$P, Q \in l_1; \ P, R \in l_2; \ Q, R \in l_3; \ P, S \in l_4.$$

We will find the equations of $l_1, l_2, l_3$ and $l_4$. We have

$$\mathbf{PQ} = (4 -_\star 2, 11 -_\star 3)$$
$$= \left(2, \frac{11}{3}\right),$$
$$\mathbf{PR} = \left(\frac{1}{8} -_\star 2, \frac{1}{2} -_\star 3\right)$$
$$= \left(\frac{1}{16}, \frac{1}{6}\right),$$
$$\mathbf{QR} = \left(\frac{1}{8} -_\star 4, \frac{1}{2} -_\star 11\right)$$
$$= \left(\frac{1}{32}, \frac{1}{22}\right),$$
$$\mathbf{PS} = \left(\frac{1}{8} -_\star 2, 5 -_\star 3\right)$$
$$= \left(\frac{1}{16}, \frac{5}{3}\right).$$

Hence, the equations of $l_1, l_2, l_3$ and $l_4$ are

$$\left(\frac{X_2}{3}\right)^{\log 2} = \left(\frac{X_1}{2}\right)^{\log \frac{11}{3}},$$
$$\left(\frac{X_2}{3}\right)^{\log \frac{1}{16}} = \left(\frac{X_1}{2}\right)^{\log \frac{1}{6}},$$
$$\left(\frac{X_2}{11}\right)^{\log \frac{1}{32}} = \left(\frac{X_1}{4}\right)^{\log \frac{1}{22}},$$
$$\left(\frac{X_2}{3}\right)^{\log \frac{1}{16}} = \left(\frac{X_1}{2}\right)^{\log \frac{5}{3}},$$

respectively.

**Exercise 1.9:** Let $P\left(\frac{1}{6}, \frac{1}{12}\right)$, $Q\left(\frac{1}{2}, \frac{1}{8}\right)$, $R\left(\frac{1}{4}, \frac{1}{21}\right)$, $S(2, 4)$. Let also, $l_1, l_2, l_3, l_4$ and $l_5$ are multiplicative lines so that

$$P, Q \in l_1; \quad P, R \in l_2; \quad P, S \in l_3; \quad S, R \in l_4; \quad Q, S \in l_5.$$

Find the equations of the multiplicative lines $l_1, l_2, l_3, l_4$ and $l_5$.

**Answer 1.19:**

1. $l_1$: $(12X_2)^{\log 3} = (6X_1)^{\log \frac{3}{2}}$.
2. $l_2$: $(12X_2)^{\log \frac{3}{2}} = (6X_1)^{\log \frac{4}{7}}$.
3. $l_3$: $(12X_2)^{\log 12} = (6X_1)^{\log 48}$.
4. $l_4$: $\left(\frac{X_2}{4}\right)^{3\log 2} = \left(\frac{X_1}{2}\right)^{\log 84}$.
5. $l_5$: $\left(\frac{X_2}{4}\right)^{2\log 2} = \left(\frac{X_1}{3}\right)^{5\log 2}$.

**Example 1.13:** Let the equation of the multiplicative line $l$ is

$$\left(\frac{X_1}{2}\right)^2 \left(\frac{X_2}{3}\right)^4 = 1.$$

By equation (1.13), it follows that the point $P(2, 3) \in l$. We will find a multiplicative normnal vector $N = (N_1, N_2)$ for the multiplicative line $l$. By equation (1.13), we find

$$\log N_1 = 2 \text{ and } \log N_2 = 4.$$

Therefore

$$N_1 = e^2 \text{ and } N_2 = e^4.$$

Therefore $N = (e^2, e^4)$ is a multiplicative normal vector for $l$. Next, the vector $v = (e^{-4}, e^2)$ is a multiplicative direction vector for $l$.

**Exercise 1.10:** Let the equation of the multiplicative line $l$ is

$$(4X_1)^{\frac{2}{3}} (3X_2)^7 = 1.$$

Find a point $P(p_1, p_2) \in l$, a multiplicative normal vector and a multiplicative direction vector for $l$.

**Answer 1.20:**

1. $P\left(\frac{1}{4}, \frac{1}{3}\right)$.
2. $N = \left(e^{\frac{2}{3}}, e^7\right)$.
3. $v = \left(e^{-7}, e^{\frac{2}{3}}\right)$.

**Example 1.14:** Let $P(p_1, p_2)$ and $Q(q_1, q_2)$ be two distinct points of $E_\star^2$. We will find the equation of the multiplicative line $l$ through the multiplicative mid point $M$ of the multiplicative segment $PQ$ with a multiplicative normal vector **PQ**. We have

$$\mathbf{PQ} = \left(\frac{q_1}{p_1}, \frac{q_2}{p_2}\right).$$

Moreover,

$$
\begin{aligned}
M &= 1_\star /_\star e^2 \cdot_\star (P +_\star Q) \\
&= e /_\star e^2 \cdot_\star (p_1 q_1, p_2 q_2) \\
&= e^{\frac{\log e}{\log e^2}} \cdot_\star (p_1 q_1, p_2 q_2) \\
&= e^{\frac{1}{2}} \cdot_\star (p_1 q_1, p_2 q_2) \\
&= \left(e^{\frac{\log(p_1 q_1)}{2}}, e^{\frac{\log(p_2 q_2)}{2}}\right).
\end{aligned}
$$

Therefore

$$
l: \left(\frac{X_1}{e^{\frac{\log(p_1 q_1)}{2}}}\right)^{\log \frac{q_1}{p_1}} \left(\frac{X_2}{e^{\frac{\log(p_2 q_2)}{2}}}\right)^{\log \frac{q_2}{p_2}} = 1.
$$

Now, suppose that $a, b, c \in \mathbb{R}_\star$, $(a, b) \neq (0_\star, 0_\star)$. Consider the equation

$$a \cdot_\star x +_\star b \cdot_\star y +_\star c = 0_\star \qquad (1.15)$$

or

$$e^{\log a \log x + \log b \log y + \log c} = e^0,$$

or

$$\log a \log x + \log b \log y + \log c = 0,$$

or

$$\log x^{\log a} + \log y^{\log b} + \log c = 0,$$

or

$$c x^{\log a} y^{\log b} = 1. \tag{1.16}$$

Let $P(p_1, p_2)$, $p_1, p_2 \in \mathbb{R}_\star$, satisfies equation (1.15). Then

$$a \cdot_\star p_1 +_\star b \cdot_\star p_2 +_\star c = 0_\star,$$

or

$$c = -_\star a \cdot_\star p_1 -_\star b \cdot_\star p_2, \tag{1.17}$$

and

$$a \cdot_\star (x -_\star p_1) +_\star b \cdot_\star (y -_\star p_2) = 0_\star.$$

Set

$$N = (a, b).$$

Then

$$\langle X -_\star P, N \rangle_\star = 0_\star.$$

Thus, (1.15) or (1.16) is an equation of a multiplicative line with multiplicative normal vector $N$ and if $c$ satisfies equation (1.17), then the point $P$ lies on it.

## 1.7 Perpendicular Multiplicative Lines

**Definition 1.19:** Two multiplicative lines $l$ and $m$ are said to be perpendicular if they have multiplicative orthogonal direction vectors. In this case, we write $l \perp_\star m$.

**Example 1.15:** Let the multiplicative line $l$ has the equation

$$\left(\frac{X_1}{2}\right)^{\log 3} = \left(\frac{X_2}{3}\right)^{\log 2}$$

and the multiplicative line $m$ has the equation

$$X_1^{\log 2} = X_2^{-\log 3}.$$

By equation (1.14), it follows that the vectors

$$v^1 = (2, 3) \text{ and } v^2 = \left(\frac{1}{3}, 2\right)$$

are multiplicative directions of $l$ and $m$, respectively. We have

$$
\begin{aligned}
\langle v^1, v^2 \rangle_\star &= e^{\log 2 \log \frac{1}{3} + \log 3 \log 2} \\
&= e^{-\log 2 \log 3 + \log 2 \log 3} \\
&= e^0 \\
&= 1 \\
&= 0_\star.
\end{aligned}
$$

Consequently $l \perp_\star m$.

**Lemma 1.1:** *For any $x, y \in E_\star^2$, we have*

$$|x +_\star y|_\star^{2_\star} = |x|_\star^{2_\star} +_\star e^{2} \cdot_\star \langle x, y \rangle_\star +_\star |y|_\star^{2_\star}.$$

*Proof.* We have

$$
\begin{aligned}
|x +_\star y|_\star^{2_\star} &= |(x_1 y_1, x_2 y_2)|_\star^{2_\star} \\
&= e^{\left(\log(x_1 y_1)\right)^2 + \left(\log(x_2 y_2)\right)^2}, \\
|x|_\star^{2_\star} &= e^{(\log x_1)^2 + (\log x_2)^2}, \\
e^{2} \cdot_\star \langle x, y \rangle_\star &= e^{2} \cdot_\star e^{\log x_1 \log y_1 + \log x_2 \log y_2} \\
&= e^{2(\log x_1 \log y_1 + \log x_2 \log y_2)}, \\
|y|_\star^{2_\star} &= e^{\left(\log y_1\right)^2 + \left(\log y_2\right)^2}.
\end{aligned}
$$

Hence,

$$|x|_{\star}^{2\star} +_{\star} e^{2\cdot}\cdot_{\star} \langle x, y \rangle_{\star} +_{\star} |y|_{\star}^{2\star}$$

$$= e^{(\log x_1)^2 + (\log x_2)^2} e^{2(\log x_1 \log y_1 + \log x_2 \log y_2)} e^{\left(\log y_1\right)^2 + \left(\log y_2\right)^2}$$

$$= e^{(\log x_1)^2 + \left(\log y_1\right)^2 + (\log x_2)^2 + \left(\log y_2\right)^2 + 2\left(\log x_1 \log y_1 + \log x_2 \log y_2\right)}$$

$$= e^{\left(\log x_1 + \log y_1\right)^2 + \left(\log x_2 + \log y_2\right)^2}$$

$$= e^{\left(\log(x_1 y_1)\right)^2 + \left(\log(x_2 y_2)\right)^2}$$

$$= |x +_{\star} y|_{\star}^{2\star}.$$

This completes the proof.

**Theorem 1.21: (Multiplicative Pythagoras).** *Let P, Q and R be three distinct points. Then*

$$|R -_{\star} P|_{\star}^{2\star} = |Q -_{\star} P|_{\star}^{2\star} +_{\star} |R -_{\star} Q|_{\star}^{2\star}$$

*if and only if* **QP** *and* **RQ** *are multiplicative perpendicular.*

*Proof.* By Lemma 1.1, it follows that

$$|x +_{\star} y|_{\star}^{2\star} = |x|_{\star}^{2\star} +_{\star} |y|_{\star}^{2\star}, \quad x, y \in E_{\star}^2,$$

if and only if $\langle x, y \rangle_{\star} = 0_{\star}$, i.e., $x \perp_{\star} y$. Now, we take

$$x = \mathbf{QP}, \quad y = \mathbf{RQ}$$

and we get the desired result. This completes the proof.

**Theorem 1.22:** *Let* $l \perp_{\star} m$. *Then l and m have a unique point in common.*

*Proof.* Let $v$ and $w$ be multiplicative direction vectors of $l$ and $m$, respectively. Without loss of generality, suppose that $v$ and $w$ are multiplicative unit vectors. Thus, $\{v, w\}$ is a multiplicative orthonormal pair. Suppose that

$$l = P +_{\star} [v]_{\star},$$
$$m = Q +_{\star} [w]_{\star}.$$

Then

$$P -_\star Q = \langle P -_\star Q, v \rangle_{\star \cdot \star} v +_\star \langle P -_\star Q, w \rangle_{\star \cdot \star} w,$$

whereupon

$$P -_\star \langle P -_\star Q, v \rangle_{\star \cdot \star} v = Q +_\star \langle P -_\star Q, v \rangle_{\star \cdot \star} w.$$

Setting

$$F = P -_\star \langle P -_\star Q, v \rangle_{\star \cdot \star} v$$
$$= Q +_\star \langle P -_\star Q, v \rangle_{\star \cdot \star} w,$$

we get that $F$ is the common point of $l$ and $m$. The point $F$ is unique because otherwise the multiplicative lines $l$ and $m$ will coincide. This completes the proof.

**Example 1.16:** Let

$$l: \left(\frac{X_1}{2}\right)^{\log 2} = \left(\frac{X_2}{4}\right)^{\log 3},$$
$$m: X_1^{\log 3} = X_2^{\log \frac{1}{2}}.$$

Set

$$v^1 = (3, 2),$$
$$v^2 = \left(\frac{1}{2}, 3\right).$$

Then $v^1$ and $v^2$ are multiplicative direction vectors for the multiplicative lines $l$ and $m$, respectively. Next,

$$\langle v^1, v^2 \rangle_\star = e^{\log 3 \log \frac{1}{2} + \log 2 \log 3}$$
$$= e^{-\log 2 \log 3 + \log 2 \log 3}$$
$$= e^0$$
$$= 1$$
$$= 0_\star.$$

Thus, $v^1 \perp_\star v^2$ and $l \perp_\star m$. Let $F = l \cap m$. Note that

$$l: X_1 = 2\left(\frac{X_2}{4}\right)^{\frac{\log 3}{\log 2}},$$

$$m: X_1 = X_2^{-\frac{\log 2}{\log 3}}.$$

Then

$$X_2^{-\frac{\log 2}{\log 3}} = 2\left(\frac{X_2}{4}\right)^{\frac{\log 3}{\log 2}}$$

or

$$1 = \frac{2}{4^{\frac{\log 3}{\log 2}}} X_2^{\frac{\log 3}{\log 2} + \frac{\log 2}{\log 3}}$$

$$= \frac{1}{2^{2\frac{\log 3}{\log 2} - 1}} X_2^{\frac{(\log 3)^2 + (\log 2)^2}{\log 2 \log 3}},$$

or

$$X_2^{\frac{(\log 3)^2 + (\log 2)^2}{\log 2 \log 3}} = 2^{2\frac{\log 3}{\log 2} - 1}.$$

Hence,

$$X_2 = 2^{\frac{(2\log 3 - \log 2)\log 3}{(\log 2)^2 + (\log 3)^2}}$$

and

$$X_1 = 2^{-\frac{(2\log 3 - \log 2)\log 2}{(\log 2)^2 + (\log 3)^2}}.$$

Consequently

$$F\left(2^{-\frac{(2\log 3 - \log 2)\log 2}{(\log 2)^2 + (\log 3)^2}}, 2^{\frac{(2\log 3 - \log 2)\log 3}{(\log 2)^2 + (\log 3)^2}}\right)$$

is the common point of the multiplicative lines $l$ and $m$.

**Exercise 1.11:** Let

$$l: \left(\frac{X_1}{4}\right)^{\log 5} = \left(\frac{X_2}{3}\right)^{\log 7},$$

$$m: \left(\frac{X_1}{11}\right)^{\log 7} = \left(\frac{X_2}{4}\right)^{\log \frac{1}{5}}.$$

Prove that $l \perp_\star m$ and find the point $F = l \cap m$.

**Answer 1.23:**

$$F = \left( \frac{(\log 7)^2}{11^{(\log 5)^2 + (\log 7)^2}} 3^{-\frac{\log 7 \log 5}{(\log 5)^2 + (\log 7)^2}} 4^{\frac{\log 5 (\log 5 + \log 7)}{(\log 5)^2 + (\log 7)^2}}, \; 11^{\frac{\log 5 \log 7}{(\log 5)^2 + (\log 7)^2}} \right.$$

$$\left. 3^{\frac{(\log 7)^2}{(\log 5)^2 + (\log 7)^2}} 4^{\frac{\log 5 (\log 5 - \log 7)}{(\log 5)^2 + (\log 7)^2}} \right).$$

**Theorem 1.24:** *Let $R$ be a point in $E_\star^2$ and let $l$ be a multiplicative line. Then there is a unique multiplicative line $m$ through $R$ perpendicular to $l$. Furthermore,*

1. *$m = R +_\star [N]_\star$, where $N$ is a multiplicative unit normal vector to $l$.*
2. *$l$ and $m$ intersect in the point*

$$F = R -_\star \langle R -_\star P, N \rangle_{\star \cdot \star} N,$$

   *where $P$ is any point on $l$.*

3. *$d_\star(R, F) = |\langle R -_\star P, N \rangle_\star|_\star$.*

*Proof.* Let $Q(Q_1, Q_2) \in l$ and the equation of $l$ is given by

$$\left(\frac{X_1}{Q_1}\right)^{\log N_1} \left(\frac{X_2}{Q_2}\right)^{\log N_2} = 1,$$

where $N = (N_1, N_2)$ is a multiplicative unit normal vector to $l$. Then

$$m = R +_\star [N]_\star$$

and

$$m: \left(\frac{X_1}{R_1}\right)^{\log N_2} = \left(\frac{X_2}{R_2}\right)^{\log N_1}.$$

We have

$$F = R +_\star \mu \cdot_\star N,$$

where $\mu \in \mathbb{R}_\star$ will be determined below. Note that

$$F -_\star P = R -_\star P +_\star \mu \cdot_\star N.$$

Hence,

$$\begin{aligned}
0_\star &= \langle F -_\star P, N \rangle_\star \\
&= \langle R -_\star P, N \rangle_\star +_\star \mu \cdot_\star \langle N, N \rangle_\star \\
&= \langle R -_\star P, N \rangle_\star +_\star \mu,
\end{aligned}$$

whereupon

$$\mu = -_\star \langle R -_\star P, N \rangle_\star.$$

Consequently

$$F = R -_\star \langle R -_\star P, N \rangle_\star \cdot_\star N$$

and hence,

$$\begin{aligned}
d_\star(F, R) &= |\langle R -_\star P, N \rangle_\star \cdot_\star N| \\
&= |\langle R -_\star P, N \rangle_\star|_\star \cdot_\star |N|_\star \\
&= |\langle R -_\star P, N \rangle_\star|_\star.
\end{aligned}$$

This completes the proof.

**Remark 1.2:** Suppose that all conditions of Theorem 1.24 hold. Let $P(P_1, P_2)$, $R(R_1, R_2)$, $F(F_1, F_2)$. Then

$$R -_\star P = \left( \frac{R_1}{P_1}, \frac{R_2}{P_2} \right),$$

$$\langle R -_\star P, N \rangle_\star = e^{\log \frac{R_1}{P_1} \log N_1 + \log \frac{R_2}{P_2} \log N_2}$$

and

$$\langle R -_\star P, N \rangle_\star \cdot_\star N = \left( e^{\left( \log \frac{R_1}{P_1} \log N_1 + \log \frac{R_2}{P_2} \log N_2 \right) \log N_1}, \right.$$
$$\left. e^{\left( \log \frac{R_1}{P_1} \log N_1 + \log \frac{R_2}{P_2} \log N_2 \right) \log N_2} \right)$$

and

$$F = (F_1, F_2)$$
$$= \left(R_1 e^{-\left(\log \frac{R_1}{P_1} \log N_1 + \log \frac{R_2}{P_2} \log N_2\right) \log N_1},\right.$$
$$\left. R_2 e^{-\left(\log \frac{R_1}{P_1} \log N_1 + \log \frac{R_2}{P_2} \log N_2\right) \log N_2}\right)$$

and

$$d_\star(R, F) = e^{\log \frac{R_1}{P_1} \log N_1 + \log \frac{R_2}{P_2} \log N_2}.$$

**Definition 1.20:** The number $d_\star(R, F)$ is called the multiplicative distance of the point $R$ to the line $l$ and it is written as $d_\star(R, l)$.

**Example 1.17:** Let

$$l: \left(\frac{X_1}{2}\right)^{\log 3} \left(\frac{X_2}{4}\right)^{\log 7} = 1$$

and $R(3, 5)$. We will find a multiplicative line $m$ so that $R \in m$ and $l \perp_\star m$. By equation (1.13), it follows that $N = (3, 7)$ is a multiplicative normal vector to $l$. Then $N$ is a multiplicative direction vector to $m$. Hence, applying equation (1.14), we find

$$m: \left(\frac{X_2}{5}\right)^{\log 3} = \left(\frac{X_1}{3}\right)^{\log 7}.$$

Let $F = m \cap l$. Consider the system

$$\left(\frac{X_1}{2}\right)^{\log 3} \left(\frac{X_2}{4}\right)^{\log 7} = 1$$
$$\left(\frac{X_2}{5}\right)^{\log 3} = \left(\frac{X_1}{3}\right)^{\log 7}.$$

We have

$$X_1 = 3\left(\frac{X_2}{5}\right)^{\frac{\log 3}{\log 7}}$$

$$\left(\frac{3}{2}\left(\frac{X_2}{5}\right)^{\frac{\log 3}{\log 7}}\right)^{\log 3} \left(\frac{X_2}{4}\right)^{\log 7} = 1,$$

whereupon

$$\frac{3^{\log 3}}{2^{\log 3} 5^{\frac{(\log 3)^2}{\log 7}}} X_2^{\frac{(\log 3)^2}{\log 7}} X_2^{\log 7} \frac{1}{4^{\log 7}} = 1,$$

or

$$X_2^{\frac{(\log 3)^2 + (\log 7)^2}{\log 7}} = \frac{2^{\log 3} 5^{\frac{(\log 3)^2}{\log 7}} 4^{\log 7}}{3^{\log 3}},$$

or

$$X_2 = \frac{2^{\frac{\log 3 \log 7}{(\log 3)^2 + (\log 7)^2}} 5^{\frac{(\log 3)^2}{(\log 3)^2 + (\log 7)^2}} 4^{\frac{(\log 7)^2}{(\log 3)^2 + (\log 7)^2}}}{3^{\frac{\log 3 \log 7}{(\log 3)^2 + (\log 7)^2}}},$$

or

$$X_2 = \frac{2^{\frac{\log 7 (\log 3 + 2\log 7)}{(\log 3)^2 + (\log 7)^2}} 5^{\frac{(\log 3)^2}{(\log 3)^2 + (\log 7)^2}}}{3^{\frac{\log 3 \log 7}{(\log 3)^2 + (\log 7)^2}}},$$

and

$$X_1 = \frac{3}{5^{\frac{\log 3}{\log 7}}} \frac{2^{\frac{\log 3 (\log 3 + 2\log 7)}{(\log 3)^2 + (\log 7)^2}} 5^{\frac{(\log 3)^3}{\log 7 \left((\log 3)^2 + (\log 7)^2\right)}}}{3^{\frac{(\log 3)^2}{(\log 3)^2 + (\log 7)^2}}}$$

$$= \frac{2^{\frac{\log 3 (\log 3 + 2\log 7)}{(\log 3)^2 + (\log 7)^2}} 5^{-\frac{\log 3 \log 7}{\left((\log 3)^2 + (\log 7)^2\right)}}}{3^{-\frac{(\log 7)^2}{(\log 3)^2 + (\log 7)^2}}}$$

$$= \frac{2^{\frac{\log 3 (\log 3 + 2\log 7)}{(\log 3)^2 + (\log 7)^2}} 3^{\frac{(\log 7)^2}{(\log 3)^2 + (\log 7)^2}}}{5^{\left(\frac{\log 3 \log 7}{(\log 3)^2 + (\log 7)^2}\right)}}.$$

Consequently

$$F\left(\frac{\frac{\log 3(\log 3+2\log 7)}{2}\frac{(\log 7)^2}{(\log 3)^2+(\log 7)^2}\frac{3(\log 3)^2+(\log 7)^2}{}}{5\left((\log 3)^2+(\log 7)^2\right)}, \frac{\frac{\log 7(\log 3+2\log 7)}{2}\frac{(\log 3)^2}{(\log 3)^2+(\log 7)^2}\frac{5(\log 3)^2+(\log 7)^2}{}}{3(\log 3)^2+(\log 7)^2}\right) = m \cap l.$$

**Exercise 1.12:** Let $P(7, 8)$ and

$$l:\ X_1^{\log 10}\left(\frac{X_2}{4}\right)^{\log 15} = 1.$$

Find a multiplicative line $m$ so that $P \in m$ and $l\perp_* m$.

**Answer 1.25:**

$$\left(\frac{X_1}{7}\right)^{\log 15} = \left(\frac{X_2}{8}\right)^{\log 10}.$$

**Definition 1.21:** Let $PQ$ be a multiplicative segment and $M$ be its multiplicative midpoint. The multiplicative line through $M$ that is multiplicative perpendicular to **PQ** is called multiplicative perpendicular bisector of the multiplicative segment $PQ$.

---

## 1.8 Parallel and Intersecting Multiplicative Lines

**Definition 1.22:** Two distinct multiplicative lines $l$ and $m$ are said to be multiplicative parallel if they have no point of intersection. In this case, we will write $l\|_* m$.

**Example 1.18:** Consider the multiplicative lines

$$l_1: X_1^2 X_2 = 1,$$
$$l_2: X_1^2\left(\frac{X_2}{2}\right) = 1.$$

These multiplicative lines have not any point of intersection. In Fig. 1.3 they are shown in graphics.

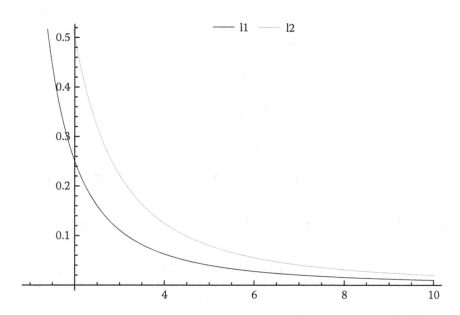

**FIGURE 1.3**
Parallel multiplicative lines.

**Theorem 1.26:** *Two distinct multiplicative lines l and m are multiplicative parallel if and only if they have the same multiplicative direction.*

*Proof.*

1. Let $l\|_\star m$. Suppose that $l$ and $m$ have different multiplicative direction vectors $v$ and $w$, respectively. Let $P \in l$ and $Q \in m$. Because $v$ and $w$ are not multiplicative proportional, there exist numbers $s, t \in \mathbb{R}_\star$ so that

$$P -_\star Q = t \cdot_\star v +_\star s \cdot_\star w,$$

i.e.,

$$P -_\star t \cdot_\star v = Q +_\star s \cdot_\star w.$$

Thus, the point

$$F = P -_\star t \cdot_\star v$$
$$= Q +_\star s \cdot_\star w$$

lies on $l$ and $m$. This is a contradiction. Therefore $l$ and $m$ have the same multiplicative direction vectors.

2. Suppose that $l$ and $m$ have the same multiplicative direction vectors. Let $l$ and $m$ have a common point $F$. Then

$$l = F +_\star [v]_\star \text{ and } m = F +_\star [w]_\star.$$

Because $l$ and $m$ are distinct, we have that

$$[v]_\star \neq [w]_\star.$$

This is a contradiction. Therefore $l$ and $m$ have not any common point. Thus, $l\|_\star m$. This completes the proof.

**Theorem 1.27:** *Let $l$, $m$ and $n$ be multiplicative lines so that $l\|_\star m$ and $l\|_\star n$. Then $m = n$ or $m\|_\star n$.*

*Proof.* Since $l\|_\star m$, we have that $l$ and $m$ are distinct and have the same multiplicative direction $v$. Because $l\|_\star n$, we have that $l$ and $n$ are distinct and have the same multiplicative direction $v$. Hence, $m$ and $n$ have the same multiplicative direction $v$. Therefore $m = n$ or $m\|_\star n$. This completes the proof.

**Theorem 1.28:** *Let $l$, $m$ and $n$ be multiplicative lines so that $l\|_\star m$ and $m\perp_\star n$. Then $l\perp_\star n$.*

*Proof.* Since $l\|_\star m$, we have that $l$ and $m$ are distinct and have the same multiplicative direction $v$ and the same multiplicative normal vector $N$. Because $m\perp_\star n$, we have that $N$ is a multiplicative direction vector for $n$. Hence, $l\perp_\star n$. This completes the proof.

**Theorem 1.29:** *Let $l$, $m$ and $n$ be multiplicative lines so that $l\perp_\star m$ and $l\perp_\star n$. Then $m\|_\star n$ or $m = n$.*

*Proof.* Let $v$ and $N$ be a multiplicative direction vector and a multiplicative normal vector, respectively, of $l$. Since $l\perp_\star m$ and $l\perp_\star n$, we have that $N$ is a multiplicative direction vector for $m$ and $n$. Hence, $m\|_\star n$ or $m = n$. This completes the proof.

---

## 1.9 Multiplicative Rays and Multiplicative Angles

**Definition 1.23:** Let $P$ be a point of $E_\star^2$ and $v$ be a multiplicative nonzero vector. Then

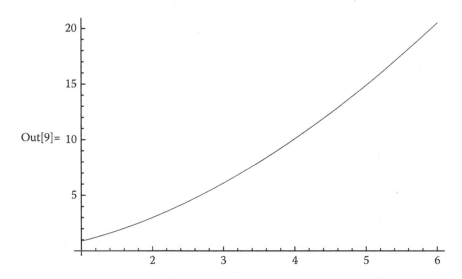

**FIGURE 1.4**
The multiplicative ray through $P(2,3)$ and a multiplicative direction $v=(e^4,e^7)$.

$$r = \{P +_\star \ t\cdot_\star v : t \geq 0_\star\}$$

is called a multiplicative ray with origin $P$ and multiplicative direction $v$. Clearly, every multiplicative line through $P$ is the union of two multiplicative rays with origin $P$. Their directions are multiplicative negative of each other. In Fig. 1.4, it is shown a multiplicative ray through $P(2, 3)$ with direction $v = (e^4, e^7)$.

**Definition 1.24:** The union of two multiplicative rays $r_1$ and $r_2$ with common origin $P$ is called a multiplicative angle with vertex $P$ and multiplicative arms $r_1$ and $r_2$. When $r_1 = r_2$, then the multiplicative angle is said to be the multiplicative zero angle. If $r_1$ and $r_2$ are two halves of the same multiplicative line, we say that they are multiplicative opposite rays and the multiplicative angle is a multiplicative straight angle. If $r_1 \perp_\star r_2$, we call the multiplicative angle a multiplicative right angle.
   Let

$$r_1 = \{(2, 3) +_\star \ t\cdot_\star(e^4, e^7), \ t \geq 0_\star\},$$

$$r_2 = \left\{(2, 3) +_\star \ t\cdot_\star\left(e, e^{\frac{1}{2}}\right), \ t \geq 0_\star\right\}.$$

In Fig. 1.5, it is shown a multiplicative angle with vertex $(2, 3)$ and multiplicative arms $r_1$ and $r_2$. For a given two distinct points $P$ and $Q$, there is a

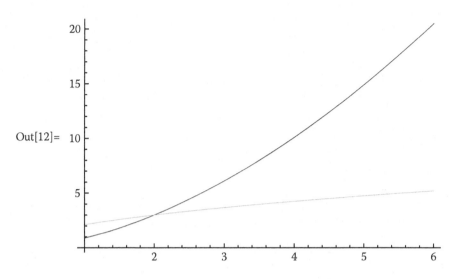

**FIGURE 1.5**
A multiplicative angle.

unique multiplicative ray with origin the point $P$ that passes through the point $Q$. We denote this ray by **PQ**. The multiplicative angle with vertex the point $Q$ and multiplicative arms **QR** and **QP** will be denoted by $\angle_\star PQR$ and $\angle_\star RQP$.

**Definition 1.25:** Let $\mathscr{A}$ be a multiplicative angle whose multiplicative arms have multiplicative unit direction vectors $u$ and $v$. The multiplicative radian measure of $\mathscr{A}$ is defined by

$$\alpha = \arccos_\star(\langle u, v \rangle_\star).$$

**Example 1.19:** Let $u = \left( e^{\frac{1}{\sqrt{2}}}, e^{\frac{1}{\sqrt{2}}} \right)$ and $v = \left( e^{\sqrt{\frac{2}{3}}}, e^{\sqrt{\frac{1}{3}}} \right)$. Then

$$|u|_\star = e^{\left( \left( \log e^{\frac{1}{\sqrt{2}}} \right)^2 + \left( \log e^{\frac{1}{\sqrt{2}}} \right)^2 \right)^{\frac{1}{2}}}$$
$$= e^{\left( \frac{1}{2} + \frac{1}{2} \right)^{\frac{1}{2}}}$$
$$= e$$
$$= 1_\star$$

and

$$|v|_\star = e^{\left(\left(\log e\sqrt{\frac{2}{3}}\right)^2 + \left(\log e\sqrt{\frac{1}{3}}\right)^2\right)^{\frac{1}{2}}}$$

$$= e^{\left(\frac{2}{3}+\frac{1}{3}\right)^{\frac{1}{2}}}$$

$$= e$$

$$= 1_\star.$$

We have

$$\langle u, v \rangle_\star = e^{\log e\frac{1}{\sqrt{2}} \log e\sqrt{\frac{2}{3}} + \log e\frac{1}{\sqrt{2}} \log e\sqrt{\frac{1}{3}}}$$

$$= e^{\frac{1}{\sqrt{2}} \cdot \frac{\sqrt{2}}{\sqrt{3}} + \frac{1}{\sqrt{2}} \cdot \frac{1}{\sqrt{3}}}$$

$$= e^{\frac{1+\sqrt{2}}{\sqrt{6}}}$$

and

$$\alpha = \arccos_\star(\langle u, v \rangle_\star)$$

$$= e^{\arccos(\log(\langle u,v \rangle_\star))}$$

$$= e^{\arccos\left(\log e^{\frac{1+\sqrt{2}}{\sqrt{6}}}\right)}$$

$$= e^{\arccos\left(\frac{1+\sqrt{2}}{\sqrt{6}}\right)}.$$

Let

$$u = (\cos_\star\theta, \sin_\star\theta), \quad v = (\cos_\star\phi, \sin_\star\phi).$$

Then

$$u = (e^{\cos(\log\theta)}, e^{\sin(\log\theta)}),$$

$$v = (e^{\cos(\log\phi)}, e^{\sin(\log\phi)})$$

and

$$\langle u, v \rangle_\star = e^{\cos(\log\theta)\cos(\log\phi) + \sin(\log\theta)\sin(\log\phi)}$$

$$= e^{\cos(\log\theta - \log\phi)}$$

$$= e^{\cos\left(\log\frac{\theta}{\phi}\right)}$$

and the multiplicative radian measure is

$$\alpha = \arccos_\star(\langle u, v \rangle_\star)$$

$$= e^{\arccos(\log(\langle u,v\rangle_\star))}$$

$$= e^{\arccos(\cos(\log\frac{\theta}{\phi}))}$$

$$= e^{\log\frac{\theta}{\phi}}$$

$$= \frac{\theta}{\phi}$$

$$= \theta -_\star \phi.$$

**Exercise 1.13:** Let $u = (2, 8)$ and $v = (3, 7)$. Find the multiplicative radian measure.

**Theorem 1.30:** *Let $\mathscr{A}$ be a multiplicative angle. Its multiplicative radian measure $\alpha$ is $0_\star$ if $\mathscr{A}$ is the multiplicative zero angle.*

*Proof.* Let $u, v$ be multiplicative unit vectors that are the multiplicative arms of $\mathscr{A}$. Then

$$\alpha = \arccos_\star(\langle u, v \rangle_\star)$$
$$= 0_\star$$

if and only if

$$e^{\arccos(\log(\langle u,v\rangle_\star))} = e^0$$

if and only if

$$\arccos(\log(\langle u, v \rangle_\star)) = 0$$

if and only if

$$\log(\langle u, v \rangle_\star) = 1$$

if and only if

$$\langle u, v \rangle_\star = e$$

if and only if $u = v$ or $u\|_\star v$. This completes the proof.

**Theorem 1.31:** *Let $\mathscr{A}$ be a multiplicative angle. Its multiplicative radian measure $\alpha$ is $e^\pi$ if and only if $\mathscr{A}$ is a multiplicative straight angle.*

*Proof.* By the proof of Theorem 1.30, it follows that

$$\alpha = e^{\pi}$$

if and only if

$$\arccos\left(\log(\langle u, v\rangle_{\star})\right) = \pi$$

if and only if

$$\log\left(\langle u, v\rangle_{\star}\right) = -1$$

if and only if

$$\langle u, v\rangle_{\star} = e^{-1}$$

if and only if

$$e^{\log u_1 \log v_1 + \log u_2 \log v_2} = e^{-1}$$

if and only if

$$\log u_1 \log v_1 + \log u_2 \log v_2 = -1$$

if and only if

$$v = \left(\frac{1}{u_1}, \frac{1}{u_2}\right)$$

if and only if $\mathscr{A}$ is a multiplicative straight angle. This completes the proof.

**Theorem 1.32:** *Let $\mathscr{A}$ be a multiplicative angle. Then its multiplicative radian measure is in the interval $[1, e^{\pi}]$.*

*Proof.* We have that

$$\alpha = e^{\arccos\left(\log(\langle u, v\rangle_{\star})\right)}.$$

Hence, $\alpha \in [1, e^{\pi}]$. This completes the proof.

**Definition 1.26:** A multiplicative angle $\mathscr{A}$ is said to be multiplicative acute if its multiplicative radiation measure is $< e^{\frac{\pi}{2}}$. If its multiplicative radian measure is $> e^{\frac{\pi}{2}}$, then it is said to be multiplicative obtuse.

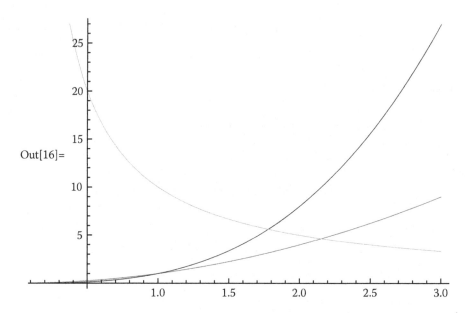

Out[16]=

**FIGURE 1.6**
A multiplicative triangle.

Let $\mathscr{A} = \angle_\star PQR$. By the definition of multiplicative radian measure, it follows that $\mathscr{A}$ is multiplicative acute(obtuse) if and only if

$$\langle P -_\star Q, R -_\star Q \rangle_\star > (<)0_\star.$$

**Definition 1.27:** Let $P, Q, R \in E_\star^2$ be multiplicative noncollinear points. The multiplicative triangle $\triangle_\star PQR$ or $PQR$ is the multiplicative rectilinear figure consisting of the multiplicative segments $PQ$, $PR$ and $RQ$. The multiplicative segments are called the multiplicative sides of the multiplicative triangle.

In Fig. 1.6 it is shown a multiplicative triangle.

**Definition 1.28:** A point $X$ is said to be in the interior of the multiplicative triangle $PQR$ if it is in the interior of all three multiplicative angles determined by the points $P$, $Q$ and $R$.

---

## 1.10 The Space $E_\star^2$

Set

$$\mathbb{R}_\star^3 = \{x = (x_1, x_2, x_3) : x_1, x_2, x_3 \in \mathbb{R}_\star\}.$$

**Definition 1.29:** For $x = (x_1, x_2, x_3)$, $y = (y_1, y_2, y_3) \in \mathbb{R}^3_*$, define

$$x +_* y = (x_1 +_* , x_2 +_* y_2, x_3 +_* y_3),$$
$$x -_* y = (x_1 -_* y_1, x_2 -_* y_2, x_3 -_* y_3),$$
$$c \cdot_* x = (c \cdot_* x_1, c \cdot_* x_2, c \cdot_* x_3),$$
$$\langle x, y \rangle_* = e^{\log x_1 \log y_1 + \log x_2 \log y_2 + \log x_3 \log y_3},$$
$$|x|_* = e^{((\log x_1)^2 + (\log x_2)^2 + (\log x_3)^2)^{\frac{1}{2}}}.$$

**Example 1.20:** Let

$$x = (1, e, 3),$$
$$y = (2, 3, 4).$$

Then

$$x +_* y = (1 +_* 2, e +_* 3, 3 +_* 4)$$
$$= (2, 3e, 12)$$

and

$$x -_* y = (1 -_* 2, e -_* 3, 3 -_* 4)$$
$$= \left( \frac{1}{2}, \frac{e}{3}, \frac{3}{4} \right),$$

and

$$3 \cdot_* x = (3 \cdot_* 1, 3 \cdot_* e, 3 \cdot_* 3)$$
$$= (e^{\log 3 \log 1}, e^{\log 3 \log e}, e^{(\log 3)^2})$$
$$= (1, 3, e^{(\log 3)^2}),$$

and

$$\langle x, y \rangle_* = e^{\log 1 \log 2 + \log e \log 3 + \log 3 \log 4}$$
$$= e^{\log 3(1 + 2\log 2)},$$

and

$$|x|_\star = e^{((\log 1)^2 + (\log e)^2 + (\log 3)^2)^{\frac{1}{2}}}$$

$$= e^{(1 + (\log 3)^2)^{\frac{1}{2}}}.$$

**Exercise 1.14:** Let

$$x = (3, e^2, 11),$$
$$y = (4, e, e^5).$$

Find

1. $x +_\star y$.
2. $x -_\star y$.
3. $4 \cdot_\star x$.
4. $5 \cdot_\star y$.
5. $\langle x, y \rangle_\star$.
6. $|x|_\star$.
7. $|y|_\star$.

**Definition 1.30:** For $x, y \in \mathbb{R}^3_\star$, define the multiplicative distance between $x$ and $y$ as follows:

$$d_\star(x, y) = |x -_\star y|_\star.$$

The symbol $E^3_\star$ will be used to denote $\mathbb{R}^3_\star$ equipped with the multiplicative distance $d_\star$.

---

## 1.11 The Multiplicative Cross Product

**Definition 1.31:** Let $u = (u_1, u_2, u_3)$, $v = (v_1, v_2, v_3) \in E^3_\star$. Then we define the multiplicative cross product of $u$ and $v$, denoted by $u \times_\star v$, to be the vector

$$u \times_\star v = \left( e^{\log u_2 \log v_3 - \log v_2 \log u_3}, e^{\log v_1 \log u_3 - \log u_1 \log v_3}, e^{\log u_1 \log v_2 - \log v_1 \log u_2} \right).$$

**Example 1.21:** Let

$$u = (e, e^2, e^3),$$
$$v = (e^2, e^4, e^6).$$

Then

$$u \times_\star v = (e^{\log e^2 \log e^6 - \log e^4 \log e^3}, e^{\log e^2 \log e^3 - \log e \log e^6}, e^{\log e \log e^4 - \log e^2 \log e^2})$$
$$= (e^{12-12}, e^{6-6}, e^{4-4})$$
$$= (1, 1, 1).$$

**Exercise 1.15:** Let

$$u = (e^3, e, e^4),$$
$$v = (e^8, e, e^{15}).$$

Find $u \times_\star v$.

Below, we will deduct some of the properties of the multiplicative cross product. Suppose that

$$u = (u_1, u_2, u_3), \quad v = (v_1, v_2, v_3), \quad w = (w_1, w_2, w_3), \quad z = (z_1, z_2, z_3) \in E_\star^3.$$

Let also, $a \in \mathbb{R}_\star$. Then we have the following.

1. $\langle u \times_\star v, u \rangle_\star = \langle u \times_\star v, v \rangle_\star = 0_\star$.

   *Proof.* We have

   $$\log u_1 \log u_2 \log v_3 - \log u_1 \log v_2 \log u_3 + \log u_2 \log v_1 \log u_3$$
   $$- \log u_2 \log u_1 \log v_3 + \log u_3 \log u_1 \log v_2 - \log u_3 \log v_1 \log u_2$$
   $$= 0$$

   and

   $$\langle u \times_\star v, u \rangle_\star = e^0$$
   $$= 1$$
   $$= 0_\star.$$

   Next,

   $$\log v_1 \log u_2 \log v_3 - \log v_1 \log v_2 \log u_3 + \log v_2 \log v_1 \log u_3$$
   $$- \log v_2 \log u_1 \log u_3 + \log v_3 \log u_1 \log v_2 - \log v_3 \log v_1 \log u_2$$
   $$= 0$$

   and

$$\langle u \times_\star v, v \rangle_\star = e^0$$
$$= 1$$
$$= 0_\star.$$

This completes the proof.

2. $u \times_\star v = -_\star v \times_\star u$.

*Proof.* We have

$$v \times_\star u = \left(e^{\log v_2 \log u_3 - \log u_2 \log v_3}, e^{\log u_1 \log v_3 - \log v_1 \log u_3}, e^{\log v_1 \log u_2 - \log u_1 \log v_2}\right)$$

$$= \left(e^{-(\log u_2 \log v_3 - \log v_2 \log u_3)}, e^{-(\log v_1 \log u_3 - \log u_1 \log v_3)}, e^{-(\log u_1 \log v_2 - \log v_1 \log u_2)}\right)$$

$$= \left(-_\star e^{\log u_2 \log v_3 - \log v_2 \log u_3}, -_\star e^{\log v_1 \log u_3 - \log u_1 \log v_3}, -_\star e^{\log u_1 \log v_2 - \log v_1 \log u_2}\right)$$

$$= -_\star \left(e^{\log u_2 \log v_3 - \log v_2 \log u_3}, e^{\log v_1 \log u_3 - \log u_1 \log v_3}, e^{\log u_1 \log v_2 - \log v_1 \log u_2}\right)$$

$$= -_\star u \times_\star v.$$

This completes the proof.

3. $\langle u \times_\star v, w \rangle_\star = \langle u, v \times_\star w \rangle_\star$.

*Proof.* We have

$$u \times_\star v = \left(e^{\log u_2 \log v_3 - \log v_2 \log u_3}, e^{\log v_1 \log u_3 - \log u_1 \log v_3}, e^{\log u_1 \log v_2 - \log v_1 \log u_2}\right)$$

and

$$\langle u \times_\star v, w \rangle_\star = \exp(\log w_1 \log u_2 \log v_3 - \log w_1 \log v_2 \log u_3$$
$$+ \log w_2 \log v_1 \log u_3 - \log w_2 \log u_1 \log v_3$$
$$+ \log w_3 \log u_1 \log v_3 - \log w_3 \log v_1 \log u_2).$$

Next,

$$v \times_\star w = \left(e^{\log v_2 \log w_3 - \log w_2 \log v_3}, e^{\log w_1 \log v_3 - \log v_1 \log w_3}, e^{\log v_1 \log w_2 - \log w_1 \log v_2}\right)$$

and

$$\langle u, v \times_\star w \rangle_\star = \exp(\log u_1 \log v_2 \log w_3 - \log u_1 \log w_2 \log v_3$$
$$+ \log u_2 \log w_1 \log v_3 - \log u_2 \log v_1 \log w_3$$
$$+ \log u_3 \log v_1 \log w_2 - \log u_3 \log w_1 \log v_2).$$

Thus,

$$\langle u \times_\star v, w \rangle_\star = \langle u, v \times_\star w \rangle_\star.$$

This completes the proof.

4. $(u \times_\star v) \times_\star w = \langle u, w \rangle_{\star \cdot_\star} v -_\star \langle v, w \rangle_{\star \cdot_\star} u.$

*Proof.* We have

$$u \times_\star v = \left( e^{\log u_2 \log v_3 - \log v_2 \log u_3}, \ e^{\log v_1 \log u_3 - \log u_1 \log v_3}, \ e^{\log u_1 \log v_2 - \log v_1 \log u_2} \right)$$

and

$$(u \times_\star v) \times_\star w = \big( e^{(\log v_1 \log u_3 - \log u_1 \log v_3)\log w_3 - \log w_2 (\log u_1 \log v_2 - \log v_1 \log u_2)}$$
$$e^{(\log u_1 \log v_2 - \log v_1 \log u_2)\log w_1 - (\log u_2 \log v_3 - \log v_2 \log u_3)\log w_3},$$
$$e^{(\log u_2 \log v_3 - \log v_2 \log u_3)\log w_2 - \log w_1 (\log v_1 \log u_3 - \log u_1 \log v_3)} \big).$$

Moreover,

$$\langle u, w \rangle_{\star \cdot_\star} v = e^{\log u_1 \log w_1 + \log u_2 \log w_2 + \log u_3 \log w_3} \cdot_\star v$$
$$= \big( e^{\log v_1 (\log u_1 \log w_1 + \log u_2 \log w_2 + \log u_3 \log w_3)},$$
$$e^{\log v_2 (\log u_1 \log w_1 + \log u_2 \log w_2 + \log u_3 \log w_3)},$$
$$e^{\log v_3 (\log u_1 \log w_1 + \log u_2 \log w_2 + \log u_3 \log w_3)} \big)$$

and

$$\langle v, w \rangle_{\star \cdot_\star} u = e^{\log v_1 \log w_1 + \log v_2 \log w_2 + \log v_3 \log w_3} \cdot_\star u$$
$$= \big( e^{\log u_1 (\log v_1 \log w_1 + \log v_2 \log w_2 + \log v_3 \log w_3)},$$
$$e^{\log u_2 (\log v_1 \log w_1 + \log v_2 \log w_2 + \log v_3 \log w_3)},$$
$$e^{\log u_3 (\log v_1 \log w_1 + \log v_2 \log w_2 + \log v_3 \log w_3)} \big).$$

Note that

$$\log v_1 (\log u_1 \log w_1 + \log u_2 \log w_2 + \log u_3 \log w_3)$$
$$- \log u_1 (\log v_1 \log w_1 + \log v_2 \log w_2 + \log v_3 \log w_3)$$
$$= (\log v_1 \log u_3 - \log u_1 \log v_3)\log w_3 + (\log u_2 \log v_1 - \log u_1 \log v_2)\log w_2$$

and

$$\log v_2 (\log u_1 \log w_1 + \log u_2 \log w_2 + \log u_3 \log w_3)$$
$$- \log u_2 (\log v_1 \log w_1 + \log v_2 \log w_2 + \log v_3 \log w_3)$$
$$= \log w_1 (\log u_1 \log v_2 - \log u_2 \log v_1) - \log w_3 (\log u_2 \log v_3 - \log u_3 \log v_2),$$

and

$$\log v_3 (\log u_1 \log w_1 + \log u_2 \log w_2 + \log u_3 \log w_3)$$
$$- \log u_3 (\log v_1 \log w_1 + \log v_2 \log w_2 + \log v_3 \log w_3)$$
$$= \log w_2 (\log u_2 \log v_3$$
$$- \log u_3 \log v_2) - \log w_1 (\log u_3 \log v_1 - \log v_3 \log u_1).$$

Therefore

$$(u \times_\star v) \times_\star w = \langle u, w \rangle_{\star \cdot \star} v -_\star \langle v, w \rangle_{\star \cdot \star} u.$$

This completes the proof.

5. $u \times_\star (a \cdot_\star u) = 0_\star$.

*Proof.* We have

$$a \cdot_\star u = \left( e^{\log a \log u_1}, e^{\log a \log u_2}, e^{\log a \log u_3} \right)$$

and

$$u \times_\star (a \cdot_\star u) = \left( e^{\log a \log u_2 \log u_3 - \log a \log u_2 \log u_3}, e^{\log a \log u_1 \log u_3 - \log a \log u_1 \log u_3}, \right.$$
$$\left. e^{\log a \log u_1 \log u_2 - \log a \log u_1 \log u_2} \right)$$
$$= (e^0, e^0, e^0)$$
$$= (1, 1, 1)$$
$$= (0_\star, 0_\star, 0_\star).$$

This completes the proof.

6.

$$\langle u \times_\star v, w \times_\star z \rangle_\star = \langle u, w \rangle_{\star \cdot \star} \langle v, z \rangle_\star$$
$$-_\star \langle v, w \rangle_{\star \cdot \star} \langle u, z \rangle_\star.$$

*Proof.* We have

$$\langle u \times_\star v, w \times_\star z \rangle_\star = \langle u, v \times_\star (w \times_\star z) \rangle_\star$$
$$= -_\star \langle u, (w \times_\star z) \times_\star v \rangle_\star$$
$$= -_\star \langle u, \langle w, v \rangle_{\star \cdot \star} z -_\star \langle z, v \rangle_{\star \cdot \star} w \rangle_\star$$
$$= -_\star \langle u, \langle w, v \rangle_{\star \cdot \star} z \rangle_\star +_\star \langle u, \langle z, v \rangle_{\star \cdot \star} w \rangle_\star$$
$$= \langle u, w \rangle_{\star \cdot \star} \langle z, v \rangle_\star -_\star \langle u, z \rangle_{\star \cdot \star} \langle w, v \rangle_\star.$$

This completes the proof.

7. $|u \times_\star v|_\star^{2\star} = |u|_\star^{2\star} \cdot_\star |v|_\star^{2\star} -_\star \langle u, v \rangle_\star^{2\star}.$

*Proof.* Take $w \times_\star z = u \times v$ in the equality in the previous point and we get the desired result. This completes the proof.

8. Let

$$e_1 = (1_\star, 0_\star, 0_\star),$$
$$e_2 = (0_\star, 1_\star, 0_\star),$$
$$e_3 = (0_\star, 0_\star, 1_\star).$$

Then

$$e_1 \times_\star e_2 = e_3,$$
$$e_2 \times e_3 = e_1,$$
$$e_3 \times_\star e_1 = e_2.$$

*Proof.* We have

$$e_1 \times_\star e_2 = (e^{\log 1 \log 1 - \log e \log 1}, e^{\log 1 \log 1 - \log e \log 1}, e^{\log e \log e - \log 1 \log 1})$$
$$= (1, 1, e)$$
$$= (0_\star, 0_\star, 1_\star).$$

The other two equalities we leave to the reader as an exercise. This completes the proof.

Let $u, v \in E_\star^2$ and $u \times_\star v \neq 0_\star$. Assume that there are $\lambda, \mu, \nu \in \mathbb{R}_\star$ so that

$$\lambda \cdot_\star u +_\star \mu \cdot_\star v +_\star \nu \cdot_\star (u \times_\star v) = 0_\star. \tag{1.18}$$

We take multiplicative inner product with $u \times_* v$ of both sides of (1.18) and we get

$$v \cdot_* | u \times_* v |_*^{2_*} = 0_*.$$

Then $v = 0_*$. Hence,

$$\lambda \cdot_* (u \times_* v) = 0_*$$

and

$$\lambda \cdot_* (u \times v) = 0_*.$$

Consequently $\lambda = \mu = 0_*$.

**Exercise 1.16:** Prove that

$$(e_1 \times_* e_2) \times_* e_3 = 0_* = \langle e_1, e_3 \rangle_{*} \cdot_* e_2 -_* \langle e_2, e_3 \rangle_{*} \cdot_* e_1,$$
$$(e_2 \times_* e_3) \times_* e_3 = -_* e_2 = \langle e_2, e_3 \rangle_{*} \cdot_* e_3 -_* \langle e_3, e_3 \rangle_{*} \cdot_* e_2,$$
$$(e_3 \times_* e_1) \times_* e_3 = e_1 = \langle e_3, e_3 \rangle_{*} \cdot_* e_1 -_* \langle e_1, e_3 \rangle_{*} \cdot_* e_3.$$

---

## 1.12 Multiplicative Orthonormal Bases

**Definition 1.32:** A triple $\{u, v, w\}$ of mutually multiplicative orthogonal multiplicative unit vectors is called a multiplicative orthonormal basis.

If $\{u, v, w\}$ is a multiplicative orthonormal basis, then any $x \in E_*^3$ can be represented in the form

$$x = \langle x, u \rangle_{*} \cdot_* u +_* \langle x, v \rangle_{*} \cdot_* v +_* \langle x, w \rangle_{*} \cdot_* w.$$

**Definition 1.33:** A multiplicative plane $\alpha$ is a set of points of $E_*^3$ with the following properties.

1. $\alpha$ is not contained in any multiplicative line.
2. The multiplicative line joining any two points of $\alpha$ lies in $\alpha$.
3. Not any point in $E_*^3$ is in $\alpha$.

**Theorem 1.33:** *If $u$ is any multiplicative unit vector, then there are vectors $v$ and $w$ so that $\{u, v, w\}$ is a multiplicative orthonormal basis.*

*Proof.* Let $\xi \in E_\star^3$ be such that $\xi \neq \pm_\star u$. Take

$$v = \frac{u \times_\star \xi}{|u \times_\star \xi|_\star} \text{ and } w = u \times_\star v.$$

We have $u \perp_\star v$, $u \perp_\star w$, $v \perp_\star w$ and

$$\begin{aligned}
|w|_\star^{2_\star} &= |u \times_\star v|_\star^{2_\star} \\
&= |u|_\star^{2_\star} \cdot_\star |v|_\star^{2_\star} -_\star \langle u, v \rangle_\star^{2_\star} \\
&= 1_\star \cdot_\star 1_\star -_\star 0_\star \\
&= 1_\star.
\end{aligned}$$

This completes the proof.

---

## 1.13 Multiplicative Planes

**Definition 1.34:** A multiplicative plane is a set $\Pi_\star$ of points of $E_\star^3$ with the following properties.

1. $\Pi_\star$ is not contained in one multiplicative line.
2. The multiplicative line joining any two points of $\Pi_\star$ lies in $\Pi_\star$.
3. Not every point of $E_\star^3$ is in $\Pi_\star$.

In Fig. 1.7 it is shown a multiplicative plane.

**Theorem 1.34:** *Let $v$ and $w$ be not multiplicative proportional and $P$ be any point of $E_\star^3$. Then*

$$P +_\star [v, w]$$

*is a multiplicative plane. We will speak of the multiplicative plane through $P$ spanned by $\{v, w\}$. Here*

$$[v, w] = \{t \cdot_\star v +_\star s \cdot_\star w : t, s \in \mathbb{R}_\star\}$$

*is called the multiplicative span of $\{v, w\}$.*

*Proof.* Let

$$\alpha = P +_\star [v, w].$$

Set

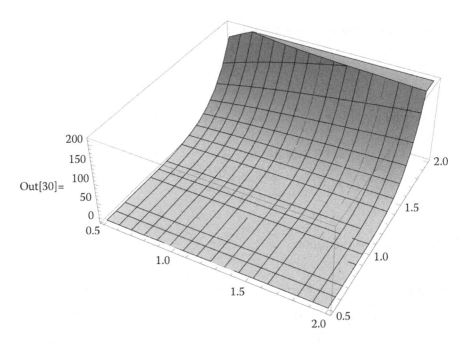

Out[30]=

**FIGURE 1.7**
A multiplicative plane.

$$Q = P +_\star v, \quad R = P +_\star w.$$

Then $Q -_\star P$ and $R -_\star P$ are not multiplicative proportional. Thus, $P$, $Q$ and $R$ are not multiplicative collinear and they do not lie on one multiplicative line. Set

$$X = P +_\star v \times_\star w.$$

Since $\{v, w, v \times_\star w\}$ is a multiplicative linear independent system, we conclude that $X \notin \alpha$. Thus, $\alpha$ does not contain any point of $E_\star^3$. Now, suppose that

$$Y = P +_\star y_1 \cdot_\star v +_\star y_2 \cdot_\star w,$$
$$Z = P +_\star z_1 \cdot_\star v +_\star z_2 \cdot_\star w$$

be two points of $\alpha$. Take $t \in \mathbb{R}_\star$ arbitrarily. Then

$$(1_\star -_\star t)\cdot_\star Y +_\star t\cdot_\star Z$$
$$= (1_\star -_\star t)\cdot_\star P +_\star y_1\cdot_\star (1_\star -_\star t)\cdot_\star v +_\star y_2\cdot_\star (1_\star -_\star t)\cdot_\star w$$
$$+_\star t\cdot_\star P +_\star t\cdot_\star z_1\cdot_\star v +_\star t\cdot_\star z_2\cdot_\star w$$
$$= P +_\star ((1_\star -_\star t)\cdot_\star y_1 +_\star t\cdot_\star z_1)\cdot_\star v$$
$$+_\star ((1_\star -_\star t)\cdot_\star y_2 +_\star t\cdot_\star z_2)\cdot_\star w.$$

Hence,

$$(1_\star -_\star t)\cdot_\star Y +_\star t\cdot_\star Z \in \alpha.$$

Therefore $\alpha$ is a multiplicative plane. This completes the proof.

**Theorem 1.35:** *Let P, Q and R be multiplicative noncollinear points. Then there exists a unique multiplicative plane through P, Q and R. In this case we speak for the multiplicative plane PQR.*

*Proof.* Let

$$v = Q -_\star P,$$
$$w = R -_\star P.$$

Then $P +_\star [v, w]$ is a multiplicative plane containg the points $P, Q$ and $R$. Assume that there is other multiplicative plane $\beta$ containing the points $P, Q$ and $R$. Then

$$P +_\star \lambda\cdot_\star v +_\star \mu\cdot_\star w = P +_\star \lambda\cdot_\star (Q -_\star P) +_\star \mu\cdot_\star (R -_\star P)$$
$$= (1_\star -_\star \lambda -_\star \mu)\cdot_\star P +_\star \lambda\cdot_\star Q +_\star \mu\cdot_\star R$$
$$= (1_\star -_\star \lambda -_\star \mu)\cdot_\star P$$
$$+_\star (\lambda +_\star \mu)\cdot_\star ((\lambda/_\star(\lambda +_\star \mu))\cdot_\star Q +_\star (\mu/_\star(\lambda +_\star \mu))\cdot_\star R).$$

Let

$$X = (\lambda/_\star(\lambda +_\star \mu))\cdot_\star Q +_\star (\mu/_\star(\lambda +_\star \mu))\cdot_\star R.$$

Thus, $X$ belongs to the multiplicative line through $Q$ and $R$. Therefore any plane containing $P, Q$ and $R$ contains $P +_\star [v, w]$. Suppose that $\beta$ contains a point $S$ that does not belong in $P +_\star [v, w]$. Then $\{S -_\star P, v, w\}$ is a multiplicative linearly independent system. Let $Z \in E_\star^3$ be arbitrarily chosen. Then, for some $\lambda, \mu, \nu \in E_\star^3$, we have

$$Z -_\star P = \lambda\cdot_\star v +_\star \mu\cdot_\star w +_\star \nu\cdot_\star (S -_\star P)$$

and

$$Z = (1_\star -_\star v)\cdot_\star P$$
$$+_\star (1_\star -_\star v)\cdot_\star ((\lambda/_\star (1_\star -_\star v))\cdot_\star v +_\star (\mu/_\star (1_\star -_\star v))\cdot_\star w).$$

Therefore any point of $E_\star^3$ is on a multiplicative line joining $S$ and a point of $\beta$, i.e., $\beta$ contains any point of $E_\star^3$. This is a contradiction. This completes the proof.

**Theorem 1.36:** *If $N$ is a multiplicative unit vector and $P \in E_\star^3$, then*

$$\{X: \langle X -_\star P, N\rangle_\star = 0_\star\}$$

*is a multiplicative plane. We speak of the multiplicative plane through $P$ and multiplicative normal $N$.*

*Proof.* By Theorem 1.33, it follows that there are vectors $v$ and $w$ so that $\{N, v, w\}$ is a multiplicative orthonormal basis. Then any point $X \in E_\star^3$ can be represented in the form

$$X -_\star P = \langle X -_\star P, N\rangle_\star \cdot_\star N +_\star \langle X -_\star P, v\rangle_\star \cdot_\star v$$
$$+_\star \langle X -_\star P, w\rangle_\star \cdot_\star w.$$

Hence, $X -_\star P$ lies in $[v, w]$ if and only if

$$\langle X -_\star P, N\rangle_\star = 0_\star.$$

This completes the proof.

---

## 1.14 Advanced Practical Problems

**Problem 1.1:** Let $x = (3, 7)$, $y = (4, 1) \in \mathbb{R}_\star^2$. Find

1. $x +_\star y$.
2. $x -_\star y$.
3. $3 \cdot_\star (x -_\star y)$.

**Answer 1.37:**

1. $(12, 7)$.

2. $\left(\frac{3}{4}, 7\right)$.

3. $\left(e^{\log 3 \log \frac{3}{4}}, e^{\log 3 \log 7}\right)$.

**Problem 1.2:** Let

$$x = (2, 3), \quad y = (4, 5).$$

Find

1. $x +_\star y$.
2. $x -_\star y$.
3. $\langle x, y \rangle_\star$.

**Answer 1.38:**

1. $(8, 15)$.
2. $\left(\frac{1}{2}, \frac{3}{5}\right)$.
3. $e^{2(\log 2)^2 + \log 3 \log 5}$.

**Problem 1.3:** Let $P = (1, 4)$, $Q = (3, 8)$ and $R = (2, 9)$. Find

1. $d_\star(P, Q)$.
2. $d_\star(P, R)$.
3. $d_\star(R, Q)$.

**Answer 1.39:**

1. $e^{\left((\log 3)^2 + (\log 2)^2\right)^{\frac{1}{2}}}$.

2. $e^{\left((\log 2)^2 + \left(\log \frac{9}{4}\right)^2\right)^{\frac{1}{2}}}$.

3. $e^{\left(\left(\log \frac{3}{2}\right)^2 + \left(\log \frac{8}{9}\right)^2\right)^{\frac{1}{2}}}$.

**Problem 1.4:** Find the equation of the multiplicative line through $P(7, 6)$ and multiplicative direction $v = (8, 5)$.

**Answer 1.40:**

$$\left(\frac{X_1}{7}\right)^{\log 5} = \left(\frac{X_2}{6}\right)^{\log 8}.$$

**Problem 1.5:** Find the equation of the multiplicative line $l$ through $P(1, 4)$ with multiplicative normal vector $N = (7, 2)$.

**Answer 1.41:**

$$X_1^{\log 7}\left(\frac{X_2}{4}\right)^{\log 2} = 1.$$

**Problem 1.6:** Let $P(6, 7)$, $Q(8, 9)$, $R(2, e)$, $S(1, 4)$, $T(2, 11)$. Let also, $l_1, l_2, l_3, l_4, l_5$ and $l_6$ are multiplicative lines so that

$$P, Q \in l_1; \ P, R \in l_2; \ P, S \in l_3; \ P, T \in l_4; \ R, S \in l_5; \ S, T \in l_6.$$

Find the equations of the multiplicative lines $l_1, l_2, l_3, l_4, l_5$ and $l_6$.

**Answer 1.42:**

1.

$$l_1: \ \left(\frac{X_1}{6}\right)^{\log \frac{9}{7}} = \left(\frac{X_2}{7}\right)^{\log \frac{4}{3}}.$$

2.

$$l_2: \ \left(\frac{X_1}{6}\right)^{1-\log 7} = \left(\frac{X_2}{7}\right)^{-\log 3}.$$

3.

$$l_3: \ \left(\frac{X_1}{6}\right)^{\log \frac{4}{7}} = \left(\frac{X_2}{7}\right)^{-\log 6}.$$

4.

$$l_4: \ \left(\frac{X_1}{6}\right)^{\log \frac{11}{7}} = \left(\frac{X_2}{7}\right)^{-\log 3}.$$

5.

$$l_5: \left(\frac{X_1}{2}\right)^{2\log 2 - 1} = \left(\frac{X_2}{e}\right)^{-\log 2}.$$

$$l_6: (X_1)^{\log \frac{11}{4}} = \left(\frac{X_2}{4}\right)^{\log 2}.$$

**Problem 1.7:** Let the equation of the multiplicative line $l$ is

$$X_1^7 \left(\frac{X_2}{5}\right)^8 = 1.$$

Find a point $P(p_1, p_2) \in l$, a multiplicative normal vector and a multiplicative direction vector for the multiplicative line $l$.

**Answer 1.43:**

1.

$$P(1, 5).$$

2.

$$N = (e^7, e^8).$$

$$v = (e^{-8}, e^7).$$

**Problem 1.8:** Let

$$l: \left(\frac{X_1}{13}\right)^{\log 4} = X_2^{\log 8},$$

$$m: \left(\frac{X_2}{5}\right)^{\log 8} = X_1^{\log \frac{3}{4}}.$$

Find the point $F = l \cap m$.

**Answer 1.44:**

$$F(13^{\frac{2\log2}{2\log2-\log\frac{3}{4}}}\cdot5^{\frac{3\log2\log\frac{3}{4}}{2\log2-\log\frac{3}{4}}},\ 13^{3\ \frac{2\log\frac{3}{4}}{2\log2-\log\frac{3}{4}}}\cdot5^{\frac{2\log2\log\frac{3}{4}}{2\log2-\log\frac{3}{4}}}).$$

**Problem 1.9:** Check if the multiplicative lines

$$l:\ X_1^{\log18} = \left(\frac{X_2}{3}\right)^{\log11},$$

$$m:\ X_1^{\log4} = X_2^{\log15}$$

are perpendicular.

**Answer 1.45:** *No.*

**Problem 1.10:** Let $P(3, 7)$ and

$$l:\ \left(\frac{X_1}{7}\right)^{\log2}\left(\frac{X_2}{3}\right)^{\log5} = 1.$$

Find a multiplicative line $m$ so that $P \in m$ and $m \perp_* l$.

**Answer 1.46:**

$$\left(\frac{X_1}{7}\right)^{\log5} = \left(\frac{X_2}{3}\right)^{\log2}.$$

**Problem 1.11:** Let

$$l_1:\ X_1 X_2^4 = 1,$$

$$l_2:\ \left(\frac{X_1}{2}\right)^{\log2} X_2^3 = 1,$$

$$l_3:\ \left(\frac{X_1}{3}\right)^{\log4}\left(\frac{X_2}{2}\right)^{\log\frac{1}{3}} = 1.$$

Check if $l_1 \perp_* l_2$, $l_2 \perp_* l_3$, $l_1 \perp_* l_3$.

**Answer 1.47:** *No.*

**Problem 1.12:** Let $P(6, 9)$, $N = \left(e^{\frac{1}{2}}, e^{\frac{\sqrt{3}}{2}}\right)$. Find the equation of the multiplicative line $l$ through $P$ and with multiplicative unit normal vector $N$.

**Answer 1.48:**

$$\left(\frac{X_1}{6}\right)^{\frac{1}{2}}\left(\frac{X_2}{9}\right)^{\frac{\sqrt{3}}{2}} = 1.$$

**Problem 1.13:** Let

$$x = \left(\frac{1}{3}, e^{\frac{1}{4}}, 18\right),$$
$$y = (e^3, e^2, e^4).$$

Find

1. $x +_\star y$.
2. $x -_\star y$.
3. $4 \cdot_\star x$.
4. $5 \cdot_\star y$.
5. $\langle x, y \rangle_\star$.
6. $|x|_\star$.
7. $|y|_\star$.

**Problem 1.14:** Let

$$u = (4, e^2, e^8),$$
$$v = (5, e^3, e^7).$$

Find $u \times_\star v$.

# 2

---

## *Multiplicative Curves in* $\mathbb{R}^n$

---

### 2.1 Multiplicative Frenet Curves in $\mathbb{R}^n$

Let $a, b \in \mathbb{R}_\star$, $a < b$.

**Definition 2.1:** A multiplicative curve in $\mathbb{R}^n$ is a continuous function $f: [a, b] \to \mathbb{R}^n$. The function $f$ that defines the multiplicative curve is called a multiplicative parameterization of the multiplicative curve and the multiplicative curve is a multiplicative parametric curve.

**Example 2.1:** Suppose that $f: [1, 5] \to \mathbb{R}^3$ is given by

$$f(t) = (t - 1, t, t^2 - 5), \ t \in [1, 5].$$

Then $f: [1, 5] \to \mathbb{R}^3$ is a multiplicative curve. Its graphic is shown in Fig. 2.1.

**Definition 2.2:** A multiplicative regular parameterized curve is a function $f: [a, b] \to \mathbb{R}^n$ such that $f \in \mathscr{C}_\star^1([a, b])$ and

$$f^\star(t) \neq (0_\star, 0_\star, ..., 0_\star), \ t \in [a, b].$$

**Definition 2.3:** Let $f: [a, b] \to \mathbb{R}^n$ be a multiplicative parameterized curve such that $f \in \mathscr{C}_\star^1([a, b])$. Then the multiplicative vector $f^\star(t_0)$ is called the multiplicative tangent vector to $f$ at $t_0$ and the multiplicative line spanned by this multiplicative vector through $f(t_0)$ is called the multiplicative tangent to $f$ at this point.

**Example 2.2:** Let $[a, b] = [2, 4]$ and $f: [2, 4] \to \mathbb{R}^2$ is given by

$$f(t) = (t, t^3 + t^2), \ t \in [2, 4].$$

DOI: 10.1201/9781003299844-2

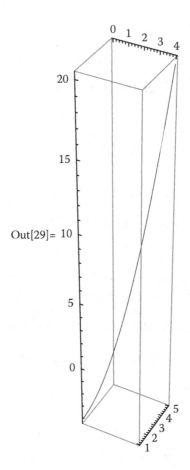

**FIGURE 2.1**
The graphic of the function $f$.

Here

$$f_1(t) = t,$$
$$f_2(t) = t^3 + t^2, \quad t \in [2, 4].$$

Hence,

$$f_1^\star(t) = e,$$
$$f_2^\star(t) = e^{\frac{3t+2}{t+1}}, \quad t \in [2, 4].$$

Thus,

$$f^\star(t) = (f_1^\star(t), f_2^\star(t))$$

$$= \left(e, e^{\frac{3t+2}{t+1}}\right)$$

$$\neq (0_\star, 0_\star), \quad t \in [2, 4].$$

Thus, the considered multiplicative curve is a multiplicative regular curve. In Fig. 2.2, they are given the graphics of $f$ and its multiplicative tangent line through $(2, 12)$.

**Example 2.3:** Let $[a, b] = \left[\frac{1}{3}, \frac{2}{3}\right]$ and $f: \left[\frac{1}{3}, \frac{2}{3}\right] \to \mathbb{R}^3$ is given by

$$f(t) = \left(t^2 - t - 4, 2t^3 - \frac{3}{2}t^2, t - t^2\right), \quad t \in \left[\frac{1}{3}, \frac{2}{3}\right].$$

Here

$$f_1(t) = t^2 - t - 4,$$
$$f_2(t) = 2t^3 - \frac{3}{2}t^2,$$
$$f_3(t) = t - t^2, \quad t \in \left[\frac{1}{3}, \frac{2}{3}\right].$$

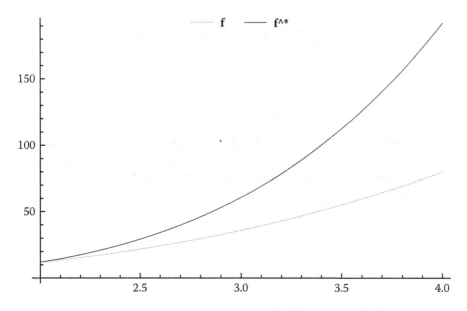

**FIGURE 2.2**
The graphic of the function $f$.

Then

$$f_1^\star(t) = e^{\frac{t(2t-1)}{t^2-t-4}},$$

$$f_2^\star(t) = e^{\frac{6t-3}{2t-\frac{3}{2}}},$$

$$f_3^\star(t) = e^{\frac{1-2t}{1-t}}, \quad t \in \left[\frac{1}{3}, \frac{2}{3}\right].$$

Hence,

$$f^\star(t) = \left(f_1^\star(t), f_2^\star(t), f_3^\star(t)\right)$$

$$= \left(e^{\frac{t(2t-1)}{t^2-t-4}}, e^{\frac{6t-3}{2t-\frac{3}{2}}}, e^{\frac{1-2t}{1-t}}\right), \quad t \in \left[\frac{1}{3}, \frac{2}{3}\right].$$

We have that

$$f^\star(t) \neq (0_\star, 0_\star, 0_\star) \text{ for } t \neq \frac{1}{2}.$$

Thus, $f$ is a multiplicative regular curve on $\left[\frac{1}{3}, \frac{1}{2}\right) \cup \left(\frac{1}{2}, \frac{2}{3}\right]$.

**Exercise 2.1:** Prove that the multiplicative curve $f: \mathbb{R}_\star \to \mathbb{R}^4$, given by

$$f(t) = \left(\frac{1}{1+t}, \frac{1}{1+2t}, \frac{1}{1+3t}, \frac{1}{1+4t}\right), \quad t \in \mathbb{R}_\star,$$

is a multiplicative regular curve.

**Definition 2.4:** Let $[\alpha, \beta] \subset \mathbb{R}_\star$. $\phi: [\alpha, \beta] \to [a, b]$ be such that $\phi \in \mathscr{C}_\star^1([\alpha, \beta])$ such that $\phi^\star(t) > 0_\star$, $t \in [\alpha, \beta]$. Let also, $f: [a, b] \to \mathbb{R}_\star$ be a multiplicative regular curve. Then the multiplicative curves $f$ and $f \circ \phi$ are said to be multiplicative equivalent.

**Example 2.4:** Let $f: \left[0, \frac{2}{3}\right] \to \mathbb{R}^2$ is given by

$$f(t) = (1+t, t^2), \quad t \in \left[0, \frac{2}{3}\right].$$

Let also, $\phi: [0, 2] \to \left[0, \frac{2}{3}\right]$ is given by

$$\phi(t) = \frac{t}{1+t}, \quad t \in [0, 2].$$

Then $\phi\colon [0, 2] \to \left[0, \frac{2}{3}\right]$ and

$$\phi^\star(t) = e^{t\left(\frac{1}{(1+t)^2}\right)\frac{1}{1+t}}$$
$$= e^{\frac{1}{1+t}}, \ t \in [0, 2].$$

Next,

$$f \circ \phi(t) = \left(1 + \frac{t}{1+t}, \frac{t^2}{(1+t)^2}\right)$$
$$= \left(\frac{2t+1}{1+t}, \frac{t^2}{(1+t)^2}\right), \ t \in [0, 2].$$

Therefore $f$ and $f \circ \phi$ are multiplicative equivalent.

**Definition 2.5:** Let $f\colon [a, b] \to \mathbb{R}^n$ be a multiplicative regular parameterized curve and

$$f(t) = (f_1(t), f_2(t), \ldots, f_n(t)), \ t \in [a, b].$$

The multiplicative arc length parameter $L_f(t, a)$, $t \in [a, b]$, is defined as follows

$$L_f(t, a) = \int_{\star a}^{t} |f^\star(s)|_\star \cdot_\star d_\star s.$$

We have

$$L_f(t, a) = \int_{\star a}^{t} |f^\star(s)|_\star \cdot_\star d_\star s$$

$$= e^{\int_a^t \frac{1}{s} \log |f^\star(s)|_\star \, ds}$$

$$= e^{\int_a^t \frac{1}{s} e^{\left((\log f_1^\star(s))^2 + (\log f_2^\star(s))^2 + \cdots + (\log f_n^\star(s))^2\right)^{\frac{1}{2}}} ds}$$

$$= e^{\int_a^t \frac{1}{s} ((\log f_1^\star(s))^2 + (\log f_2^\star(s))^2 + \cdots + (\log f_n^\star(s))^2)^{\frac{1}{2}} ds}$$

$$= e^{\int_a^t \frac{1}{s} \left(\left(\log e^{\frac{sf_1'(s)}{f_1(s)}}\right)^2 + \left(\log e^{\frac{sf_2'(s)}{f_2(s)}}\right)^2 + \cdots + \left(\log e^{\frac{sf_n'(s)}{f_n(s)}}\right)^2\right)^{\frac{1}{2}} ds}$$

$$= e^{\int_a^t \frac{1}{s} \left(\frac{s^2 (f_1'(s))^2}{(f_1(s))^2} + \frac{s^2 (f_2'(s))^2}{(f_2(s))^2} + \cdots + \frac{s^2 (f_n'(s))^2}{(f_n(s))^2}\right)^{\frac{1}{2}} ds}$$

$$= e^{\int_a^t \left(\left(\frac{f_1'(s)}{f_1(s)}\right)^2 + \left(\frac{f_2'(s)}{f_2(s)}\right)^2 + \cdots + \left(\frac{f_n'(s)}{f_n(s)}\right)^2\right)^{\frac{1}{2}} ds},$$

$t \in [a, b]$.

**Example 2.5:** Let $f\colon [1, 10] \to \mathbb{R}^3$ be defined as follows

$$f(t) = \left(e^{2t^2}, e^{t^2}, e^{t^3}\right), \quad t \in [1, 10].$$

Here

$$f_1(t) = e^{2t^2},$$
$$f_2(t) = e^{t^2},$$
$$f_3(t) = e^{t^3}, \quad t \in [1, 10],$$

and

$$f_1'(t) = 4te^{2t^2},$$
$$f_2'(t) = 2te^{t^2},$$
$$f_3'(t) = 3t^2 e^{t^3},$$
$$\frac{f_1'(t)}{f_1(t)} = 4t,$$
$$\frac{f_2'(t)}{f_2(t)} = 2t,$$
$$\frac{f_3'(t)}{f_3(t)} = 3t^2, \quad t \in [1, 10].$$

Hence,

$$
\begin{aligned}
L_f(t, a) &= e^{\int_1^t (16s^2 + 4s^2 + 9s^4)^{\frac{1}{2}} ds} \\
&= e^{\int_1^t (20s^2 + 9s^4)^{\frac{1}{2}} ds} \\
&= e^{\int_1^t s(20 + 9s^2)^{\frac{1}{2}} ds} \\
&= e^{\frac{1}{18} \int_1^t (20 + 9s^2)^{\frac{1}{2}} d(20 + 9s^2)} \\
&= e^{\left. \frac{1}{18} \frac{(20 + 9s^2)^{\frac{3}{2}}}{\frac{3}{2}} \right|_{s=1}^{s=t}} \\
&= e^{\left. \frac{(20 + 9s^2)^{\frac{3}{2}}}{27} \right|_{s=1}^{s=t}} \\
&= e^{\frac{(20 + 9t^2)^{\frac{3}{2}} - 29^{\frac{3}{2}}}{27}}, \quad t \in [1, 10].
\end{aligned}
$$

**Definition 2.6:** We will say that a multiplicative regular parameterized curve is multiplicative naturally parameterized if

$$|f^\star(s)|_\star = 1_\star, \quad s \in [a, b].$$

Usually, the natural parameter is denoted by $s$.

**Theorem 2.1:** *The multiplicative arc length of any two multiplicative equivalent curves are equal.*

*Proof.* Let $f: [a, b] \to \mathbb{R}^n$ be a multiplicative regular curve, $\phi: [\alpha, \beta] \to [a, b]$, $\phi(\alpha) = a$, $\phi(\beta) = b$, $\phi \in \mathscr{C}^1_\star([\alpha, \beta])$, $\phi^\star > 0_\star$ on $[\alpha, \beta]$. Then $f$ and $f \circ \phi$ are multiplicative equivalent. Set

$$g = f \circ \phi.$$

Then

$$
\begin{aligned}
L_g(\beta, \alpha) &= \int_{\star\alpha}^\beta |g^\star(\tau)|_\star \cdot_\star d_\star\tau \\
&= \int_{\star\alpha}^\beta |f^\star(\phi(\tau))\cdot_\star \phi^\star(\tau)|_\star \cdot_\star d_\star\tau \\
&= \int_{\star\alpha}^\beta |f^\star(\phi(\tau))|_\star \cdot_\star \phi^\star(\tau)\cdot_\star d_\star\tau \\
&= \int_{\star\alpha}^\beta |f^\star(\phi(\tau))|_\star \cdot_\star d_\star\phi(\tau) \\
&= \int_{\star a}^b |f^\star(t)|_\star \cdot_\star d_\star t \\
&= L_f(b, a).
\end{aligned}
$$

This completes the proof.

**Theorem 2.2:** *For any multiplicative regular parameterized curve there is a multiplicative naturally parameterized curve that is multiplicative equivalent to it.*

*Proof.* Let $f: [a, b] \to \mathbb{R}^n$ be a multiplicative regular parameterized curve. Define

$$\phi(t) = \int_{\star a}^t |f^\star(\tau)|_\star \cdot_\star d_\star\tau, \quad t \in [a, b].$$

We have that $\phi \in \mathscr{C}^1_\star([a, b])$, $\phi > 0_\star > 0$ on $[a, b]$. and

$$
\begin{aligned}
\phi^\star(t) &= |f^\star(t)|_\star \\
&\not\leq 0_\star, \quad t \in [a, b].
\end{aligned}
$$

Hence, $\phi' > 0$ on $[a, b]$ and $\phi: [a, b] \to [\alpha, \beta]$ is a diffeomorphism for some interval $[\alpha, \beta] \subset \mathbb{R}_\star$. Note that

$$(\phi^{-1})^\star (s) = 1_\star /_\star \phi^\star (\phi^{-1}(s))$$
$$= 1_\star /_\star |f^\star (\phi^{-1}(s))|_\star , \quad s \in [\alpha, \beta].$$

Let

$$g(t) = f(\phi^{-1}(t)), \quad t \in [a, b].$$

Then

$$g^\star (t) = f^\star (\phi^{-1}(t)) \cdot_\star (\phi^{-1})^\star (t)$$
$$= f^\star (\phi^{-1}(t)) /_\star |f^\star (\phi^{-1}(t))|_\star , \quad t \in [a, b],$$

and

$$|g^\star (t)|_\star = 1_\star, \quad t \in [a, b].$$

This completes the proof.

We have

$$f(t) = g(s(t)), \quad t \in [a, b],$$

and

$$f^\star (t) = g^\star (s(t)) \cdot_\star s^\star (t)$$
$$= g^\star (s(t)) \cdot_\star |f^\star (t)|_\star , \quad t \in [a, b].$$

---

## 2.2 Analytical Representations of Curves

Let $I \subset \mathbb{R}_\star$.

**Definition 2.7:** A subset $M \subset \mathbb{R}_\star^n$ is called a multiplicative regular curve or a multiplicative one-dimensional multiplicative smooth manifold of $\mathbb{R}_\star^n$ if for each point $t_0 \in M$ there is a multiplicative regular parameterized curve

$f: I \to \mathbb{R}^n_\star$ whose support $f(I)$ is an open neighbourhood in $M$ of the point $t_0$, i.e., is a set of the form $M \cap U$, where $U$ is an neighbourhood of $t_0$ in $\mathbb{R}^n_\star$, while the map $f: I \to f(I)$ is a homeomorphism with respect to the topology of subspace of $f(I)$. A multiplicative parameterized curve with these properties is called a local parametrization of the multiplicative curve $M$ around the point $t_0$. If for a multiplicative curve $M$ there is a local parametrization which is global, i.e., $f(I) = M$, the multiplicative curve is called multiplicative simple curve.

### 2.2.1 Multiplicative Plane Curves

**Definition 2.8:** A multiplicative regular curve $M \subset \mathbb{R}^3_\star$ is called a multiplicative plane curve if it is contained in a multiplicative plane $\pi$. We shall usually assume that the multiplicative plane $\pi$ coincides with the coordinate multiplicative plane $xO_\star y$. Here $O_\star = (0_\star, 0_\star) = (1, 1)$.

#### 2.2.1.1 Parametric Representation

We choose an arbitrary local parametrization $(I, f(t)) = (f_1(t), f_2(t), f_3(t))$ of the multiplicative curve. Then the support $f(t)$ of this local parametrization is an open subset of the multiplicative curve. For a global parametrization of a multiplicative simple curve, $f(I)$ is the entire multiplicative curve. Thus, any point $t_0$ of the multiplicative curve has an open neighbourhood which is the support of the parameterized multiplicative curve

$$x = f_1(t)$$
$$y = f_2(t). \tag{2.1}$$

**Definition 2.9:** The equations (2.1) are called the parametric equations of the multiplicative curve in the neighbourhood of the point $t_0$. Usually, unless the multiplicative curve is simple, we cannot use the same set of equations to describe the points of an entire multiplicative curve.

**Example 2.6:** Let $I = [e, 20]$. Then

$$f_1(t) = t^2 + t + 1$$
$$f_2(t) = 10 + \sin t + \cos_\star t +_\star \frac{1 - t + t^2}{1 + t^2 + t^4}, \ t \in I,$$

is a parametric representation of a multiplicative plane curve.

**Example 2.7:** Let $I = [1, 4]$,

$$f_1(t) = (1 +_\star t^2 -_\star t^4)/_\star (3 +_\star t +_\star +_\star t^2_\star)$$
$$f_2(t) = (1 + t + t^2) -_\star (1 + t^8), \quad t \in I,$$

is a parametric representation of a multiplicative plane curve.

**Example 2.8:** Let $I = [3, 243]$. Then

$$f_1(t) = t,$$
$$f_2(t) = \sinh_\star t +_\star 2\cdot_\star t^{3\star}, \ t \in I,$$

is a parametric representation of a multiplicative plane curve.

### 2.2.1.2 Explicit Representation

Suppose that $I$ is an open interval in $\mathbb{R}_\star$ and $f: I \to \mathbb{R}_\star$ is a smooth function, i.e., $f \in \mathscr{C}^1_\star(I)$. Then its graph is

$$C = \{(t, f(t)): t \in I\} \qquad\qquad (2.2)$$

is a multiplicative simple curve, which has the global representation

$$x = t$$
$$y = f(t), \ t \in I.$$

**Definition 2.10:** The equation

$$y = f(x)$$

is called the explicit equation of the multiplicative curve (2.2). Sometimes for the explicit representation of a multiplicative plane curve it is, also, used the term nonparametric form.

**Example 2.9:** Let $I = (1, 128)$. Then

$$y = \frac{x + 1}{1 + x + x^2}, \ x \in I,$$

is an explicit representation of a multiplicative plane curve.

**Example 2.10:** Let $I = (0, 15)$. Then

$$y = x + 1 + \sinh_\star(x, 1), \quad x \in I,$$

is an explicit representation of a multiplicative plane curve.

**Example 2.11:** Let $I = (1, 24)$. Then

$$y = \frac{1}{1 + x^2} + \cosh_\star x, \quad x \in I,$$

is an explicit equation of a multiplicative plane curve.

### 2.2.1.3 Implicit Representation

Let $D \subset \mathbb{R}^2_\star$. Let $F: D \to \mathbb{R}_\star$ be a multiplicative smooth function and

$$C = \{(x, y) \in D: F(x, y) = 0_\star\}$$

is the $0_\star$-level set of the function $F$. In the general case, $C$ is not a multiplicative regular curve. Nevertheless, if at the point $(x_0, y_0) \in C$ the multiplicative vector gradient

$$\text{grad}_\star F(x_0, y_0) = \left( F_x^\star(x_0, y_0), F_y^\star(x_0, y_0) \right)$$

is not vanishing, then there exists an open neighbourhood $U$ of the point $(x_0, y_0)$ and a multiplicative smooth function $y = f(x)$ defined on an open neighbourhood $I \subset \mathbb{R}_\star$ of the point $x_0$ such that

$$C \cap U = \{(x, f(x)): x \in I\}.$$

If $\text{grad}\, F \neq 0_\star$ in all points of $C$, then $C$ is a multiplicative regular curve.

**Example 2.12:** Let

$$C = \{(x, y) \in [2, 4] \times [2, 4]: x^{3_\star} -_\star y^{3_\star} = 0_\star\}.$$

We have that

$$F(x, y) = x^{3_\star} -_\star y^{3_\star}$$
$$= e^{(\log x)^3 - (\log y)^3}, \quad (x, y) \in [2, 4] \times [2, 4].$$

Then

$$F_x(x, y) = \frac{3(\log x)^2}{x} e^{(\log x)^3 - (\log y)^3},$$

$$F_y(x, y) = -\frac{3(\log y)^2}{y} e^{(\log x)^3 - (\log y)^3}, \quad (x, y) \in [2, 4] \times [2, 4],$$

and

$$F_x^\star(x, y) = e^{x \frac{F_x(x,y)}{F(x,y)}}$$

$$= e^{3(\log x)^2},$$

$$F_x^\star = e^{y \frac{F_y(x,y)}{F(x,y)}}$$

$$= e^{-3(\log y)^2}, \quad (x, y) \in [2, 4] \times [2, 4].$$

Therefore

$$\mathrm{grad}_\star F(x, y) = (F_x^\star(x, y), F_y^\star(x, y))$$

$$= \left( e^{3(\log x)^2}, e^{-3(\log y)^2} \right)$$

$$\neq (0_\star, 0_\star), \quad (x, y) \in [2, 4] \times [2, 4].$$

Thus, $C$ is a multiplicative regular curve.

**Remark 2.1:** Note that the condition for multiplicative nonsingularity of the $\mathrm{grad}_\star F$ is only a sufficient condition for the equation $F(x, y) = 0$ to represent a multiplicative curve. If $\mathrm{grad}_\star F(x_0, y_0) = 0_\star$ for some $(x_0, y_0) \in D$, then we cannot claim that the equation represent a multiplicative curve in the neighbourhood of that point and the opposite.

**Example 2.13:** Let $F: \mathbb{R}_\star^2 \to \mathbb{R}_\star$ is given by

$$F(x, y) = (x -_\star y)^{2_\star}, \quad (x, y) \in \mathbb{R}_\star^2.$$

We have

$$\mathrm{grad}_\star F(x, y) = \left( e^{\frac{2\log\frac{x}{y}}{x}}, e^{-\frac{2\log\frac{x}{y}}{y}} \right), \quad (x, y) \in \mathbb{R}_\star^2.$$

Hence,

$$\mathrm{grad}_\star F(x, y) = 0_\star \text{ for } x = y, \quad x, y \in \mathbb{R}_\star.$$

If we denote

$$C = \{(x, y) \in \mathbb{R}^2 \colon F(x, y) = 0_\star\},$$

then $\mathrm{grad}_\star F(x, y) = 0_\star$ at any point $(x, y) \in C$. But, clearly, $C$ is a multiplicative curve.

### 2.2.2 Multiplicative Space Curves

#### 2.2.2.1 Parametric Representation

As in the case of plane curves, with local parametrization

$$x = f_1(t)$$
$$y = f_2(t)$$
$$z = f_3(t), \ t \in I,$$

we can represent either the entire curve, or only a neighbourhood of one of its points.

**Example 2.14:** Let $I = \mathbb{R}_\star$. Then

$$x = t + 1$$
$$y = t^2 + 1$$
$$z = \frac{1}{1+t+t^2}, \ t \in \mathbb{R}_\star,$$

is a parametric representation of a multiplicative space curve.

#### 2.2.2.2 Explicit Representation

Let $f, g \colon I \to \mathbb{R}$ be two multiplicative smooth functions on an open interval $I \subset \mathbb{R}_\star$. Then the set

$$C = \{(x, f(x), g(x)) \subset \mathbb{R}_\star^3 \colon x \in I\}$$

is a multiplicative simple curve with a global representation given by

$$x = t$$
$$y = f(t)$$
$$z = g(t), \ t \in I.$$

**Definition 2.11:** The equations
$$y = f(x)$$
$$z = g(x), \ x \in I$$

are called the explicit equations of the multiplicative curve.

**Example 2.15:** Let $I = [1, 20]$. Then

$$y = x + 1$$
$$z = x^2 + x + 1, \quad x \in I,$$

is an explicit representation of a multiplicative space curve.

### 2.2.2.3 Implicit Representation

Let $D \subset \mathbb{R}^3_*$ and $F, G: D \to \mathbb{R}$ are multiplicative smooth functions. Consider the set

$$C = \{(x, y, z) \in D: F(x, y, z) = 0_*, \ G(x, y, z) = 0_*\},$$

i.e., the set of solutions of the system

$$F(x, y, z) = 0_*$$
$$G(x, y, z) = 0_*.$$

In the general case, the set $C$ is not a multiplicative regular curve. Nevertheless, if $a = (x_0, y_0, z_0) \in C$ and

$$\mathrm{rank} \begin{pmatrix} F_x^\star(a) & F_y^\star(a) & F_z^\star(a) \\ G_x^\star(a) & G_y^\star(a) & G_z^\star(a) \end{pmatrix} = 2,$$

then there is an open neighbourhood $U \subset D$ of the point $a$ such that $C \cap U$ is a multiplicative curve. If the rank of the matrix

$$\begin{pmatrix} F_x^\star & F_y^\star & F_z^\star \\ G_x^\star & G_y^\star & G_z^\star \end{pmatrix}$$

is equal to two, then $C$ is a multiplicative curve.

**Example 2.16:** Let

$$F(x, y, z) = e^{x+y+z},$$
$$G(x, y, z) = e^{2x+3y+z}, \quad (x, y, z) \in \mathbb{R}^3_*.$$

Then

$$F_x^\star(x, y, z) = e,$$
$$F_y^\star(x, y, z) = e,$$
$$F_z^\star(x, y, z) = e, \quad (x, y, z) \in \mathbb{R}_\star^3,$$

and

$$G_x^\star(x, y, z) = e^2,$$
$$G_y^\star(x, y, z) = e^3,$$
$$G_z^\star(x, y, z) = e, \quad (x, y, z) \in \mathbb{R}^3.$$

Hence,

$$\text{rank} \begin{pmatrix} F_x^\star & F_y^\star & F_z^\star \\ G_x^\star & G_y^\star & G_z^\star \end{pmatrix} = \begin{pmatrix} e & e & e \\ e^2 & e^3 & e \end{pmatrix}$$
$$= 2.$$

Therefore

$$C = \{(x, y, z) \in \mathbb{R}_\star^2 : F(x, y, z) = 0_\star, \ G(x, y, z) = 0_\star\}$$

is a multiplicative curve.

---

## 2.3 The Multiplicative Tangent and the Multiplicative Normal Plane. The Multiplicative Normal at a Multiplicative Plane Curve

Suppose that $I \subseteq \mathbb{R}_\star$, $t_0 \in I$ and $f: I \to \mathbb{R}_\star^n$ is a multiplicative curve so that $f^\star(t_0) \neq 0_\star$.

**Definition 2.12:** The multiplicative line passing through $f(t_0)$ and having multiplicative direction of the vector $f^\star(t_0)$ is called multiplicative tangent of the multiplicative curve at the point $f(t_0)$ (or at the point $t_0$).
    The equations of the multiplicative tangent line read as follows:

$$F(\tau) = f(t_0) +_\star \ \tau \cdot_\star f^\star(t_0)$$
$$= (f_1(t_0), f_2(t_0), ..., f_n(t_0)) +_\star \ \tau \cdot_\star (f_1^\star(t_0), f_2^\star(t_0), ..., f_n^\star(t_0))$$
$$= (f_1(t_0), f_2(t_0), ..., f_n(t_0)) +_\star (\tau \cdot_\star f_1^\star(t_0), \tau \cdot_\star f_2^\star(t_0), ..., \tau \cdot_\star f_n^\star(t_0))$$
$$= (f_1(t_0) +_\star \ \tau \cdot_\star f_1^\star(t_0), f_2(t_0) +_\star \ \tau \cdot_\star f_2^\star(t_0), ..., f_n(t_0) +_\star \ \tau \cdot_\star f_n^\star(t_0))$$
$$= (f_1(t_0)e^{\log\tau \log f_1^\star(t_0)}, f_2(t_0)e^{\log\tau \log f_2^\star(t_0)}, ..., f_n(t_0)e^{\log\tau \log f_n^\star(t_0)}),$$

$\tau \in \mathbb{R}_\star.$

**Example 2.17:** Let

$$f(t) = (t^2, t^3 + t, t), \ t \in \mathbb{R}_\star.$$

Here

$$f_1(t) = t^2,$$
$$f_2(t) = t^3 + t,$$
$$f_3(t) = t, \ t \in \mathbb{R}_\star.$$

Hence,

$$f_1'(t) = 2t,$$
$$f_2'(t) = 3t^2 + 1,$$
$$f_3'(t) = 1, \quad t \in \mathbb{R}_\star,$$

and

$$f_1^\star(t) = e^{t\frac{f_1'(t)}{f_1(t)}}$$
$$= e^{t\frac{2t}{t^2}}$$
$$= e^2,$$

$$f_2^\star(t) = e^{t\frac{f_2'(t)}{f_2(t)}}$$
$$= e^{t\frac{3t^2+1}{t^3+t}}$$
$$= e^{\frac{3t^2+1}{t^2+1}},$$

$$f_3^\star(t) = e^{t\frac{f_3'(t)}{f_3(t)}}$$
$$= e^{t\frac{1}{t}}$$
$$= e, \quad t \in \mathbb{R}_\star.$$

Therefore the equation of the multiplicative line to $f$ at arbitrary point is given by

$$F(\tau) = \left( t^2 e^{2\log \tau},\ (t^3 + t)e^{\frac{3t^2+1}{t^2+1}\log \tau},\ e^{\log \tau} \right),\ \tau, t \in \mathbb{R}_\star.$$

**Example 2.18:** Let $f: \mathbb{R}_\star \to \mathbb{R}_\star^4$ be given by

$$f(t) = \left( e^{-t^3-t},\ e^{-3t},\ e^{-t+1},\ e \right),\ t \in \mathbb{R}_\star.$$

Here

$$f_1(t) = e^{-t^2-t},$$
$$f_2(t) = e^{-3t},$$
$$f_3(t) = e^{-t+1},$$
$$f_4(t) = e,\ t \in \mathbb{R}_\star.$$

We have

$$f_1'(t) = (-2t-1)e^{-t^2-t},$$
$$f_2'(t) = -3e^{-3t},$$
$$f_3'(t) = -e^{-t+1},$$
$$f_4'(t) = 0,\ t \in \mathbb{R}_\star.$$

Hence,

$$f_1^\star(t) = e^{t\frac{f_1'(t)}{f_1(t)}}$$
$$= e^{-2t^2-t},$$
$$f_2^\star(t) = e^{t\frac{f_2'(t)}{f_2(t)}}$$
$$= e^{-3t},$$
$$f_3^\star(t) = e^{t\frac{f_3'(t)}{f_3(t)}}$$
$$= e^{-t},$$
$$f_4^\star(t) = e^{t\frac{f_4'(t)}{f_4(t)}}$$
$$= 1,\quad t \in \mathbb{R}_\star.$$

Thus, the equation of the multiplicative tangent line to $f$ at arbitrary point is given by

$$F(\tau) = \left(e^{-t^2-t-t(2t+1)\log\tau}, e^{-3t-3t\log\tau}, e^{-t+1-t\log\tau}, e\right),$$

$t, \tau \in \mathbb{R}_\star.$

**Exercise 2.2:** Let $f: \mathbb{R}_\star \to \mathbb{R}_\star^2$ be given by

$$f(t) = \left(\frac{t+1}{t+2}, t^2\right), \quad t \in \mathbb{R}_\star.$$

Find the equation of the multiplicative tangent line at arbitrary point.

**Answer 2.3:**

$$F(\tau) = \left(\frac{t+1}{t+2}e^{\log\tau\log\frac{t}{(t+1)(t+2)}}, t^2e^{2\log\tau}\right), \quad t, \tau \in \mathbb{R}_\star.$$

**Theorem 2.4:** *The multiplicative tangent vectors of two multiplicative equivalent curves are multiplicative collinear at corresponding points and the multiplicative tangent lines coincide.*

*Proof.* Let $(I, f)$ and $(J, g)$ be two multiplicative curves that are multiplicative equivalent. Let $s: I \to J$ be the parameter change. Then

$$f(t) = g(s(t)), \quad t \in I,$$

and

$$f^\star(t) = g^\star(s(t)) \cdot_\star s^\star(t), \quad t \in I.$$

This completes the proof.

Now, suppose that $f: I \to \mathbb{R}_\star^n$ is a multiplicative regular curve. For $h$, enough close to 1, or $h \to 0_\star$, by Taylor's formula, we have

$$f(t_0 +_\star h) = f(t_0) +_\star h \cdot_\star f^\star(t_0) +_\star h \cdot_\star \varepsilon,$$

where $\varepsilon \to 0_\star$, as $h \to 0_\star$. Let $l$ be an arbitrary multiplicative line passing through $f(t_0)$ and having multiplicative unit direction vector $m$. Set

$$d_\star(h) = d_\star(f(t_0 +_\star h), l).$$

**Theorem 2.5:** *The multiplicative line $l$ is the multiplicative tangent line to the multiplicative regular parameterized curve $f$ at the point $t_0$ if and only if*

$$\lim_{h \to 0_\star} (d_\star(h))/_\star |h|_\star = 0_\star.$$

*Proof.* We have

$$\begin{aligned}
d_\star(h) &= |(f(t_0 +_\star h) -_\star f(t_0)) \times_\star m|_\star \\
&= |(h\cdot_\star f^\star(t_0) +_\star h\cdot_\star \varepsilon) \times_\star m|_\star \\
&= |h\cdot_\star (f^\star(t_0) \times_\star m) +_\star h\cdot_\star (\varepsilon \times_\star m)|_\star \\
&= |h|_\star \cdot_\star |(f^\star(t_0) \times_\star m) +_\star (\varepsilon \times_\star m)|_\star
\end{aligned}$$

and

$$d_\star(h)/_\star |h|_\star = |(f^\star(t_0) \times_\star m) +_\star (\varepsilon \times_\star m)|_\star.$$

Hence, using that $\varepsilon \to 0_\star$, $|(f^\star(t_0) \times_\star m) +_\star (\varepsilon \times_\star m)|_\star \to |f^\star(t_0) \times_\star m|_\star$, as $h \to 0_\star$, we find

$$\begin{aligned}
\lim_{h \to 0_\star} d_\star(h)/_\star |h|_\star &= \lim_{h \to 0_\star} |(f^\star(t_0) \times_\star m) +_\star (\varepsilon \times_\star m)|_\star \\
&= |f^\star(t_0) \times_\star m|_\star.
\end{aligned}$$

1. Let $l$ be the multiplicative tangent line to $f$ at $t_0$. Then $f^\star(t_0)$ and $m$ are multiplicative collinear and

$$|f^\star(t_0) \times_\star m|_\star = 0_\star.$$

Hence,

$$\lim_{h \to 0_\star} d_\star(h)/_\star |h|_\star = 0_\star.$$

2. Now, suppose that (2.3) holds. Then

$$|f^\star(t_0) \times_\star m|_\star = 0_\star$$

and $f^\star(t_0)$ and $m$ are multiplicative collinear. Thus, $l$ is a multiplicative tangent line to $f$ at $t_0$. This completes the proof.

**Definition 2.13:** Let $(I, f)$ be a multiplicative parametric curve and $t_0 \in I$. The multiplicative normal plane at $f(t_0)$ is the multiplicative plane through

$f(t_0)$ that is multiplicative perpendicular to the multiplicative tangent line to the multiplicative curve at the point $f(t_0)$.

The equation for the multiplicative normal plane is as follows

$$(R(t) -_\star f(t_0)) \cdot_\star f^\star(t_0) = 0_\star, \ t \in I.$$

We have

$$0_\star = e^0$$
$$= ((R_1(t), R_2(t), \ldots, R_n(t)) -_\star (f_1(t_0), f_2(t_0), \ldots, f_n(t_0)))$$
$$\cdot_\star (f_1^\star(t_0), f_2^\star(t_0), \ldots, f_n^\star(t_0))$$
$$= (R_1(t) -_\star f_1(t_0), R_2(t) -_\star f_2(t_0), \ldots, R_n(t) -_\star f_n(t_0))$$
$$\cdot_\star (f_1^\star(t_0), f_2^\star(t_0), \ldots, f_n^\star(t_0))$$
$$= \left( \frac{R_1(t)}{f_1(t_0)}, \frac{R_2(t)}{f_2(t_0)}, \ldots, \frac{R_n(t)}{f_n(t_0)} \right) \cdot_\star (f_1^\star(t_0), f_2^\star(t_0), \ldots, f_n^\star(t_0))$$
$$= e^{\log \frac{R_1(t)}{f_1(t_0)} \log f_1^\star(t_0) + \log \frac{R_2(t)}{f_2(t_0)} \log f_2^\star(t_0) + \cdots + \log \frac{R_n(t)}{f_n(t_0)} \log f_n^\star(t_0)},$$

$t \in I$, i.e.,

$$\log \frac{R_1(t)}{f_1(t_0)} \log f_1^\star(t_0) + \log \frac{R_2(t)}{f_2(t_0)} \log f_2^\star(t_0) + \cdots + \log \frac{R_n(t)}{f_n(t_0)} \log f_n^\star(t_0) = 0,$$

$$t \in I.$$

**Example 2.19:** Let $f \colon \mathbb{R}_\star \to \mathbb{R}_\star^3$ is defined by

$$f(t) = \left( t^2, e^{-t^2+1}, e^{-t+2} \right), \ t \in \mathbb{R}_\star.$$

Here

$$f_1(t) = t^2,$$
$$f_2(t) = e^{-t^2+1},$$
$$f_3(t) = e^{-t+2}, \ t \in \mathbb{R}_\star.$$

Hence,

$$f_1'(t) = 2t,$$
$$f_2'(t) = -2te^{-t^2+1},$$
$$f_3'(t) = -e^{-t+2}, \quad t \in \mathbb{R}_\star,$$

and

$$f_1^\star(t) = e^{t\frac{f_1'(t)}{f_1(t)}}$$
$$= e^2,$$
$$f_2^\star(t) = e^{t\frac{f_2'(t)}{f_2(t)}}$$
$$= e^{-2t^2},$$
$$f_3^\star(t) = e^{t\frac{f_3'(t)}{f_3(t)}}$$
$$= e^{-t}, \quad t \in \mathbb{R}_\star.$$

Take $t_0 \in I$ arbitrarily. Then the equation of the multiplicative normal plane to $f$ at $f(t_0)$ is

$$\log\frac{R_1(t)}{t_0^2}\log e^2 + \log\frac{R_2(t)}{e^{-t_0^2+1}}\log e^{-2t_0^2} + \log\frac{R_3(t)}{e^{-t_0+2}}\log e^{-t_0} = 0, \quad t \in \mathbb{R}_\star,$$

or

$$2\log\frac{R_1(t)}{t_0^2} - 2t_0^2\log\frac{R_2(t)}{e^{-t_0^2+1}} - t_0\log\frac{R_3(t)}{e^{-t_0+2}} = 0, \quad t \in \mathbb{R}_\star.$$

**Example 2.20:** Let $f\colon \mathbb{R}_\star \to \mathbb{R}_\star^4$ is defined by

$$f(t) = \left(t, e^{-t}, e^{-t^2}, e^{-t^3}\right), \quad t \in \mathbb{R}_\star.$$

Here

$$f_1(t) = t,$$
$$f_2(t) = e^{-t},$$
$$f_3(t) = e^{-t^2},$$
$$f_4(t) = e^{-t^3}, \quad t \in \mathbb{R}_\star.$$

Hence,

$$f_1'(t) = 1,$$
$$f_2'(t) = -e^{-t},$$
$$f_3'(t) = -2te^{-t^2},$$
$$f_4'(t) = -3t^2e^{-t^3}, \quad t \in \mathbb{R}_\star,$$

and

$$f_1^\star(t) = e^{t\frac{f_1'(t)}{f_1(t)}}$$
$$= e,$$
$$f_2^\star(t) = e^{t\frac{f_2'(t)}{f_2(t)}}$$
$$= e^{-t},$$
$$f_3^\star(t) = e^{t\frac{f_3'(t)}{f_3(t)}}$$
$$= e^{-2t^2},$$
$$f_4^\star(t) = e^{t\frac{f_4'(t)}{f_4(t)}}$$
$$= e^{-3t^3}, \quad t \in \mathbb{R}_\star.$$

Then the equation of the multiplicative normal plane to $f$ at the point $f(t_0)$, where $t_0 \in \mathbb{R}_\star$ is arbitrarily chosen, is

$$\log \frac{R_1(t)}{t_0} \log e + \log \frac{R_2(t)}{e^{-t_0}} \log e^{-t_0} + \log \frac{R_3(t)}{e^{-t_0^2}} \log e^{-2t_0^2}$$
$$+ \log \frac{R_4(t)}{e^{-t_0^3}} \log e^{-3t_0^3} = 0,$$

$t \in \mathbb{R}_\star$, or

$$\log \frac{R_1(t)}{t_0} - t_0 \log \frac{R_2(t)}{e^{-t_0}} - 2t_0^2 \log \frac{R_3(t)}{e^{-t_0^2}} - 3t_0^3 \log \frac{R_4(t)}{e^{-t_0^3}} = 0,$$

$t \in I$.

**Exercise 2.3:** Let $f: \mathbb{R}_\star \to \mathbb{R}_\star^3$ is given by

$$f(t) = (e^t, e^{2t}, e^{3t}), \quad t \in \mathbb{R}_\star.$$

Find the equation of the multiplicative normal plane to $f$ at arbitrarily chosen point $f(t_0)$, $t_0 \in \mathbb{R}_\star$.

**Answer 2.6:**

$$\log\frac{R_1(t)}{e} + 2\log\frac{R_2(t)}{e^2} + 3\log\frac{R_3(t)}{e^3} = 0, \ t \in \mathbb{R}_\star.$$

1. Let $l$ be the multiplicative tangent line to $f$ at $t_0$. Then $f^\star(t_0)$ and $m$ are multiplicative collinear and

$$|f^\star(t_0) \times_\star m|_\star = 0_\star.$$

Hence,

$$\lim_{h\to 0_\star} d_\star(h)/_\star |h|_\star = 0_\star. \tag{2.3}$$

2. Now, suppose that (2.3) holds. Then

---

## 2.4 Multiplicative Osculating Plane

**Definition 2.14:** A multiplicative parameterized curve $f = f(t)$, $t \in I$, is said to be multiplicative biregular at the point $t_0$ if the multiplicative vectors $f^\star(t_0)$ and $f^{\star\star}(t_0)$ are not multiplicative collinear, i.e., if

$$f^\star(t_0) \times_\star f^{\star\star}(t_0) \neq 0_\star.$$

**Example 2.21:** Let $f\colon \mathbb{R}_\star \to \mathbb{R}_\star^3$ is given by

$$f(t) = (t^2, e^t, e^{2t}), \ t \in \mathbb{R}_\star.$$

Here

$$f_1(t) = t^2,$$
$$f_2(t) = e^t,$$
$$f_3(t) = e^{2t}, \ t \in \mathbb{R}_\star.$$

Hence,

$$f_1'(t) = 2t,$$
$$f_2'(t) = e^t,$$
$$f_3'(t) = 2e^{2t}, \ t \in \mathbb{R}_\star,$$

and

$$f_1^\star(t) = e^{t\frac{f_1'(t)}{f_1(t)}}$$
$$= e^2,$$
$$f_2^\star(t) = e^{t\frac{f_2'(t)}{f_2(t)}}$$
$$= e^t,$$
$$f_3^\star(t) = e^{t\frac{f_3'(t)}{f_3(t)}}$$
$$= e^{2t},$$
$$f_1^{\star\star}(t) = 1,$$
$$f_2^{\star\star}(t) = e^t,$$
$$f_3^{\star\star}(t) = e^{2t}, \ t \in \mathbb{R}_\star.$$

Therefore

$$f^\star(t) = (e^2, e^t, e^{2t}),$$
$$f^{\star\star}(t) = (1, e^t, e^{2t}), \ t \in \mathbb{R}_\star,$$

and

$$f^\star(t) \times_\star f^{\star\star}(t) = (e^{\log e^t \log e^{2t} - \log e^t \log e^{2t}},$$
$$e^{\log e^{2t} \log 1 - \log e^2 \log e^{2t}},$$
$$e^{\log e^2 \log e^t - \log 1 \log e^t})$$
$$= (e^0, e^{-4t}, e^{2t})$$
$$\neq (0_\star, 0_\star, 0_\star)$$

for any $t \in \mathbb{R}_\star$. Thus, the considered multiplicative curve is a multiplicative biregular curve.

**Definition 2.15:** Let $(I, f)$ be a multiplicative parameterized curve and $t_0 \in I$. The multiplicative osculating plane through $f(t_0)$ that is multiplicative parallel to $f^\star(t_0)$ and $f^{\star\star}(t_0)$ is defined by

$$0_\star = \langle F(t) - f(t_0), f^\star(t_0) \times_\star f^{\star\star}(t_0) \rangle_\star, \ t \in I.$$

We have

$$
\begin{aligned}
F(t) \to_\star f(t_0) &= (F_1(t), F_2(t), F_3(t)) \to_\star (f_1(t_0), f_2(t_0), f_3(t_0)) \\
&= (F_1(t) \to_\star f_1(t_0), F_2(t) \to_\star f_2(t_0), F_3(t) \to_\star f_3(t_0)) \\
&= \left( \frac{F_1(t)}{f_1(t_0)}, \frac{F_2(t)}{f_2(t_0)}, \frac{F_3(t)}{f_3(t_0)} \right), \quad t \in I,
\end{aligned}
$$

and

$$
f^\star(t_0) = \left( f_1^\star(t_0), f_2^\star(t_0), f_3^\star(t_0) \right),
$$

$$
f^{\star\star}(t_0) = \left( f_1^{\star\star}(t_0), f_2^{\star\star}(t_0), f_3^{\star\star}(t_0) \right),
$$

and

$$
\begin{aligned}
f^\star(t_0) \times_\star f^{\star\star}(t_0) = \big( &e^{\log f_2^\star(t_0)\log f_3^{\star\star}(t_0) - \log f_2^{\star\star}(t_0)\log f_3^\star(t_0)}, \\
&e^{\log f_3^\star(t_0)\log f_1^{\star\star}(t_0) - \log f_3^{\star\star}(t_0)\log f_1^\star(t_0)}, \\
&e^{\log f_1^\star(t_0)\log f_2^{\star\star}(t_0) - \log f_1^{\star\star}(t_0)\log f_2^\star(t_0)} \big).
\end{aligned}
$$

Hence,

$$
\begin{aligned}
0 = \ &\log \tfrac{F_1(t)}{f_1(t_0)} \left( \log f_2^\star(t_0)\log f_3^{\star\star}(t_0) - \log f_2^{\star\star}(t_0)\log f_3^\star(t_0) \right) \\
&+ \log \tfrac{F_2(t)}{f_2(t_0)} \left( \log f_3^\star(t_0)\log f_1^{\star\star}(t_0) - \log f_3^{\star\star}(t_0)\log f_1^\star(t_0) \right) \\
&+ \log \tfrac{F_3(t)}{f_3(t_0)} \left( \log f_1^\star(t_0)\log f_2^{\star\star}(t_0) - \log f_1^{\star\star}(t_0)\log f_2^\star(t_0) \right),
\end{aligned}
$$

$t \in I$, or

$$
0_\star =
\begin{vmatrix}
F_1(t) \to_\star f_1(t_0) & F_2(t) \to_\star f_2(t_0) & F_3(t) \to_\star f_3(t_0) \\
f_1^\star(t_0) & f_2^\star(t_0) & f_3^\star(t_0) \\
f_1^{\star\star}(t_0) & f_2^{\star\star}(t_0) & f_3^{\star\star}(t_0)
\end{vmatrix}, \quad t \in I,
$$

is the equation of the multiplicative osculating plane.

**Example 2.22:** Let $f\colon \mathbb{R}_\star \to \mathbb{R}^3_\star$, is given by

$$f(t) = (t, t^2, t + t^2),\ t \in \mathbb{R}_\star.$$

Here

$$f_1(t) = t,$$
$$f_2(t) = t^2,$$
$$f_3(t) = t + t^2,\ t \in \mathbb{R}_\star.$$

Then

$$f'_1(t) = 1,$$
$$f'_2(t) = 2t,$$
$$f'_3(t) = 1 + 2t,\quad t \in \mathbb{R}_\star,$$

and

$$f^\star_1(t) = e^{t\frac{f'_1(t)}{f_1(t)}}$$
$$= e,$$
$$f^\star_2(t) = e^{t\frac{f'_2(t)}{f_2(t)}}$$
$$= e^2,$$
$$f^\star_3(t) = e^{t\frac{f'_3(t)}{f_3(t)}}$$
$$= e^{t\frac{(1+2t)}{t+t^2}}$$
$$= e^{\frac{1+2t}{1+t}},\ t \in \mathbb{R}_\star,$$

and

$$(f^\star_1)'(t) = 0,$$
$$(f^\star_2)'(t) = 0,$$
$$(f^\star_3)'(t) = \frac{2(1+t)-(1+2t)}{(1+t)^2}e^{\frac{1+2t}{1+t}}$$
$$= \frac{1}{(1+t)^2}e^{\frac{1+2t}{1+t}},\ t \in \mathbb{R}_\star,$$

and

$$f_1^{\star\star}(t) = e^{t\frac{(f_1^\star)'(t)}{f_1^\star(t)}}$$
$$= 1,$$
$$f_2^{\star\star}(t) = e^{t\frac{(f_2^\star)'(t)}{f_2^\star(t)}}$$
$$= 1,$$
$$f_3^{\star\star}(t) = e^{t\frac{(f_3^\star)'(t)}{f_3^\star(t)}}$$
$$= e^{\frac{t}{(1+t)^2}}, \quad t \in \mathbb{R}_\star.$$

Therefore

$$f^\star(t) = (f_1^\star(t), f_2^\star(t), f_3^\star(t))$$
$$= \left( e, e^2, e^{\frac{1+2t}{1+t}} \right),$$
$$f^{\star\star}(t) = (f_1^{\star\star}(t), f_2^{\star\star}(t), f_3^{\star\star}(t))$$
$$= \left( 1, 1, e^{\frac{t}{(1+t)^2}} \right), \quad t \in \mathbb{R}_\star,$$

and the equation of the multiplicative osculating plane is

$$0 = \log\frac{F_1(t)}{t_0}\left( \frac{2t_0}{(1+t_0)^2} - 0 \right) + \log\frac{F_2(t)}{t_0^2}\left( 0 - \frac{t_0}{(1+t_0)^2} \right)$$
$$+ \log\frac{F_3(t)}{t_0 + t_0^2}(0 - 0)$$
$$= \frac{2t_0}{(1+t_0)^2}\log\frac{F_1(t)}{t_0} - \frac{t_0}{(1+t_0)^2}\log\frac{F_2(t)}{t_0^2}, \quad t, t_0 \in \mathbb{R}_\star.$$

**Example 2.23:** Let $f\colon \mathbb{R}_\star \to \mathbb{R}_\star^3$ is given by

$$f(t) = \left( e^t, e^{t^2}, e^{t^3} \right), \quad t \in \mathbb{R}_\star.$$

Here

$$f_1(t) = e^t,$$
$$f_2(t) = e^{t^2},$$
$$f_3(t) = e^{t^3}, \quad t \in \mathbb{R}_\star.$$

Hence,

$$f_1'(t) = e^t,$$
$$f_2'(t) = 2te^{t^2},$$
$$f_3'(t) = 3t^2e^{t^3}, \quad t \in \mathbb{R}_\star,$$

and

$$f_1^\star(t) = e^{t\frac{f_1'(t)}{f_1(t)}}$$
$$= e^t,$$
$$f_2^\star(t) = e^{t\frac{f_2'(t)}{f_2(t)}}$$
$$= e^{2t^2},$$
$$f_3^\star(t) = e^{t\frac{f_3'(t)}{f_3(t)}}$$
$$= e^{3t^3}, \quad t \in \mathbb{R}_\star,$$

and

$$\left(f_1^\star\right)'(t) = e^t,$$
$$\left(f_2^\star\right)'(t) = 4te^{2t^2},$$
$$\left(f_3^\star\right)'(t) = 9t^2e^{3t^3}, \quad t \in \mathbb{R}_\star,$$

and

$$f_1^{\star\star}(t) = e^{t\frac{(f_1^\star)'(t)}{f_1^\star(t)}}$$
$$= e^t,$$
$$f_2^{\star\star}(t) = e^{t\frac{(f_2^\star)'(t)}{f_2^\star(t)}}$$
$$= e^{4t^2},$$
$$f_3^{\star\star}(t) = e^{t\frac{(f_3^\star)'(t)}{f_3^\star(t)}}$$
$$= e^{9t^3}, \quad t \in \mathbb{R}_\star.$$

Therefore

$$f^\star(t) = \left(f_1^\star(t), f_2^\star(t), f_3^\star(t)\right)$$

$$= (e^t, e^{2t^2}, e^{3t^3}),$$

$$f^{\star\star}(t) = \left(f_1^{\star\star}(t), f_2^{\star\star}(t), f_3^{\star\star}(t)\right)$$

$$= (e^t, e^{4t^2}, e^{9t^3}), \quad t \in \mathbb{R}_\star,$$

and the equation of the multiplicative osculating plane is

$$0 = \frac{F_1(t)}{e^{t_0}} (18t_0^5 - 12t_0^5) + \log\frac{F_2(t)}{e^{t_0^2}} (3t_0^4 - 9t_0^4)$$

$$+ \log\frac{F_3(t)}{e^{t_0^3}} (4t_0^3 - 2t_0^3)$$

$$= 6t_0^5 \log\frac{F_1(t)}{e^{t_0}} - 6t_0^4 \log\frac{F_2(t)}{e^{t_0^2}} + 2t_0^3 \log\frac{F_3(t)}{e^{t_0^3}}, \quad t, t_0 \in \mathbb{R}_\star,$$

or

$$0 = 3t_0^2 \log\frac{F_1(t)}{e^{t_0}} - 3t_0^2 \log\frac{F_2(t)}{e^{t_0^2}} + \log\frac{F_3(t)}{e^{t_0^3}}, \quad t, t_0 \in \mathbb{R}_\star.$$

**Exercise 2.4:** Let $f\colon \mathbb{R}_\star \to \mathbb{R}_\star^3$ is given by

$$f(t) = \left(e^{t^2}, e^{t^3}, e^{t^4}\right), \quad t \in \mathbb{R}_\star.$$

Find the equation of the multiplicative osculating plane at arbitrary point $t_0 \in \mathbb{R}_\star$.

**Answer 2.7:**

$$0 = 6t_0^2 \log\frac{F_1(t)}{e^{t_0^2}} - 8t_0 \log\frac{F_2(t)}{e^{t_0^3}} + 3\log\frac{F_3(t)}{e^{t_0^4}}, \quad t, t_0 \in \mathbb{R}_\star.$$

**Theorem 2.8:** *The multiplicative osculating planes of two multiplicative equivalent curves coincide at the biregular points.*

*Proof.* Let $(I, f)$ and $(J, g)$ be two multiplicative equivalent curves. Let also, $s\colon I \to J$ be the parameter change. Then

$$f(t) = g(s(t)),$$
$$f^\star(t) = g^\star(s(t)) \cdot_\star s^\star(t),$$
$$f^{\star\star}(t) = f^{\star\star}(t) \cdot_\star (s^\star(t))^{2\star} +_\star g^\star(s(t)) \cdot_\star s^{\star\star}(t), \quad t \in I.$$

Thus, the systems $\{f^\star(t), f^{\star\star}(t)\}$ and $\{g^\star(s(t)), g^{\star\star}(s(t))\}$ are multiplicative equivalent. This completes the proof.

Let $(I, f)$ be a multiplicative parameterized curve, $t_0 \in I$, and $f(t_0)$ is multiplicative biregular. Suppose that $\alpha$ is a multiplicative plane with a multiplicative unit normal vector $e$ and $\alpha$ passes through $f(t_0)$. Denote

$$d_\star(h) = d_\star(f(t_0 +_\star h), \alpha).$$

**Theorem 2.9:** *The multiplicative plane $\alpha$ is a multiplicative osculating plane to the multiplicative curve $f = f(t)$ at the multiplicative biregular point $f(t_0)$ if and only if*

$$\lim_{h \to 0_\star} d_\star(h) /_\star |h|_\star^{2\star} = 0_\star. \tag{2.4}$$

*Proof.* By the Taylor formula, we have

$$f(t_0 +_\star h) = f(t_0) +_\star h \cdot_\star f^\star(t_0) +_\star (1_\star /_\star e^2) \cdot_\star h^{2\star} \cdot_\star f^{\star\star}(t_0)$$
$$+_\star h^{2\star} \cdot_\star \epsilon,$$

where $\epsilon \to 0_\star$, as $h \to 0_\star$. From here,

$$\begin{aligned}
d_\star(h) &= |\langle e, f(t_0 +_\star h) -_\star f(t_0) \rangle_\star |_\star \\
&= |\langle e, h \cdot_\star f^\star(t_0) +_\star (1_\star /_\star e^2) \cdot_\star h^{2\star} \cdot_\star f^{\star\star}(t_0) +_\star h^{2\star} \cdot_\star \epsilon \rangle_\star |_\star \\
&= |\langle e, h \cdot_\star f^\star(t_0) \rangle_\star +_\star \langle e, (1_\star /_\star e^2) \cdot_\star h^{2\star} \cdot_\star f^{\star\star}(t_0) \rangle_\star \\
&\quad +_\star \langle e, h^{2\star} \cdot_\star \epsilon \rangle_\star |_\star ,
\end{aligned}$$

whereupon

$$\begin{aligned}
d_\star(h) /_\star h^{2\star} &= |(1_\star /_\star h) \cdot_\star \langle e, f^\star(t_0) \rangle_\star +_\star (1_\star /_\star e^2) \cdot_\star \langle e, f^{\star\star}(t_0) \rangle_\star \\
&\quad +_\star \langle e, \epsilon \rangle_\star |_\star
\end{aligned}$$

and

$$\lim_{h \to 0_\star} d_\star (h) /_\star h^2_\star = \lim_{h \to 0_\star} |\, (1_\star /_\star h) \cdot_\star \langle e, f^\star (t_0) \rangle_\star +_\star (1_\star /_\star e^2) \cdot_\star \langle e, f^{\star\star} (t_0) \rangle_\star \,|_\star \, .$$

$$(2.5)$$

1. Let (2.4) holds. Then

$$\langle e, f^\star (t_0) \rangle_\star = 0_\star,$$
$$\langle e, f^{\star\star} (t_0) \rangle_\star = 0_\star. \qquad (2.6)$$

Thus,

$$e \|_\star f^\star (t_0) \times f^{\star\star} (t_0)$$

and $\alpha$ is a multiplicative osculating plane.

2. $\alpha$ is a multiplicative osculating plane. Then, we have (2.6) and using (2.5), we get (2.4). This completes the proof.

---

## 2.5 Multiplicative Curvature of a Multiplicative Curve

Let $(I, f)$ and $(J, g)$ be two multiplicative curves that are multiplicative equivalent with the parameter change $s$. Then

$$f (t) = g (s(t)), \ t \in I.$$

We have $s: J \to I$, $s^\star > 0_\star$ on $I$ and

$$|\, g^\star (s) \,|_\star = 1_\star.$$

**Theorem 2.10:** *The vector* $g^{\star\star} (s)$ *does not depend on the parameter change.*

*Proof.* Let $(J_1, g_1)$ be another multiplicative naturally parameterized curve with the parameter change $s_1$, i.e.,

$$g (s) = g_1 (s_1(s)), \ s \in J,$$

$s_1: J \to J_1$. Then

$$g^\star (s) = g_1^\star (s_1(s)) \cdot_\star s_1^\star (s), \ s \in J.$$

Hence,

$$|s_1^\star(s)|_\star = 1_\star$$

and

$$s_1^\star(s) = \pm_\star 1_\star, \quad s \in J.$$

Therefore

$$s_1(s) = \pm_\star s +_\star s_0,$$

for some constant $s_0 \in \mathbb{R}_\star$. Then

$$s_1^{\star\star}(s) = 0_\star, \quad s \in J,$$

and

$$
\begin{aligned}
g^{\star\star}(s) &= g_1^{\star\star}(s_1(s)) \cdot_\star (s_1^\star(s))^{2_\star} +_\star g^\star(s_1(s)) \cdot_\star s_1^{\star\star}(s) \\
&= g_1^{\star\star}(s_1(s)) \cdot_\star (s_1^\star(s))^{2_\star} \\
&= g_1^{\star\star}(s_1(s)), \quad s \in J.
\end{aligned}
$$

This completes the proof.

**Definition 2.16:** The multiplicative vector

$$\mathbf{k}(t) = g^{\star\star}(s(t))$$

will be called the multiplicative curvature of the multiplicative curve $f = f(t)$ at the point $t$ and

$$k(t) = |g^{\star\star}(s(t))|_\star$$

will be called the multiplicative curvature of $f$ at the point $t$.

In fact, we have

$$f^\star(t) = g^\star(s(t)) \cdot_\star s^\star(t), \quad t \in I,$$

and

$$s^\star(t) = |f^\star(t)|_\star, \quad t \in I.$$

Therefore

$$g^\star(s(t)) = f^\star(t)/_\star \, s^\star(t)$$
$$= f^\star(t)/_\star \, |f^\star(t)|_\star \,, \quad t \in I.$$

Next,

$$s^{\star\star}(t) = \, |f^\star|^\star_\star(t)$$
$$= \, d_\star((\langle f^\star(t), f^\star(t)\rangle_\star))^{\frac{1}{2}\star}/_\star \, d_\star t$$
$$= (\langle f^\star(t), f^{\star\star}(t)\rangle_\star)/_\star(|f^\star(t)|_\star), \quad t \in I.$$

Since

$$f^{\star\star}(t) = g^{\star\star}(s(t))\cdot_\star(s^\star(t))^{2\star} +_\star \, g^\star(s(t))\cdot_\star s^{\star\star}(t), \quad t \in I,$$

we get

$$g^{\star\star}(s(t))\cdot_\star(s^\star(t))^{2\star} = \, f^{\star\star}(t)$$
$$-_\star \, (\langle f^\star(t), f^{\star\star}(t)\rangle_\star \cdot_\star f^\star(t))/_\star\left(|f^\star(t)|^{2\star}_\star\right), \quad t \in I,$$

or

$$g^{\star\star}(s(t))\cdot_\star |f^\star(t)|^{2\star}_\star = \, f^{\star\star}(t)$$
$$-_\star \, (\langle f^\star(t), f^{\star\star}(t)\rangle_\star \cdot_\star f^\star(t))/_\star\left(|f^\star(t)|^{2\star}_\star\right), \quad t \in I,$$

or

$$g^{\star\star}(s(t)) = \, f^{\star\star}(t)/_\star |f^\star(t)|^{2\star}_\star$$
$$-_\star \, (\langle f^\star(t), f^{\star\star}(t)\rangle_\star \cdot_\star f^\star(t))/_\star\left(|f^\star(t)|^{4\star}_\star\right), \quad t \in I.$$

Consequently

$$\mathbf{k}(t) = \, f^{\star\star}(t)/_\star |f^\star(t)|^{2\star}_\star$$
$$-_\star \, (\langle f^\star(t), f^{\star\star}(t)\rangle_\star \cdot_\star f^\star(t))/_\star\left(|f^\star(t)|^{4\star}_\star\right), \quad t \in I,$$

and

$$
\begin{aligned}
k(t) &= |\mathbf{k}(t)|_\star \\
&= |g^{\star\star}(s(t))|_\star \\
&= |g^{\star\star}(s(t)) \times_\star g^\star(s(t))|_\star \\
&= |f^\star(t) \times_\star f^{\star\star}(t)|_\star /_\star \left(|f^\star(t)|_\star^{3\star}\right), \quad t \in I.
\end{aligned}
$$

**Example 2.24:** Let $f\colon \mathbb{R}_\star \to \mathbb{R}_\star^3$ be as in Example 2.22. Then

$$
f^\star(t) \times_\star f^{\star\star}(t) = \left(e^{\frac{2t}{(1+t)^2}}, \, e^{-\frac{t}{(1+t)^2}}, \, 1\right),
$$

$$
\begin{aligned}
|f^\star(t) \times_\star f^{\star\star}(t)|_\star &= e^{\left(\left(\frac{2t}{(1+t)^2}\right)^2 + \left(-\frac{t}{(1+t)^2}\right)^2\right)^{\frac{1}{2}}} \\
&= e^{\frac{\sqrt{5}\,t}{(1+t)^2}},
\end{aligned}
$$

$$
\begin{aligned}
|f^\star(t)|_\star &= e^{\left(1 + 4 + \left(\frac{1+2t}{1+t}\right)^2\right)^{\frac{1}{2}}} \\
&= e^{\left(5 + \frac{(1+2t)^2}{(1+t)^2}\right)^{\frac{1}{2}}} \\
&= e^{\frac{1}{1+t}(5 + 10t + 5t^2 + 1 + 4t + 4t^2)^{\frac{1}{2}}} \\
&= e^{\frac{1}{1+t}(6 + 14t + 9t^2)^{\frac{1}{2}}}, \quad t \in I,
\end{aligned}
$$

and

$$
\begin{aligned}
k(t) &= \left(e^{\frac{\sqrt{5}\,t}{(1+t)^2}}\right) \Big/ \left(e^{\frac{1}{(1+t)^3}(6 + 14t + 9t^2)^{\frac{3}{2}}}\right)_\star \\
&= e^{\dfrac{\frac{\sqrt{5}\,t}{(1+t)^2}}{\frac{1}{(1+t)^3}(6 + 14t + 9t^2)^{\frac{3}{2}}}} \\
&= e^{\frac{\sqrt{5}\,t(1+t)}{(6 + 14t + 9t^2)^{\frac{3}{2}}}}, \quad t \in I.
\end{aligned}
$$

**Example 2.25:** Let $f\colon \mathbb{R}_\star \to \mathbb{R}_\star^3$ be as in Example 2.23. Then

$$f^\star(t) \times_\star f^{\star\star}(t) = \left(e^{18t^5-12t^5}, e^{3t^4-9t^4}, e^{4t^3-2t^3}\right)$$
$$= \left(e^{6t^5}, e^{-6t^4}, e^{2t^3}\right),$$
$$|f^\star(t) \times_\star f^{\star\star}(t)|_\star = e^{((6t^5)^2+(-6t^4)^2+(2t^3)^2)^{\frac{1}{2}}}$$
$$= e^{(36t^{10}+36t^8+4t^6)^{\frac{1}{2}}}$$
$$= e^{2t^3(9t^4+9t^2+1)^{\frac{1}{2}}},$$
$$|f^\star(t)|_\star = e^{(t^2+4t^4+9t^6)^{\frac{1}{2}}}$$
$$= e^{t(1+4t^2+9t^4)^{\frac{1}{2}}}, \quad t \in I.$$

Consequently

$$k(t) = \left(e^{2t^3(9t^4+9t^2+1)^{\frac{1}{2}}}\right) \Big/_\star \left(e^{t^3(1+4t^2+9t^4)^{\frac{3}{2}}}\right)$$

$$= e^{\frac{2t^3(9t^4+9t^2+1)^{\frac{1}{2}}}{t^3(1+4t^2+9t^4)^{\frac{3}{2}}}}$$

$$= e^{\frac{2(9t^4+9t^2+1)^{\frac{1}{2}}}{(1+4t^2+9t^4)^{\frac{3}{2}}}}, \quad t \in \mathbb{R}_\star.$$

**Exercise 2.5:** Let $f\colon \mathbb{R}_\star \to \mathbb{R}^3_\star$ be defined by

$$f(t) = \left(e^{t^2}, e^{2t^2}, e^{t^3}\right), \quad t \in I.$$

Find the multiplicative curvature of $f$ at arbitrary point $t$.

**Answer 2.11:**

$$e^{\frac{6\sqrt{5}t^2}{(20+9t^2)^{\frac{3}{2}}}}, \quad t \in \mathbb{R}_\star.$$

---

## 2.6 The Multiplicative Frenet Frame

Suppose that $(I, f)$, $f\colon \mathbb{R}_\star \to \mathbb{R}^3_\star$ is a multiplicative biregular curve.

**Definition 2.17:** The multiplicative Frenet frame or the multiplicative moving frame at the point $t_0$ is a multiplicative orthonormal frame in $\mathbb{R}^3_\star$

with the origin the point $f(t_0)$ and the multiplicative coordinate vectors $\{\tau(t_0), \nu(t_0), \beta(t_0)\}$, where

1.

$$\tau(t_0) = f^\star(t_0)/_\star |f^\star(t_0)|_\star$$

and it is called the multiplicative unit tangent at $t_0$.

2.

$$\nu(t_0) = \mathbf{k}(t_0)/_\star |k(t_0)|_\star$$

and it is called the multiplicative unit principal normal at the point $t_0$.

3.

$$\beta(t_0) = \tau(t_0) \times_\star \nu(t_0)$$

and it is called the multiplicative binormal.

For a multiplicative naturally parameterized curve $(J, g)$, we have

$$\tau(s_0) = g^\star(s_0),$$
$$\nu(s_0) = g^{\star\star}(s_0)/_\star |g^{\star\star}(s_0)|_\star,$$
$$\beta(s_0) = (g^\star(s_0) \times_\star g^{\star\star}(s_0))/_\star |g^{\star\star}(s_0)|_\star.$$

We have

$$\tau(t) = f^\star(t)/_\star |f^\star(t)|_\star, \quad t \in I,$$

and

$$\mathbf{k}(t) = f^{\star\star}(t)/_\star |f^\star(t)|_\star^{2\star}$$
$$-_\star (\langle f^\star(t), f^{\star\star}(t) \rangle_\star \cdot_\star f^\star(t))/_\star \left( |f^\star(t)|_\star^{4\star} \right), \quad t \in I,$$

and

$$k(t) = |f^\star(t) \times_\star f^{\star\star}(t)|_\star /_\star \left( |f^\star(t)|_\star^{3\star} \right), \quad t \in I.$$

Hence,

$$\nu(t) = \mathbf{k(t)}/_\star k(t)$$
$$= (|f^\star(t)|_\star /_\star |f^\star \times_\star f^{\star\star}(t)|_\star)\cdot_\star f^{\star\star}(t)$$
$$\rightarrow_\star (((\langle f^\star(t), f^{\star\star}(t)\rangle_\star)/_\star(|f^\star(t) \times_\star f^{\star\star}(t)|_\star \cdot_\star |f^\star(t)|_\star))$$
$$\cdot_\star f^\star(t), \quad t \in I,$$

and

$$\beta(t) = \tau(t) \times_\star \nu(t)$$
$$= (f^\star(t) \times_\star f^{\star\star}(t))/_\star |f^\star(t) \times_\star f^{\star\star}(t)|_\star,$$

$t \in I$. Next,

$$\beta(t) \times_\star \tau(t) = (\tau(t) \times_\star \nu(t)) \times_\star \tau(t)$$
$$= \langle \tau(t), \tau(t)\rangle_\star \cdot_\star \nu(t)$$
$$\rightarrow_\star \langle \nu(t), \tau(t)\rangle_\star \cdot_\star \tau(t)$$
$$= \nu(t), \quad t \in I.$$

## 2.7 Multiplicative Oriented Space Curves. The Multiplicative Frenet Frame of an Multiplicative Oriented Space Curve

**Definition 2.18:** A multiplicative orientation of a multiplicative regular curve $C \subset \mathbb{R}^3$ is a family of multiplicative local parameterizations $\{(I_\alpha, f_\alpha)\}_{\alpha \in A}$ such that

1. $C = \cup_{\alpha \in A} f_\alpha(I_\alpha)$.
2. For any connected component $C_{\alpha\beta}^b$ of

$$C_{\alpha\beta} = f_\alpha(I_\alpha) \cap f_\beta(I_\beta), \quad \alpha, \beta \in A,$$

the multiplicative parameterized curves $(I_\alpha^b, f_\alpha^b)$, $(I_\beta^b, f_\beta^b)$ with

$$I_\alpha^b = f_\alpha^{-1}\left(C_{\alpha\beta}^b\right), \quad f_\alpha^b = f_\alpha\Big|_{I_\alpha^b},$$

$$I_\beta^b = f_\beta^{-1}\left(C_{\alpha\beta}^b\right), \quad f_\beta^b = f_\beta\Big|_{I_\beta^b},$$

are positively equivalent.

**Definition 2.19:** A multiplicative regular curve $C \subset \mathbb{R}^3$ with a multiplicative orientation is called a multiplicative oriented regular curve.

**Definition 2.20:** A multiplicative local parameterization $(I, f)$ of a multiplicative oriented regular curve $C$ is called compatible with the multiplicative orientation defined by the family $\{(I_\alpha, f_\alpha)\}_{\alpha \in A}$ if on the intersections $f(I) \cap f_\alpha(I_\alpha)$ the multiplicative parameterized curves $(I, f)$ and $(I_\alpha, f_\alpha)$ are positively oriented.

**Definition 2.21:** The multiplicative Frenet frame of a multiplicative oriented biregular curve $C$ at a point $x \in C$ is the multiplicative Frenet frame of a multiplicative biregular parameterized curve $f = f(t)$ at $t_0$, where $f = f(t)$ is a multiplicative local parameterization of the multiplicative curve $C$, multiplicative compatible with the multiplicative orientation such that $f(t_0) = x$.

---

## 2.8 The Multiplicative Frenet Formulae. The Multiplicative Torsion

Let $(I, f)$ be a multiplicative biregular curve. Let also, $\{\tau, \nu, \beta\}$ be the multiplicative Frenet frame. Then

$$\tau(t) = f^\star(t) /_\star |f^\star(t)|_\star, \quad t \in I,$$

and

$$
\begin{aligned}
\tau^\star(t) &= (f^{\star\star}(t) \cdot_\star |f^\star(t)|_\star \\
&\quad -_\star f^\star(t) \cdot_\star (\langle f^\star(t), f^{\star\star}(t) \rangle_\star /_\star |f^\star(t)|_\star))) /_\star |f^\star(t)|_\star^{2_\star} \\
&= (f^{\star\star}(t) \cdot_\star |f^\star(t)|_\star^{2_\star} -_\star f^\star(t) \cdot_\star \langle f^\star(t), f^{\star\star}(t) \rangle_\star) /_\star |f^\star(t)|_\star^{3_\star} \\
&= |f^\star(t)|_\star \cdot_\star (|f^\star(t)| \times_\star f^{\star\star}(t)|_\star /_\star |f^\star(t)|_\star^{3_\star}) \\
&\quad \cdot_\star ((|f^\star(t)|_\star /_\star |f^\star(t)| \times_\star f^{\star\star}(t)|_\star) \cdot_\star f^{\star\star}(t) \\
&\quad -_\star (\langle f^\star(t), f^{\star\star}(t) \rangle_\star) /_\star (|f^\star(t)|_\star \cdot_\star |f^\star(t)| \times_\star f^{\star\star}(t)|_\star)) \\
&= |f^\star(t)|_\star \cdot_\star k(t) \cdot_\star \nu(t), \quad t \in I.
\end{aligned}
$$

Further,

$$\beta(t) = \tau(t) \times_\star \nu(t), \ t \in I,$$

whereupon

$$\begin{aligned} \beta^\star(t) &= \tau^\star(t) \times_\star \nu(t) +_\star \tau(t) \times_\star \nu^\star(t) \\ &= |f^\star(t)|_\star \cdot_\star k(t) \cdot_\star (\nu(t) \times_\star \nu(t)) +_\star \tau(t) \times_\star \nu^\star(t) \\ &= \tau(t) \times_\star \nu^\star(t), \ t \in I. \end{aligned}$$

Therefore

$$\beta^\star(t) \perp_\star \tau(t), \ t \in I,$$

and by the equality

$$|\beta(t)|_\star^{2\star} = 1_\star, \quad t \in I.$$

we get

$$\langle \beta^\star(t), \beta(t) \rangle_\star = 0_\star, \ t \in I,$$

and then

$$\beta^\star(t) \perp_\star \beta(t), \ t \in I.$$

Therefore $\beta^\star(t), \ t \in I$, is multiplicative collinear with

$$\beta(t) \times_\star \tau(t), \ t \in I$$

or $\beta^\star(t), \ t \in I$, is multiplicative collinear with $\nu(t), \ t \in I$. We can write

$$\beta^\star(t) = -_\star |f^\star(t)|_\star \cdot_\star \kappa(t) \cdot_\star \nu(t), \ t \in I.$$

By the equation

$$\nu(t) = \beta(t) \times_\star \tau(t), \ t \in I,$$

we find

$$\begin{aligned}
\nu^\star(t) &= \beta^\star(t) \times_\star \tau(t) +_\star \beta(t) \times_\star \tau^\star(t) \\
&= (-_\star |f^\star(t)|_\star \cdot_\star \kappa(t) \cdot_\star \nu(t)) \times_\star \tau(t) \\
&\quad +_\star \beta(t) \times_\star (|f^\star(t)|_\star \cdot_\star k(t) \cdot_\star \nu(t)) \\
&= -_\star |f^\star(t)|_\star \cdot_\star \kappa(t) \cdot_\star (\nu(t) \times_\star \tau(t)) \\
&\quad +_\star |f^\star(t)|_\star \cdot_\star k(t) \cdot_\star (\beta(t) \times_\star \nu(t)) \\
&= |f^\star(t)|_\star \cdot_\star \kappa(t) \cdot_\star \beta(t) -_\star |f^\star(t)|_\star \cdot_\star k(t) \cdot_\star \tau(t) \\
&= |f^\star(t)|_\star \cdot_\star (-_\star k(t) \cdot_\star \tau(t) +_\star \kappa(t) \cdot_\star \beta(t)), \quad t \in I.
\end{aligned}$$

Therefore

$$\begin{aligned}
\tau^\star(t) &= |f^\star(t)|_\star \cdot_\star k(t) \cdot_\star \nu(t) \\
\beta^\star(t) &= -_\star |f^\star(t)|_\star \cdot_\star \kappa(t) \cdot_\star \nu(t) \\
\nu^\star(t) &= |f^\star(t)|_\star \cdot_\star (-_\star k(t) \cdot_\star \tau(t) +_\star \kappa(t) \cdot_\star \beta(t)), \quad t \in I.
\end{aligned} \tag{2.7}$$

**Definition 2.22:** The formulae (2.7) will be called multiplicative Frenet formulae.

**Definition 2.23:** The quantity $\kappa(t)$, $t \in I$ will be called multiplicative torsion or second multiplicative curvature.

Now, we will deal with a multiplicative naturally parameterized curve $(J, g = g(s))$. Then the multiplicative Frenet formulae take the form

$$\begin{aligned}
\tau^\star(s) &= k(s) \cdot_\star \nu(s) \\
\beta^\star(s) &= -_\star \kappa(s) \cdot_\star \nu(s) \\
\nu^\star(s) &= -_\star k(s) \cdot_\star \tau(s) +_\star \kappa(s) \cdot_\star \beta(s), \quad s \in J.
\end{aligned}$$

The multiplicative Frenet frame is as follows:

$$\begin{aligned}
\tau &= g^\star \\
\nu &= (1_\star /_\star k) \cdot_\star g^{\star\star} \\
\beta &= (1_\star /_\star k) \cdot_\star (g^\star \times g^{\star\star}).
\end{aligned} \tag{2.8}$$

Then

$$\begin{aligned}
\langle \beta^\star(s), \nu(s) \rangle_\star &= \langle -_\star \kappa(s) \cdot_\star \nu(s), \nu(s) \rangle_\star \\
&= -_\star \kappa(s) \cdot_\star \langle \nu(s), \nu(s) \rangle_\star \\
&= -_\star \kappa(s), \quad s \in J.
\end{aligned}$$

By the third equation of (2.8), we find

$$\beta^\star(s) = (1_\star/_\star k(s))^\star \cdot_\star (g^\star(s) \times_\star g^{\star\star}(s))$$
$$+_\star (1_\star/_\star k(s)) \cdot_\star (g^{\star\star}(s) \times_\star g^{\star\star}(s))$$
$$+_\star (1_\star/_\star k(s)) \cdot_\star (g^\star(s) \times g^{\star\star\star}(s))$$
$$= (1_\star/_\star k(s))^\star \cdot_\star (g^\star(s) \times_\star g^{\star\star}(s))$$
$$+_\star (1_\star/_\star k(s)) \cdot_\star (g^\star(s) \times g^{\star\star\star}(s)), \quad s \in J.$$

Hence,

$$-_\star \kappa(s) = \langle \beta^\star(s), \nu(s) \rangle_\star$$
$$= (1_\star/_\star k(s))^\star \cdot_\star \langle g^\star(s) \times g^{\star\star}(s), \nu(s) \rangle_\star$$
$$+_\star (1_\star/_\star k(s)) \cdot_\star \langle g^\star(s) \times_\star g^{\star\star\star}(s), \nu(s) \rangle_\star$$
$$= (1_\star/_\star k(s)) \cdot_\star \langle g^\star(s) \times_\star g^{\star\star\star}(s), \nu(s) \rangle_\star$$
$$= (1_\star/_\star k(s))^{2_\star} \cdot_\star \langle g^\star(s) \times_\star g^{\star\star\star}(s), g^{\star\star}(s) \rangle_\star$$
$$= -_\star (1_\star/_\star k(s))^{2_\star} \cdot_\star \langle g^{\star\star\star}(s) \times_\star g^\star(s), g^{\star\star}(s) \rangle_\star$$
$$= -_\star (1_\star/_\star k(s))^{2_\star} \cdot_\star \langle g^{\star\star\star}(s), g^\star(s) \times_\star g^{\star\star}(s) \rangle_\star$$
$$= -_\star (1_\star/_\star k(s))^{2_\star} \cdot_\star \langle g^\star(s) \times_\star g^{\star\star}(s), g^{\star\star\star}(s) \rangle_\star, \quad s \in J.$$

Therefore

$$\kappa(s) = (1_\star/_\star k(s))^{2_\star} \cdot_\star \langle g^\star(s) \times_\star g^{\star\star}(s), g^{\star\star\star}(s) \rangle_\star, \quad s \in J.$$

**Theorem 2.12:** *Let* $(I, f = f(t))$ *and* $(J, g = g(s))$ *be two multiplicative positive oriented equivalent parametric curves with the parameter change* $s \to J, s^\star > 0_\star$ *on* $I$. *Then they have the same multiplicative torsions at the corresponding points* $t$ *and* $s(t)$.

*Proof.* Let $\{\tau, \nu, \beta\}$ and $\{\tau_1, \nu_1, \beta_1\}$ be the multiplicative Frenet frames for $(I, f)$ and $(J, g)$, respectively. Then

$$\beta_1(s(t)) = \beta(t),$$
$$\nu_1(s(t)) = \nu(t),$$
$$f^\star(t) = g^\star(s(t)) \cdot_\star s^\star(t), \quad t \in I.$$

From the second equation of the multiplicative Frenet formulae, we find

$$\langle \beta^\star(t), \nu(t) \rangle_\star = -_\star |f^\star(t)|_\star \cdot_\star \kappa(t), \quad t \in I.$$

Hence,

$$\kappa(t) = -_\star (1_\star /_\star |f^\star(t)|_\star)\cdot_\star \langle \beta^\star(t), \nu(t) \rangle_\star$$

$$= -_\star (1_\star /_\star (|g^\star(s(t))|_\star \cdot s^\star(t)))$$

$$\cdot_\star \langle \beta_1^\star(s(t))\cdot_\star s^\star(t), \nu(t) \rangle_\star$$

$$= -_\star (1_\star /_\star |g^\star(s(t))|_\star)\cdot_\star \langle \beta_1^\star(s(t)), \nu_1(s(t)) \rangle_\star$$

$$= -_\star \langle \beta_1^\star(s(t)), \nu_1(s(t)) \rangle_\star$$

$$= \kappa_1(s(t)), \quad t \in I.$$

This completes the proof.

Now, suppose that $(I, f)$ and $(J, g)$ are multiplicative equivalent with the parameter change $s: I \to J$, $s^\star > 0_\star$ on $I$. Then

$$f(t) = g(s(t)),$$

$$f^\star(t) = g^\star(s(t))\cdot_\star s^\star(t),$$

$$f^{\star\star}(t) = g^{\star\star}(s(t))\cdot_\star (s^\star(t))^{2\star} +_\star g^\star(s(t))\cdot_\star s^{\star\star}(t),$$

$$f^{\star\star\star}(t) = g^{\star\star\star}(s(t))\cdot_\star (s^\star(t))^{3\star} +_\star e^{2\star}\cdot_\star g^{\star\star}(s(t))\cdot_\star s^\star(t)\cdot_\star s^{\star\star}(t)$$

$$+_\star g^{\star\star}(s(t))\cdot_\star s^\star(t)\cdot_\star s^{\star\star}(t) +_\star g^\star(s(t))\cdot_\star s^{\star\star\star}(t)$$

$$= g^{\star\star\star}(s(t))\cdot_\star (s^\star(t))^{3\star} +_\star e^{3\star}\cdot_\star g^{\star\star}(s(t))\cdot_\star s^\star(t)\cdot_\star s^{\star\star}(t)$$

$$+_\star g^\star(s(t))\cdot_\star s^{\star\star\star}(t), \quad t \in I.$$

Hence,

$$\langle f^\star(t) \times_\star f^{\star\star}(t), f^{\star\star\star}(t) \rangle_\star = \langle (g^\star(s(t))\cdot_\star s^\star(t)) \times_\star (g^{\star\star}(s(t))\cdot_\star (s^\star(t))^{2\star}$$

$$+_\star g^\star(s(t))\cdot_\star s^{\star\star}(t)), g^{\star\star\star}(s(t))\cdot_\star (s^\star(t))^{3\star}$$

$$+_\star e^{3\star}\cdot_\star g^{\star\star}(s(t))\cdot_\star s^\star(t)\cdot_\star s^{\star\star}(t) +_\star g^\star(s(t))\cdot_\star s^{\star\star\star}(t) \rangle_\star$$

$$= (s^\star(t))^{3\star}\cdot_\star \langle g^\star(s(t)) \times_\star g^{\star\star}(s(t)), g^{\star\star\star}(s(t))\cdot_\star (s^\star(t))^{3\star}$$

$$+_\star e^{3\star}\cdot_\star g^{\star\star}(s(t))\cdot_\star s^\star(t)\cdot_\star s^{\star\star}(t) +_\star g^\star(s(t))\cdot_\star s^{\star\star\star}(t) \rangle_\star$$

$$= (s^\star(t))^{6\star}\cdot_\star \langle g^\star(s(t)) \times_\star g^{\star\star}(s(t)), g^{\star\star\star}(s(t)) \rangle_\star$$

$$= (s^\star(t))^{6\star}\cdot_\star (k(s(t)))^{2\star}\cdot_\star \kappa(s(t))$$

$$= (s^\star(t))^{6\star}\cdot_\star \left(|f^\star(t) \times_\star f^{\star\star}(t)|_\star^{2\star}\right)/_\star \left(|f^\star(t)|_\star^{6\star}\right)\cdot_\star \kappa(t)$$

$$= |f^\star(t) \times_\star f^{\star\star}(t)|_\star^{2\star}\cdot_\star \kappa(t), \quad t \in I,$$

whereupon

$$\kappa(t) = \langle f^\star(t) \times_\star f^{\star\star}(t), f^{\star\star\star}(t) \rangle_\star /_\star |f^\star(t) \times_\star f^{\star\star}(t)|_\star^{2\star},$$

$t \in I.$

**Example 2.26:** Let $f: \mathbb{R}_\star \to \mathbb{R}^3_\star$ be as in Example 2.22. Then

$$f^\star(t) \times_\star f^{\star\star}(t) = \left( e^{\frac{2t}{(1+t)^2}}, e^{-\frac{t}{(1+t)^2}}, 1 \right),$$

$$|f^\star(t) \times_\star f^{\star\star}(t)|_\star = e^{\frac{\sqrt{5}t}{(1+t)^2}}, \quad t \in \mathbb{R}_\star,$$

and

$$\left( f_1^{\star\star} \right)'(t) = 0,$$

$$\left( f_2^{\star\star} \right)'(t) = 0,$$

$$\left( f_3^{\star\star} \right)'(t) = \frac{(1+t)^2 - 2t(1+t)}{(1+t)^4} e^{\frac{t}{(1+t)^2}}$$

$$= \frac{1-t}{(1+t)^3} e^{\frac{t}{(1+t)^3}}, \quad t \in \mathbb{R}_\star,$$

and

$$f_1^{\star\star\star}(t) = 1,$$

$$f_2^{\star\star\star}(t) = 1,$$

$$f_3^{\star\star\star}(t) = e^{t \frac{\left( f_3^{\star\star} \right)'(t)}{f_3^{\star\star}(t)}}$$

$$= e^{\frac{t(1-t)}{(1+t)^3}}, \quad t \in \mathbb{R}_\star.$$

Therefore

$$f^{\star\star\star}(t) = \left( f_1^{\star\star\star}(t), f_2^{\star\star\star}(t), f_3^{\star\star\star}(t) \right)$$

$$= \left( 1, 1, e^{\frac{t(1-t)}{(1+t)^3}} \right), \quad t \in \mathbb{R}_\star,$$

and

$$\langle f^\star(t) \times_\star f^{\star\star}(t), f^{\star\star\star}(t) \rangle_\star = 0_\star, \quad t \in \mathbb{R}_\star,$$

and

$$\kappa(t) = 0_\star, \quad t \in \mathbb{R}_\star.$$

**Example 2.27:** Let $f: \mathbb{R}_\star \to \mathbb{R}_\star^3$ be as in Example 2.23. Then

$$f^\star(t) \times_\star f^{\star\star}(t) = \left(e^{6t^5}, e^{-6t^4}, e^{2t^3}\right),$$

$$|f^\star(t) \times_\star f^{\star\star}(t)|_\star = e^{2t^3(9t^4+9t^2+1)^{\frac{1}{2}}}, \quad t \in \mathbb{R}_\star,$$

and

$$(f_1^{\star\star})'(t) = e^t,$$

$$(f_2^{\star\star})'(t) = 8te^{4t^2},$$

$$(f_3^{\star\star})'(t) = 27t^2e^{9t^3}, \quad t \in \mathbb{R}_\star,$$

and

$$f_1^{\star\star\star}(t) = e^{t\frac{(f_1^{\star\star})'(t)}{f_1^{\star\star}(t)}}$$

$$= e^t,$$

$$f_2^{\star\star\star}(t) = e^{t\frac{(f_2^{\star\star})'(t)}{f_2^{\star\star}(t)}}$$

$$= e^{8t^2},$$

$$f_3^{\star\star\star}(t) = e^{t\frac{(f_3^{\star\star})'(t)}{f_3^{\star\star}(t)}}$$

$$= e^{27t^3}, \quad t \in \mathbb{R}_\star.$$

Therefore

$$f^{\star\star\star}(t) = \left(f_1^{\star\star\star}(t), f_2^{\star\star\star}(t), f_3^{\star\star\star}(t)\right)$$

$$= (e^t, e^{8t^2}, e^{27t^3}), \quad t \in \mathbb{R}_\star,$$

and

$$\langle f^\star(t) \times_\star f^{\star\star}(t), f^{\star\star\star}(t)\rangle_\star = e^{6t^6 - 48t^6 + 54t^6}$$
$$= e^{12t^6}, \quad t \in \mathbb{R}_\star,$$

and

$$\kappa(t) = e^{12t^6} /_\star \left( e^{2t^3(9t^4 + 9t^2 + 1)^{\frac{1}{2}}} \right)^{2_\star}$$

$$= e^{12t^6} /_\star e^{4t^6(9t^4 + 9t^2 + 1)}$$

$$= e^{\frac{3}{9t^4 + 9t^2 + 1}}, \quad t \in \mathbb{R}_\star.$$

**Theorem 2.13:** *Let*

$$w = |f^\star|_\star \cdot_\star ((\kappa \cdot_\star \tau +_\star k \cdot_\star \beta).$$

*Then the multiplicative Frenet formulae can be written in the form*

$$\tau^\star = w \times_\star \tau$$
$$\nu^\star = w \times_\star \nu$$
$$\beta^\star = w \times_\star \beta.$$

*Proof.* We have

$$w \times_\star \tau = |f^\star|_\star \cdot_\star (\kappa \cdot_\star \tau +_\star k \cdot_\star \beta) \times_\star \tau$$
$$= |f^\star|_\star \cdot_\star (k \cdot_\star (\beta \times_\star \tau))$$
$$= |f^\star|_\star \cdot_\star (k \cdot_\star \nu)$$
$$= \tau^\star$$

and

$$w \times_\star \nu = (|f^\star|_\star \cdot_\star (\kappa \cdot_\star \tau +_\star k \cdot_\star \beta)) \times_\star \nu$$
$$= |f^\star|_\star \cdot_\star \kappa \cdot_\star (\tau \times_\star \nu) +_\star |f^\star|_\star \cdot_\star k \cdot_\star (\beta \times_\star \nu)$$
$$= |f^\star|_\star \cdot_\star \kappa \cdot_\star \beta -_\star |f^\star|_\star \cdot_\star k \cdot_\star \tau$$
$$= \nu^\star,$$

and

$$w \times_\star \beta = (|f^\star|_\star \cdot_\star (\kappa \cdot_\star \tau +_\star k \cdot_\star \beta)) \times_\star \beta$$
$$= (|f^\star|_\star \cdot_\star \kappa) \cdot_\star (\tau \times_\star \beta) +_\star (|f^\star|_\star \cdot_\star k) \cdot_\star (\beta \times_\star \beta)$$
$$= -_\star (|f^\star|_\star \cdot_\star \kappa) \cdot_\star \nu$$
$$= \beta^\star.$$

This completes the proof.

**Definition 2.24:** The vector $w$ will be called multiplicative Darboux vector.

---

## 2.9 Applications of the Multiplicative Frenet Formulae

**Theorem 2.14:** *The support of the multiplicative biregular curve lies in a multiplicative plane if and only if the multiplicative torsion of the multiplicative curve multiplicative vanishes identically.*

*Proof.* Let $(I, f)$ be a multiplicative biregular curve.

1. Suppose that $(I, f)$ lies in a multiplicative plane $\pi$, i.e., $f(I) \subset \pi$. Then $f^\star$ and $f^{\star\star}$ are multiplicative parallel to this multiplicative plane. Thus, $\pi$ is the multiplicative osculating plane and then

$$\beta(t) = \text{const}, \ t \in I.$$

Hence,

$$0_\star = \beta^\star(t)$$
$$= -_\star |f^\star(t)|_\star \cdot_\star \kappa(t) \cdot_\star \nu(t), \ t \in I.$$

Therefore

$$\kappa(t) = 0_\star, \ t \in I.$$

2. Let

$$\kappa(t) = 0_\star, \ t \in I.$$

Then, by the Frenet formulae it follows that $\beta(t) = \text{const} = \beta_0, t \in I$. On the other hand,

$$\beta(t) = \tau(t) \times_\star \nu(t), \ t \in I.$$

Hence, $f^\star(t) \perp_\star \beta_0, \ t \in I$, and then

$$0_\star = \langle f^\star(t), \beta_0 \rangle_\star, \ t \in I.$$

Therefore

$$\langle f(\cdot), \beta_0 \rangle_\star^\star(t) = 0_\star$$

and

$$\langle f(t) -_\star f(t_0), \beta_0 \rangle_\star = 0_\star.$$

Thus, the support of $(I, f = f(t))$ is contained in a multiplicative plane that is multiplicative perpendicular to $\beta_0$. This completes the proof.

**Definition 2.25:** Let $(x_{10}, x_{20}) \in \mathbb{R}^2_\star$ and $a \in \mathbb{R}_\star$. The set

$$\{(x_1, x_2) \in \mathbb{R}^2_\star : (x_1 -_\star x_{10})^{2\star} +_\star (x_2 -_\star x_{20})^{2\star} = a^{2\star}\}$$

will be called the multiplicative circle with centre $(x_{10}, x_{20})$ and radius $a$.

We have

$$\left(\frac{x_1}{x_{10}}\right)^{2\star} +_\star \left(\frac{x_2}{x_{20}}\right)^{2\star} = a^{2\star}$$

or

$$e^{\left(\log \frac{x_1}{x_{10}}\right)^2 + \left(\log \frac{x_2}{x_{20}}\right)^2} = e^{(\log a)^2},$$

or

$$\left(\log \frac{x_1}{x_{10}}\right)^2 + \left(\log \frac{x_2}{x_{20}}\right)^2 = (\log a)^2.$$

**Theorem 2.15:** Let $(I, f = f(s))$ be a multiplicative naturally parameterized curve with a constant multiplicative curvature $k_0 > 0_\star$ and let its multiplicative torsion is $0_\star$. Then the support of $f$ lies on a multiplicative circle of radius $1_\star /_\star k_0$.

*Proof.* Since the multiplicative torsion of the multiplicative curve $f$ is $0_\star$, we have that the multiplicative curve $f$ is a multiplicative plane curve. Consider the multiplicative curve

$$f_1 = f +_\star (1_\star /_\star k_0)\cdot_\star \nu.$$

Hence,

$$f_1^\star = f^\star +_\star (1_\star /_\star k_0)\cdot_\star \nu^\star$$
$$= \tau +_\star (1_\star /_\star k_0)\cdot_\star (-_\star k_0 \cdot_\star \tau)$$
$$= \tau -_\star \tau$$
$$= 0_\star \text{ on } I.$$

Thus, $f_1$ is a constant $c$ on $I$ and

$$f -_\star c = -_\star (1_\star /_\star k_0)\cdot_\star \nu$$

and

$$|f -_\star c|_\star = (1_\star /_\star k_0)\cdot_\star |\nu|_\star = 1_\star /_\star k_0.$$

This completes the proof.

**Definition 2.26:** Let $(x_{10}, x_{20}, x_{30}) \in \mathbb{R}^3_\star$ and $a \in \mathbb{R}_\star$. The set

$$(x_1 -_\star x_{10})^{2_\star} +_\star (x_2 -_\star x_{20})^{2_\star} +_\star (x_3 -_\star x_{30})^{2_\star} = a^{2_\star}$$

will be called multiplicative sphere with centre $(x_{10}, x_{20}, x_{30})$ and radius $a$.

The equation of the multiplicative sphere with centre $(x_{10}, x_{20}, x_{30})$ and centre $a$ can be written as follows:

$$\left(\log \frac{x_1}{x_{10}}\right)^2 + \left(\log \frac{x_2}{x_{20}}\right)^2 + \left(\log \frac{x_3}{x_{30}}\right)^2 = (\log a)^2.$$

In Fig. 2.3 is shown the multiplicative sphere with centre $(0_\star, 0_\star, 0_\star)$ and radius $\sqrt{e}$.

**Theorem 2.16:** *Let the support of the multiplicative naturally parameterized curve $(I, f = f(s))$ lies on a multiplicative sphere with centre $(0_\star, 0_\star, 0_\star)$ and radius $a > 0_\star$. Then*

$$k > 1_\star /_\star a.$$

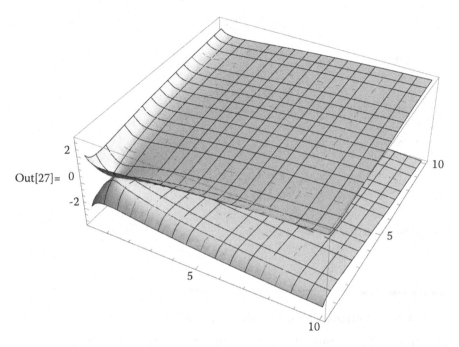

Out[27]=

**FIGURE 2.3**
The multiplicative sphere.

*Proof.* We have

$$|f|^{2_\star}_\star = a^{2_\star},$$

whereupon

$$\langle f, f^\star \rangle_\star = 0_\star$$

or

$$\langle f, \tau \rangle_\star = 0_\star.$$

We differentiate the last equation and we find

$$\begin{aligned}
0_\star &= \langle f^\star, \tau \rangle_\star +_\star \langle f, \tau^\star \rangle_\star \\
&= \langle \tau, \tau \rangle_\star +_\star \langle f, \tau^\star \rangle_\star \\
&= 1_\star +_\star \langle f, k \cdot_\star \nu \rangle_\star \\
&= 1_\star +_\star k \cdot_\star \langle f, \nu \rangle_\star,
\end{aligned}$$

whereupon

$$-_\star 1_\star = k \cdot_\star \langle f, \nu \rangle_\star.$$

Note that

$$| \langle f, \nu \rangle_\star |_\star \leq |f|_\star \cdot_\star |\nu|_\star$$
$$= a.$$

Therefore

$$k = 1_\star /_\star | \langle f, \nu \rangle_\star |_\star$$
$$\geq 1_\star /_\star a.$$

This completes the proof.

## 2.10  The Multiplicative General Helices

**Definition 2.27:** A multiplicative parameterized curve $(I, f)$ is said to be a multiplicative general helix if its multiplicative tangents make a constant multiplicative angle with a fixed multiplicative vector in the space $\mathbb{R}^3_\star$.

**Theorem 2.17: (The Multiplicative Lancret Theorem).** *A multiplicative space curve $(I, f)$ with the multiplicative curvature $k > 0_\star$ is a multiplicative general helix if and only if the multiplicative ratio of its multiplicative torsion and its multiplicative curvature is a constant.*

*Proof.* Without loss of generality, suppose that $(I, f)$ is a multiplicative naturally parameterized curve.

1. Let $(I, f)$ be a multiplicative general helix and $c$ be the fixed multiplicative direction that makes a constant multiplicative angle with its multiplicative tangents. Then

$$\langle \tau, c \rangle_\star = \cos_\star \alpha_0 = \text{const.}$$

Hence, multiplicative differentiating, we arrive at

$$\langle \tau^\star, c \rangle_\star = 0_\star,$$

whereupon

$$k \cdot_\star \langle \nu, c \rangle_\star = 0_\star.$$

Since $k > 0_\star$, we find

$$\langle \nu, c \rangle_\star = 0_\star. \tag{2.9}$$

Thus, $c \perp_\star \nu$ and

$$\langle \beta, c \rangle_\star = \sin_\star \alpha_0.$$

Now, we multiplicative differentiate the equation (2.9) and we obtain

$$\begin{aligned}
0_\star &= \langle \nu^\star, c \rangle_\star \\
&= \langle -_\star k \cdot_\star \tau +_\star \kappa \cdot_\star \beta, c \rangle_\star \\
&= -_\star k \cdot_\star \langle \tau, c \rangle_\star +_\star \kappa \cdot_\star \langle \beta, c \rangle_\star \\
&= -_\star k \cdot_\star \cos_\star \alpha_0 +_\star \kappa \cdot_\star \sin_\star \alpha_0,
\end{aligned}$$

from where

$$\begin{aligned}
\kappa /_\star k &= \cos_\star \alpha_0 /_\star \sin_\star \alpha_0 \\
&= \cot_\star \alpha_0 \\
&= \text{const.}
\end{aligned}$$

2. Let

$$\kappa /_\star k = c_0,$$

where $c_0$ is a constant. Then

$$\kappa = c_0 \cdot_\star k$$

or

$$c_0 \cdot_\star k -_\star \kappa = 0_\star.$$

Hence,

$$\begin{aligned}
0_\star &= (c_0 \cdot_\star k -_\star \kappa) \cdot_\star \nu \\
&= c_0 \cdot_\star (k \cdot_\star \nu) -_\star \kappa \cdot_\star \nu \\
&= c_0 \cdot_\star \tau^\star +_\star \beta^\star,
\end{aligned}$$

whereupon,

$$c_{0\cdot\star} \tau +_\star \beta = c_1,$$

$c_1 \neq 0_\star$. Let

$$c = c_1 /_\star |c_1|_\star .$$

Then

$$c = (c_{0\cdot\star} \tau +_\star \beta) /_\star |c_{0\cdot\star} \tau +_\star \beta|_\star$$
$$= (c_{0\cdot\star} \tau +_\star \beta) /_\star (1_\star +_\star c^{2\star})^{\frac{1}{2}\star}.$$

Hence,

$$\langle c, \tau \rangle_\star = c_0 /_\star (1_\star +_\star c^{2\star})^{\frac{1}{2}\star}.$$

Thus, $c$ and $\tau$ make a constant multiplicative angle. Therefore $(I, f)$ is a multiplicative general helix. This completes the proof.

## 2.11 The Multiplicative Bertrand Curves

**Definition 2.28:** Multiplicative curves that have the same multiplicative principal normals are said to be multiplicative Bertrand curves. Usually, for a multiplicative Bertrand curve there is only one multiplicative curve having the same multiplicative principal normals. The two multiplicative curves are said to be multiplicative Bertrand mates, or multiplicative associated Bertrand curves, or multiplicative conjugated Bertrand curves. If a multiplicative Bertrand curve has more than one multiplicative mates, then it has infinitely many and the multiplicative curve and its multiplicative Bertrand mates are said to be multiplicative circular cylindrical helix.

Let $f_1$ and $f_2$ be multiplicative Bertrand mates and $f_1$ be multiplicative naturally parameterized with the parameter change $s$. Then $f_2$ depends on $s$ and we assume that $f_2(s)$ and $f_1(s)$ have the same multiplicative principal normals. Both points will be called corresponding points.

**Theorem 2.18: (The Multiplicative Shell Theorem).** *Suppose that $f_1$ and $f_2$ are two multiplicative associated Bertrand curves and $f_1$ is multiplicative naturally* *parameterized with the parameter changes. Then the multiplicative angle of the multiplicative tangents at the corresponding points is a constant.*

*Proof.* Let $I \subseteq \mathbb{R}_\star$ is the range of $s$. Let also $\{\tau_1, \nu_1, \beta_1\}$ and $\{\tau_2, \nu_2, \beta_2\}$ be the multiplicative Frenet frames of $f_1$ and $f_2$, respectively. Then

$$f_2(s) = f_1(s) +_\star a(s) \cdot_\star \nu_1(s), \quad s \in I, \tag{2.10}$$

for some $a \in \mathscr{C}^1_\star(I)$. For the multiplicative principal normals $\nu_1$ and $\nu_2$, we have the relations

$$\nu_2(s) = \pm_\star \nu_1(s), \quad s \in I. \tag{2.11}$$

We multiplicative differentiate equation (2.10) with respect to $s$ and we find

$$
\begin{aligned}
(d^\star f_2 /_\star d_\star s)(s) &= (d^\star f_1 /_\star d_\star s)(s) \cdot_\star (d^\star a /_\star d_\star s)(s) \cdot_\star \nu_1(s) \\
&\quad +_\star a(s) \cdot_\star (d^\star \nu_1 /_\star d_\star s)(s) \\
&= \tau_1(s) +_\star (d^\star a /_\star d_\star s)(s) \cdot_\star \nu_1(s) \\
&\quad +_\star a(s) \cdot_\star (-_\star k_1(s) \cdot_\star \tau_1(s) +_\star \kappa_1(s) \cdot_\star \beta_1(s)) \\
&= (1_\star -_\star a(s) \cdot_\star k_1(s)) \cdot_\star \tau_1(s) \\
&\quad +_\star (d^\star a /_\star d_\star s)(s) \cdot_\star \nu_1(s) \\
&\quad +_\star a(s) \cdot_\star \kappa_1(s) \cdot_\star \beta_1(s), \quad s \in I.
\end{aligned}
$$

Since

$$(d^\star f_2 /_\star d_\star s)(s), \quad s \in I$$

is a multiplicative tangent vector to $f_2$ and $s \in I$, then it is multiplicative perpendicular to $\nu_1(s)$ and $\nu_2(s)$, $s \in I$. Hence and the last equations, we find

$$
\begin{aligned}
0_\star &= \langle (d^\star f_2 /_\star d_\star s)(s), \nu_1(s) \rangle_\star \\
&= (d^\star a /_\star d_\star s)(s), \quad s \in I.
\end{aligned}
$$

Therefore $a$ is a constant on $I$ and (2.10) can be written in the form

$$f_2(s) = f_1(s) +_\star a \cdot_\star \nu_1(s), \quad s \in I, \tag{2.12}$$

and

$$(d^\star f_2 /_\star d_\star s)(s) = (1_\star -_\star a\cdot_\star k_1(s))\cdot_\star \tau_1(s)$$
$$+_\star a\cdot_\star \kappa_1(s)\cdot_\star \beta_1(s), \quad s \in I. \tag{2.13}$$

Let now, $s_2$ be the arc length parameter of $f_2$. Then

$$
\begin{aligned}
\tau_2(s) &= (d^\star f_2 /_\star d_\star s_2)(s) \\
&= (d^\star f_2 /_\star d_\star s)(s)\cdot_\star (d^\star s /_\star d_\star s_2)(s) \\
&= (1_\star -_\star a\cdot_\star k_1(s))\cdot_\star \tau_1(s)\cdot_\star (d^\star s /_\star d_\star s_2)(s) \\
&\quad +_\star a\cdot_\star \kappa_1(s)\cdot_\star \beta_1(s)\cdot_\star (d^\star s /_\star d_\star s_2)(s), \quad s \in I.
\end{aligned} \tag{2.14}
$$

Let

$$w(s) = \angle_\star(\tau_1(s), \tau_2(s)), \quad s \in I.$$

Then

$$\langle \tau_1(s), \tau_2(s) \rangle_\star = \cos_\star w(s), \quad s \in I.$$

Hence,

$$\tau_2(s) = \cos_\star w(s)\cdot_\star \tau_1(s) +_\star \sin_\star w(s)\cdot_\star \beta_1(s), \quad s \in I, \tag{2.15}$$

and

$$\beta_2(s) = \varepsilon\cdot_\star (-_\star \sin_\star w(s)\cdot_\star \tau_1(s) +_\star \cos_\star w(s)\cdot_\star \beta_1(s)),$$

$s \in I$, where $\varepsilon = \pm_\star 1_\star$. We multiplicative differentiate with respect to $s$ equation (2.15) and we find

$$
\begin{aligned}
(d^\star \tau_2 /_\star d_\star s)(s) &= -_\star \sin_\star w(s)\cdot_\star (d^\star w /_\star d_\star s)(s)\cdot_\star \tau_1(s) \\
&\quad +_\star \cos_\star w(s)\cdot_\star (d^\star \tau_1 /_\star d_\star s) \\
&\quad +_\star \cos_\star w(s)\cdot_\star (d^\star w /_\star d_\star s)(s)\cdot_\star \beta_1(s) \\
&\quad +_\star \sin_\star w(s)\cdot_\star (d^\star \beta_1 /_\star d_\star s)(s) \\
&= -_\star \sin_\star w(s)\cdot_\star (d^\star w /_\star d_\star s)(s)\cdot_\star \tau_1(s) \\
&\quad +_\star \cos_\star w(s)\cdot_\star k_1(s)\cdot_\star \nu_1(s) \\
&\quad +_\star \cos_\star w(s)\cdot_\star (d^\star w /_\star d_\star s)(s)\cdot_\star \beta_1(s) \\
&\quad -_\star \sin_\star w(s)\cdot_\star \kappa_1(s)\cdot_\star \nu_1(s), \quad s \in I.
\end{aligned}
$$

Therefore

$$\langle (d^\star \tau_2 /_\star d_\star s)(s),\ \tau_1(s)\rangle_\star = -_\star \sin_\star w(s)\cdot_\star (d^\star w /_\star d_\star s)(s)$$
$$\langle (d^\star \tau_2 /_\star d_\star s)(s),\ \beta_1(s)\rangle_\star = \cos_\star w(s)\cdot_\star (d^\star w /d_\star s)(s),\ \ s \in I.$$

We have that

$$(d^\star \tau_2 /_\star d_\star s)(s),\ s \in I,$$

is multiplicative collinear with $\nu_2(s),\ s \in I$. Therefore

$$0_\star = \langle (d^\star \tau_2 /_\star d_\star s)(s),\ \tau_1(s)\rangle_\star$$
$$0_\star = \langle (d^\star \tau_2 /_\star d_\star s)(s),\ \beta_1(s)\rangle_\star,\ s \in I,$$

and then

$$(d^\star w /_\star d_\star s)(s) = 0_\star,\ s \in I.$$

Thus, $w$ is a constant on $I$. This completes the proof.

**Theorem 2.19: (The Multiplicative Bertrand Theorem).** *Let $f_1$ be a multiplicative naturally parameterized curve on $I$ with parameter change $s$. Then $f_1$ is a multiplicative Bertrand curve if and only if its multiplicative torsion and its multiplicative curvature satisfy*

$$a\cdot_\star k_1(s) +_\star b\cdot_\star \varkappa_1(s) = 0_\star,\ s \in I, \tag{2.16}$$

*where $a$ and $b$ are multiplicative constants.*

*Proof.* We will use the notations used in the proof of Theorem 2.18.

1. Let $f_1$ and $f_2$ be multiplicative Bertrand mates. Thus, (2.14) and (2.15), we find

$$\cos_\star w = (1_\star -_\star a\cdot_\star k_1(s))\cdot_\star (d^\star s /_\star d_\star s_2)(s)$$
$$\sin_\star w = a\cdot_\star \varkappa_1(s)\cdot_\star (d^\star s /_\star d_\star s_2)(s),\ \ s \in I,$$

whereupon

$$\cot_\star w = (1_\star -_\star a\cdot_\star k_1(s)) /_\star (a\cdot_\star \varkappa_1(s)),\ s \in I,$$

or

$$(a \cdot_\star \varkappa_1(s)) \cdot_\star \cot_\star w = 1_\star -_\star a \cdot_\star k_1(s), \quad s \in I.$$

Let

$$b = a \cdot_\star \cot_\star w.$$

Hence,

$$a \cdot_\star k_1(s) +_\star b \cdot_\star \varkappa_1(s) = 1_\star, \quad s \in I,$$

i.e., we get (2.16).

2. Suppose that (2.16) holds. Then

$$(d^\star f_2 /_\star d_\star s)(s) = (1_\star -_\star a \cdot_\star k_1(s)) \cdot_\star \tau_1(s)$$
$$+_\star a \cdot_\star \varkappa_1(s) \cdot_\star \beta_1(s), \quad s \in I.$$

Therefore

$$\langle (d^\star f_2 /_\star d_\star s)(s), (d^\star f_2 /_\star d_\star s)(s) \rangle_\star = (a^{2\star} +_\star b^{2\star}) \cdot_\star |\varkappa_1(s)|_\star^{2\star},$$

$s \in I$. Note that

$$\langle (d^\star f_2 /_\star d_\star s)(s), (d^\star f_2 /_\star d_\star s)(s) \rangle_\star = \langle (d^\star f_2 /_\star d_\star s_2)(s) \cdot_\star (d^\star s_2 /_\star d_\star s)(s),$$
$$(d^\star f_2 /_\star d_\star s)(s) \cdot_\star (d^\star s_2 /_\star d_\star s)(s) \rangle_\star$$
$$= ((d^\star s_2 /_\star d_\star s)(s))^{2\star}$$
$$= 1_\star /_\star (\cdot_\star (d^\star s /_\star d_\star s_2)(s))^{2\star}, \quad s \in I.$$

Consequently

$$1_\star = (a^{2\star} +_\star b^{2\star}) \cdot_\star |\varkappa_1(s)|_\star^{2\star} \cdot_\star ((d^\star s /_\star d_\star s_2)(s))^{2\star},$$

$s \in I$. From here, we conclude that

$$\kappa_1(s)\cdot_\star\left((d^\star s/_\star d_\star s_2)(s)\right) = \text{const}, \quad s \in I.$$

As above,

$$(1_\star -_\star a\cdot_\star k_1(s))\cdot_\star\left((d^\star s/_\star d_\star s_2)(s)\right) = \text{const},$$

$s \in I$. Now, we multiplicative differentiate equation (2.14) with respect to the parameter change $s$ and we find

$$
\begin{aligned}
k_2(s)\cdot_\star \nu_2(s) =\ & (d^\star\tau_2/_\star d_\star s)(s) \\
=\ & (1_\star -_\star a\cdot_\star k_1(s))\cdot_\star(d^\star s/_\star d_\star s_2)(s) \\
=\ & (d^\star\tau_1/_\star d_\star s)(s) \\
& +_\star a\cdot_\star \kappa_1(s)\cdot_\star(d^\star s/_\star d_\star s_2)(s) \\
& \cdot_\star (d^\star\beta_1/_\star d_\star s)(s) \\
=\ & (1_\star -_\star a\cdot_\star k_1(s))\cdot_\star(d^\star s/_\star d_\star s_2)(s)\cdot_\star k_1(s)\cdot_\star \nu_1(s) \\
& -_\star a\cdot_\star \kappa_1(s)\cdot_\star(d^\star s/_\star d_\star s_2)(s)\cdot_\star \kappa_1(s)\cdot_\star \nu_1(s) \\
=\ & \left((1_\star -_\star a\cdot_\star k_1(s)\cdot_\star k_1(s) -_\star a\cdot_\star(\kappa_1(s))^2{}_\star)\right) \\
& \cdot_\star(d^\star s/_\star d_\star s_2)(s)\cdot_\star \nu_1(s), \quad s \in I.
\end{aligned}
$$

Thus, $\nu_1$ and $\nu_2$ are multiplicative collinear on $I$. Hence, we conclude that $f_1$ and $f_2$ are multiplicative Bertrand mates. This completes the proof.

---

## 2.12 The Local Behaviour of a Multiplicative Parameterized Curve

Suppose that $I \subseteq \mathbb{R}_\star$, $0_\star \in I$, $(I,f)$ is a multiplicative naturally parameterized curve and the point $f(0_\star)$ is a multiplicative biregular point. Set

$$
\begin{aligned}
\tau_0 &= \tau(0_\star), \\
\nu_0 &= \nu(0_\star), \\
\beta_0 &= \beta(0_\star).
\end{aligned}
$$

By the multiplicative Taylor formula, we get

$$f(s) = f(0_\star) +_\star s \cdot_\star f^\star(0_\star) +_\star (1_\star/_\star e^2) \cdot_\star s^{2\star} \cdot_\star f^{\star\star}(0_\star)$$
$$+_\star (1_\star/_\star e^6) \cdot_\star s^{3\star} \cdot_\star f^{\star\star\star}(0_\star) + o_\star(s^{3\star}).$$

We have

$$f^\star(0_\star) = \tau_0,$$
$$f^{\star\star}(0_\star) = k(0_\star) \cdot_\star \nu_0.$$

We multiplicative differentiate with respect to $s$ the equation

$$f^{\star\star}(s) = k(s) \cdot_\star \nu(s), \quad s \in I,$$

and we get

$$f^{\star\star\star}(s) = k^\star(s) \cdot_\star \nu(s) +_\star k(s) \cdot_\star \nu^\star(s)$$
$$= k^\star(s) \cdot_\star \nu(s)$$
$$\quad +_\star k(s) \cdot_\star (-_\star k(s) \cdot_\star \tau(s) +_\star \varkappa(s) \cdot_\star \beta(s))$$
$$= -_\star (k(s))^{2\star} \cdot_\star \tau(s) +_\star k^\star(s) \cdot_\star \nu(s)$$
$$\quad +_\star k(s) \cdot_\star \varkappa(s) \cdot_\star \beta(s), \quad s \in I.$$

Hence,

$$f^{\star\star\star}(0_\star) = -_\star (k(0_\star))^{2\star} \cdot_\star \tau_0 +_\star k^\star(0_\star) \cdot_\star \nu_0$$
$$\quad +_\star k(0_\star) \cdot_\star \varkappa(0_\star) \cdot_\star \beta_0, \quad s \in I.$$

Therefore

$$f(s) -_\star f(0_\star) = s \cdot_\star \tau_0 +_\star (1_\star/_\star e^2) \cdot_\star s^{2\star} \cdot_\star k(0_\star) \cdot_\star \nu_0$$
$$\quad +_\star (1_\star/_\star e^6) \cdot_\star s^{3\star} \cdot_\star (-_\star (k(0_\star))^{2\star} \cdot_\star \tau_0 +_\star k^\star(0_\star) \cdot_\star \nu_0$$
$$\quad +_\star k(0_\star) \cdot_\star \varkappa(0_\star) \cdot_\star \beta_0) + o_\star(s^{3\star})$$
$$= \left(s -_\star \left(1_\star/_\star e^6\right) \cdot_\star (k(0_\star))^{2\star} \cdot_\star s^{3\star} +_\star o_\star(s^{3\star})\right) \cdot_\star \tau_0$$
$$\quad +_\star \left(\left(1_\star/_\star e^2\right) \cdot_\star k(0_\star) \cdot_\star s^{2\star} +_\star \left(1_\star/_\star e^6\right) \cdot_\star k^\star(0_\star)\right.$$
$$\quad \left. \cdot_\star s^{3\star} +_\star o_\star(s^{3\star})\right) \cdot_\star \nu_0$$
$$\quad +_\star \left(\left(1_\star/_\star e^6\right) \cdot_\star k(0_\star) \cdot_\star \varkappa(0_\star) \cdot_\star s^{3\star} +_\star o_\star(s^{3\star})\right) \cdot_\star \beta_0.$$

$s \in I$. If

$$f(s) \to_{\star} f(0_{\star}) = (x_1(s), x_2(s), x_3(s)), \ s \in I,$$

then

$$x_1(s) = s \to_{\star} (1_{\star}/_{\star} e^6)\cdot_{\star}(k(0_{\star}))^{2_{\star}}\cdot_{\star} s^{3_{\star}} +_{\star} o_{\star}(s^{3_{\star}})$$
$$x_2(s) = (1_{\star}/_{\star} e^2)\cdot_{\star} k(0_{\star})\cdot_{\star} s^{2_{\star}} +_{\star} (1_{\star}/_{\star} e^6)\cdot_{\star} k^{\star}(0_{\star})\cdot_{\star} s^{3_{\star}} +_{\star} o_{\star}(s^{3_{\star}})$$
$$x_3(s) = (1_{\star}/_{\star} e^6)\cdot_{\star} k(0_{\star})\cdot_{\star} \kappa(0_{\star})\cdot_{\star} s^{3_{\star}} +_{\star} o_{\star}(s^{3_{\star}}),$$

$s \in I$.

---

## 2.13 The Multiplicative Rigid Motion

Suppose that $I \subseteq \mathbb{R}_{\star}$.

**Definition 2.29:** A multiplicative rigid motion of $\mathbb{R}_{\star}^3$ is a map $D \colon \mathbb{R}_{\star}^3 \to \mathbb{R}_{\star}^3$,

$$Dx = A\cdot_{\star} x +_{\star} b,$$

where $A \in M_{\star 3 \times 3}$, $A^T \cdot_{\star} A = I_{\star}$, $\det_{\star} A = 1_{\star}$, $b \in \mathbb{R}_{\star}^3$. The map

$$x \to A\cdot_{\star} x, \ x \in \mathbb{R}_{\star}^3$$

is said to be homogeneous part of the multiplicative motion.

**Theorem 2.20:** *Let* $(I, f = f(t))$ *be a multiplicative biregular curve and* $\{\tau(t), \nu(t), \beta(t)\}$ *be its Frenet frame at* $t \in I$. *Let also,* $D \colon \mathbb{R}_{\star}^3 \to \mathbb{R}_{\star}^3$ *be a multiplicative rigid motion with homogeneous part A. Then*

$$\{A\cdot_{\star} \tau(t), \ A\cdot_{\star} \nu(t), \ A\cdot_{\star} \beta(t)\}$$

*is the multiplicative Frenet frame of* $(I, f_1 = Df)$ *at* $t$. *Moreover, $f$ and $f_1$ have the same multiplicative curvature and multiplicative torsion.*

*Proof.* We have

$$f_1(t) = A\cdot_{\star} f(t) +_{\star} b.$$

Then

$$f_1^\star(t) = A \cdot_\star f^\star(t),$$
$$f_1^{\star\star}(t) = A \cdot_\star f^{\star\star}(t),$$
$$f_1^{\star\star\star}(t) = A \cdot_\star f^{\star\star\star}(t).$$

Therefore

$$
\begin{aligned}
\tau_1(t) &= f_1^\star(t)/_\star |f_1^\star(t)|_\star \\
&= (A \cdot_\star f^\star(t))/_\star |A \cdot_\star f^\star(t)|_\star \\
&= (A \cdot_\star f^\star(t))/_\star |f^\star(t)|_\star \\
&= A \cdot_\star (f^\star(t)/_\star |f^\star(t)|_\star) \\
&= A \cdot_\star \tau(t), \quad t \in I,
\end{aligned}
$$

and

$$
\begin{aligned}
k_1(t) &= |f_1^\star(t) \times_\star f_1^{\star\star}(t)|_\star /_\star \left( |f_1^\star(t)|_\star^{3_\star} \right) \\[2mm]
&= |A \cdot_\star f^\star(t) \times_\star A \cdot_\star f^{\star\star}(t)|_\star /_\star \left( |A \cdot_\star f^\star(t)|_\star^{3_\star} \right) \\[2mm]
&= |f^\star(t) \times_\star f^{\star\star}(t)|_\star /_\star |f^\star(t)|_\star^{3_\star} \\
&= k(t), \quad t \in I,
\end{aligned}
$$

and

$$
\begin{aligned}
\mathbf{k}_1(t) &= \left( f_1^{\star\star}(t) \right)/_\star |f_1^\star(t)|_\star^{3_\star} \\
&\quad -_\star \left( \langle f_1^\star(t), f_1^{\star\star}(t) \rangle_\star \cdot_\star f_1^\star(t) \right)/_\star |f_1^\star|_\star^{4_\star} \\
&= (A \cdot_\star f^{\star\star}(t))/_\star |A \cdot_\star f^\star(t)|_\star^{2_\star} \\
&\quad -_\star (\langle A \cdot_\star f^\star, A \cdot_\star f^{\star\star}(t) \rangle_\star \cdot_\star (A \cdot_\star f^\star(t)))/_\star |A \cdot_\star f^\star(t)|_\star^{4_\star} \\
&= A \cdot_\star (f^{\star\star}(t)/_\star |f^\star(t)|_\star^{2_\star}) \\
&\quad -_\star A \cdot_\star ((\langle f^\star(t), f^{\star\star}(t) \rangle_\star \cdot_\star f^\star(t))/_\star |f^\star(t)|_\star^{4_\star}) \\
&= A \cdot_\star (f^{\star\star}(t)/_\star |f^\star(t)|_\star^{2_\star} -_\star (\langle f^\star(t), f^{\star\star}(t) \rangle_\star \cdot_\star f^\star(t))/_\star |f^\star(t)|_\star^{4_\star}) \\
&= A \cdot_\star \mathbf{k}(t), \quad t \in I,
\end{aligned}
$$

and

$$
\begin{aligned}
\nu_1(t) &= \mathbf{k}_1(t)/_\star k_1(t) \\
&= A \cdot_\star \mathbf{k}(t)/_\star k(t) \\
&= A \cdot_\star \nu(t), \quad t \in I,
\end{aligned}
$$

and

$$
\begin{aligned}
\beta_1(t) &= \left( f_1^\star(t) \times_\star f_1^{\star\star}(t) \right)/_\star |f_1^\star(t) \times_\star f_1^{\star\star}(t)|_\star \\
&= ((A \cdot_\star f^\star(t)) \times_\star (A \cdot_\star f^{\star\star}(t)))/_\star |(A \cdot_\star f^\star(t)) \times_\star (A \cdot_\star f^{\star\star}(t))|_\star \\
&= (A \cdot_\star (f^\star(t) \times_\star f^{\star\star}(t))/_\star |f^\star(t) \times_\star f^{\star\star}(t)|_\star) \\
&= A \cdot_\star ((f^\star(t) \times_\star f^{\star\star}(t))/_\star |f^\star(t) \times_\star f^{\star\star}(t)|_\star) \\
&= A \cdot_\star \beta(t), \quad t \in I,
\end{aligned}
$$

and

$$
\begin{aligned}
\varkappa_1(t) &= \langle f_1^\star(t) \times_\star f_1^{\star\star}(t), f_1^{\star\star\star}(t) \rangle_\star /_\star |f_1^\star(t) \times_\star f_1^{\star\star}(t)|_\star^{2\star} \\
&= \langle (A \cdot_\star f^\star(t)) \times_\star (A \cdot_\star f^{\star\star}(t)). A \times_\star f^{\star\star\star}(t) \rangle_\star \\
&\quad /_\star |(A \cdot_\star f^\star(t)) \times_\star (A \cdot_\star f^{\star\star}(t))|_\star^{2\star} \\
&= (\langle f^\star(t) \times_\star f^{\star\star}(t), f^{\star\star\star}(t) \rangle_\star)/_\star |f^\star(t) \times_\star f^{\star\star}(t)|_\star^{2\star} \\
&= \varkappa(t), \quad t \in I.
\end{aligned}
$$

This completes the proof.

---

## 2.14 The Existence Theorem

Suppose that $I \subseteq \mathbb{R}_\star$.

**Theorem 2.21:** *Let* $f, g \in \mathscr{C}_\star^1(I)$, $f > 0_\star$ *on* $I$. *Then there is a unique multiplicative parameterized curve* $(I, \phi = \phi(s))$ *for which*

$$
k(s) = f(s) \text{ and } \varkappa(s) = g(s), \quad s \in I.
$$

*Proof.* Let $\{e_1, e_2, e_3\}$ be a multiplicative frame in $\mathbb{R}^3_\star$ at $t_0 \in \mathbb{R}_\star$. Consider the system

$$x_1^\star = f(s) \cdot_\star x_2$$
$$x_2^\star = -_\star f(s) \cdot_\star x_1 +_\star g(s) \cdot_\star x_3$$
$$x_3^\star = -_\star g(s) \cdot_\star x_2, \quad s \in I.$$

Define the matrix

$$A(s) = \begin{pmatrix} 0_\star & f(s) & 0_\star \\ -_\star f(s) & 0_\star & g(s) \\ 0_\star & -_\star g(s) & 0_\star \end{pmatrix}, \quad s \in I.$$

Then we get the multiplicative Cauchy problem

$$x^\star = A(s) \cdot_\star x, \quad s \in I,$$
$$x(s_0) = (e_1, e_2, e_3).$$

The last Cauchy problem has a unique solution $x$. For it, we have

$$(x^T \cdot_\star x)^\star = (x^T)^\star \cdot_\star x +_\star x^T \cdot_\star x^\star$$
$$= x^T \cdot_\star A^T \cdot_\star x +_\star x^T \cdot_\star A \cdot_\star x$$
$$= x^T \cdot_\star (A^T +_\star A) \cdot_\star x$$
$$= 0_\star.$$

Thus,

$$(x^T \cdot_\star x) = \text{const.}$$

Since

$$x^T(s_0) \cdot_\star x(s_0) = I_\star,$$

we get

$$x^T \cdot_\star x = I_\star.$$

Define

$$\phi(s) = f_0 +_\star \int_{\star s_0}^s x_1(t)\cdot_\star d_\star t, \ s \in I.$$

We have

$$\phi^\star(s) = x_1(s),$$
$$|\phi^\star(s)|_\star = |x_1(s)|_\star$$
$$= 1_\star,$$
$$\phi^{\star\star}(s) = x_1^\star(s)$$
$$= f(s)\cdot_\star x_2(s), \ s \in I.$$

Hence,

$$\phi^\star(s) \times_\star \phi^{\star\star}(s) = f(s)\cdot_\star (x_1(s) \times_\star x_2(s))$$
$$\neq 0_\star, \ s \in I.$$

Moreover,

$$\phi^{\star\star\star}(s) = f^\star(s)\cdot_\star x_2(s) +_\star f(s)\cdot_\star x_2^\star(s)$$
$$= f^\star(s)\cdot_\star x_2(s) +_\star f(s)\cdot_\star (-_\star f(s)\cdot_\star x_1(s) +_\star g(s)\cdot_\star x_3(s))$$
$$= -_\star (f(s))^{2_\star}\cdot_\star x_1(s) +_\star f^\star(s)\cdot_\star x_2(s)$$
$$+_\star (f(s)\cdot_\star g(s))\cdot_\star x_3(s), \quad s \in I.$$

From here,

$$\langle \phi^\star(s) \times_\star \phi^{\star\star}(s), \phi^{\star\star\star}(s)\rangle_\star = \langle f(s)\cdot_\star (x_1(s) \times_\star x_2(s)), -_\star (f(s))^{2_\star}\cdot_\star x_1(s)$$
$$+_\star f^\star(s)\cdot_\star x_2(s) +_\star (f(s)\cdot_\star g(s))\cdot_\star x_3(s)\rangle_\star$$
$$= (f(s))^{2_\star}\cdot_\star g(s)\cdot_\star \langle x_1(s) \times_\star x_2(s), x_3(s)\rangle_\star$$
$$= (f(s))^{2_\star}\cdot_\star g(s), \quad s \in I.$$

Therefore

$$k(s) = |\phi^\star(s) \times_\star \phi^{\star\star}(s)|_\star /_\star (|\phi^\star(s)|_\star^{3_\star})$$
$$= f(s), \ s \in I,$$

and

$$\kappa(s) = \langle \phi^\star(s) \times_\star \phi^{\star\star}(s), \phi^{\star\star\star}(s) \rangle_\star /_\star |\phi^\star(s) \times_\star \phi^{\star\star}(s)|^{2\cdot}_\star$$
$$= ((f(s))^{2\cdot}\cdot_\star g(s))/_\star (f(s))^{2\cdot}_\star$$
$$= g(s), \quad s \in I.$$

This completes the proof.

---

## 2.15 The Uniqueness Theorem

**Theorem 2.22:** *Let $(I, f)$ and $(I, g)$ be two multiplicative biregular parameterized curves. If*

$$k(t) = k_1(t), \quad \kappa(t) = \kappa_1(t), \quad |f^\star(t)|_\star = |g^\star(t)|_\star, \quad t \in I.$$

*Then there is a multiplicative rigid motion $D: \mathbb{R}^3_\star \to \mathbb{R}^3_\star$ so that*

$$g = Df.$$

*Proof.* Let $t_0 \in I$ and $\{\tau, \nu, \beta\}$ and $\{\tau_1, \nu_1, \beta_1\}$ be the multiplicative Frenet frames of $f$ and $g$, respectively, at $t \in I$. Let also $D: \mathbb{R}^3_\star \to \mathbb{R}^3_\star$ be the multiplicative rigid motion so that

$$D\{\tau, \nu, \beta\} = \{\tau_1, \nu_1, \beta_1\}.$$

Let $f_2 = Df$ and $\{\tau_2, \nu_2, \beta_2\}$ be the multiplicative Frenet frame of $f_2$. We have

$$k_2(t) = k(t)$$
$$= k_1(t),$$
$$\kappa_2(t) = \kappa(t)$$
$$= \kappa_1(t),$$
$$|f_2^\star(t)|_\star = |g^\star(t)|_\star, \quad t \in I.$$

Thus, $\{\tau_2, \nu_2, \beta_2\}$ and $\{\tau_1, \nu_1, \beta_1\}$ satisfy the system

$$\tau^\star = |f^\star|_\star \cdot_\star k \cdot_\star \nu$$
$$\nu^\star = -_\star |f^\star|_\star \cdot_\star k \cdot_\star \tau +_\star |f|_\star \cdot_\star \kappa \cdot_\star \beta$$
$$\beta^\star = -_\star |f^\star|_\star \cdot_\star \kappa \cdot_\star \nu.$$

Since for $f = f_0$, these solutions coincide, i.e.,

$$\tau_2 = \tau_1$$
$$\nu_2 = \nu_1$$
$$\beta_2 = \beta_1 \text{ on } I.$$

Hence,

$$f_2^\star (t) /_\star |f_2^\star (t)|_\star = g^\star (t) /_\star |g^\star (t)|_\star , \quad t \in I,$$

whereupon

$$f_2^\star (t) = g^\star (t), \quad t \in I,$$

and

$$f_2 (t) -_\star g (t) = \text{const}, \quad t \in I,$$

and

$$f_2 (t) -_\star g (t) = f_2 (t_0) -_\star g (t_0)$$
$$= 0_\star, \quad t \in I.$$

This completes the proof.

## 2.16 Advanced Practical Problems

**Problem 2.1:** Prove that the multiplicative curve $f \colon \mathbb{R}_\star \to \mathbb{R}^3$, given by

$$f(t) = \left( \frac{t+1}{t+2}, \frac{1}{2+t}, t^3 + t^2 + t \right), \quad t \in \mathbb{R}_\star,$$

is a multiplicative regular curve.

**Problem 2.2:** Let $f: \mathbb{R}_\star \to \mathbb{R}_\star^3$ be given by

$$f(t) = \left( \frac{t+2}{t+3}, t^3, \frac{t+4}{t+5} \right), \quad t \in \mathbb{R}_\star.$$

Find the equation of the multiplicative tangent line at arbitrary point.

**Answer 2.23:**

$$F(\tau) = \left( \frac{t+2}{t+3} e^{\log \tau \log \frac{t}{(t+2)(t+3)}}, t^3 e^{3\log \tau}, \frac{t+4}{t+5} e^{\log \tau \log \frac{t}{(t+4)(t+5)}} \right),$$

$t, \tau \in \mathbb{R}_\star$.

**Problem 2.3:** Let $f: \mathbb{R}_\star \to \mathbb{R}_\star^3$ is given by

$$f(t) = (e^{2t}, e^{-3t}, e^{-4t}), \quad t \in \mathbb{R}_\star.$$

Find the equation of the multiplicative normal plane to $f$ at arbitrarily chosen point $f(t_0)$, $t_0 \in \mathbb{R}_\star$.

**Answer 2.24:**

$$2\log \frac{R_1(t)}{e^{2t_0}} - 3\log \frac{R_2(t)}{e^{-3t_0}} - 4\log \frac{R_3(t)}{e^{-4t_0}} = 0, \quad t \in \mathbb{R}_\star.$$

**Problem 2.4:** Let $f: \mathbb{R}_\star \to \mathbb{R}_\star^3$ is given by

$$f(t) = \left( e^{t^4}, e^{t^2}, e^{t^5} \right), \quad t \in \mathbb{R}_\star.$$

Find the equation of the multiplicative osculating plane at arbitrary point $t_0$.

**Answer 2.25:**

$$0 = 15 t_0 \log \frac{F_1(t)}{e^{t_0^4}} - 10 t_0^3 \log \frac{F_2(t)}{e^{t_0^2}} - 8 \log \frac{F_3(t)}{e^{t_0^5}}, \quad t, t_0 \in \mathbb{R}_\star.$$

**Problem 2.5:** Let $f: \mathbb{R}_\star \to \mathbb{R}_\star^3$ be defined by

$$f(t) = \left(e^t, e^t, e^{t^2}\right), \ t \in \mathbb{R}_\star.$$

Find the multiplicative curvature of $f$ at arbitrary point $t$.

**Answer 2.26:**

$$\frac{2\sqrt{2}}{e^{(2+4t^2)^{\frac{3}{2}}}}, \ t \in \mathbb{R}_\star.$$

# 3

## Multiplicative Plane Curves

### 3.1 Multiplicative Envelopes of Multiplicative Plane Curves

Suppose that $I, J, A \subseteq \mathbb{R}_\star$ and

$$f = f(t, \lambda), \quad t \in I, \quad \lambda \in A. \tag{3.1}$$

**Definition 3.1:** The multiplicative envelope of the family (3.1) is a multiplicative parameterized curve that at each point is a multiplicative tangent of a multiplicative curve of the family (3.1).

**Theorem 3.1:** *The multiplicative envelope of the family (3.1) is subject to the equations*

$$f = f(t, \lambda)$$

*and*

$$f_t^\star \times_\star f_\lambda^\star = 0_\star.$$

*Moreover, the multiplicative envelope and the classical envelope for the family (3.1) coincide over $R_\star^2$.*

*Proof.* Let $(J, g)$ be the multiplicative envelope of the family (3.1) and $P \in g$. Then $P$ is a multiplicative tangency point between $g$ and a multiplicative curve of the family (3.1). Thus, its equation depends on $\lambda$, i.e.,

$$g = g(\lambda), \quad \lambda \in A.$$

Since $P$ lies on a multiplicative curve of the family (3.1), then

$$g = f(t(\lambda), \lambda).$$

DOI: 10.1201/9781003299844-3

The multiplicative tangency condition between $g$ and $f(t, \lambda)$ is as follows:

$$g_\lambda^\star \parallel_\star f_t^\star,$$

whereupon

$$g_\lambda^\star \times_\star f_t^\star = 0_\star.$$

Hence, using that

$$g_\lambda^\star = f_t^\star \cdot_\star t^\star +_\star f_\lambda^\star,$$

we get

$$0_\star = \left(f_t^\star \cdot_\star t^\star +_\star f_\lambda^\star\right) \times_\star f_t^\star$$
$$= f_\lambda^\star \times_\star f_t^\star.$$

Let $f = (f_1, f_2)$. In fact, we have

$$f_t^\star \times_\star f_\lambda^\star = (0_\star, 0_\star, f_{1\lambda}^\star \cdot_\star f_{2t}^\star -_\star f_{1t}^\star \cdot_\star f_{2\lambda}^\star).$$

Hence,

$$0_\star = f_\lambda^\star \times_\star f_t^\star$$

is equivalent to the condition

$$f_{1\lambda}^\star \cdot_\star f_{2t}^\star -_\star f_{2t}^\star \cdot_\star f_{1\lambda}^\star = 0_\star$$

or

$$0_\star = e^0$$
$$= e^{\log f_{1\lambda}^\star \log f_{2t}^\star} -_\star e^{\log f_{1t}^\star \log f_{2\lambda}^\star}$$
$$= e^{\log f_{1\lambda}^\star \log f_{2t}^\star - \log f_{1t}^\star \log f_{2\lambda}^\star},$$

or

$$\log f_{1\lambda}^\star \log f_{2t}^\star - \log f_{1t}^\star \log f_{2\lambda}^\star = 0,$$

or

$$\log e^{\frac{\lambda f_{1\lambda}}{f_1}} \log e^{\frac{t f_{2t}}{f_2}} - \log e^{\frac{\lambda f_{2\lambda}}{f_2}} \log e^{\frac{t f_{1t}}{f_1}} = 0,$$

or

$$t\lambda \frac{f_{1\lambda} f_{2t}}{f_1 f_2} - t\lambda \frac{f_{2\lambda} f_{1t}}{f_2 f_1} = 0,$$

or

$$f_{1\lambda} f_{2t} - f_{2\lambda} f_{1t} = 0,$$

which is the equation of the classical envelope of the family (3.1). This completes the proof.

**Example 3.1:** Let $f \colon \mathbb{R}_\star^2 \to \mathbb{R}_\star^2$ be given by

$$f(t, \lambda) = (t^2 + \lambda t + \lambda^2, t^2 - \lambda t + \lambda^2), \quad (t, \lambda) \in \mathbb{R}_\star^2.$$

We have

$$f_1(t, \lambda) = t^2 + \lambda t + \lambda^2,$$
$$f_2(t, \lambda) = t^2 - \lambda t + \lambda^2, \quad (t, \lambda) \in \mathbb{R}_\star^2.$$

Then

$$f_{1\lambda}(t, \lambda) = t + 2\lambda,$$
$$f_{2\lambda}(t, \lambda) = -t + 2\lambda,$$
$$f_{1t}(t, \lambda) = 2t + \lambda,$$
$$f_{2t}(t, \lambda) = 2t - \lambda, \quad (t, \lambda) \in \mathbb{R}_\star^2.$$

Hence,

$$f_t(t, \lambda) = (f_{1t}(t, \lambda), f_{2t}(t, \lambda))$$
$$= (2t + \lambda, 2t - \lambda),$$
$$f_\lambda(t, \lambda) = (f_{1\lambda}(t, \lambda), f_{2\lambda}(t, \lambda))$$
$$= (t + 2\lambda, -t + 2\lambda), \quad (t, \lambda) \in \mathbb{R}_\star^2,$$

and the equation of the multiplicative envelope for the considered family is

$$
\begin{aligned}
0 &= (t + 2\lambda)(2t - \lambda) - (2t + \lambda)(-t + 2\lambda) \\
&= 2t^2 - t\lambda + 4\lambda t - 2\lambda^2 - (-2t^2 + 4\lambda t - \lambda t + 2\lambda^2) \\
&= 2t^2 + 3\lambda t - 2\lambda^2 + 2t^2 - 3\lambda t - 2\lambda^2 \\
&= 4t^2 - 4\lambda^2 \\
&= 4(t - \lambda)(t + \lambda), \quad (t, \lambda) \in \mathbb{R}_\star^2,
\end{aligned}
$$

or

$$
t = \lambda, \ (t, \lambda) \in \mathbb{R}_\star^2.
$$

**Example 3.2:** Let $f \colon \mathbb{R}_\star^2 \to \mathbb{R}_\star^2$ be given by

$$
f(t, \lambda) = \left( t^{2\star} +_\star \lambda \cdot_\star t +_\star \lambda^{2\star}, t^{2\star} -_\star \lambda \cdot_\star t +_\star \lambda^{2\star} \right),
$$

$(t, \lambda) \in \mathbb{R}_\star^2$. We have

$$
\begin{aligned}
f_1(t, \lambda) &= t^{2\star} +_\star \lambda \cdot_\star t +_\star \lambda^{2\star} \\
&= e^{(\log t)^2} +_\star e^{\log \lambda \log t} +_\star e^{(\log \lambda)^2} \\
&= e^{(\log t)^2 + \log \lambda \log t + (\log \lambda)^2}, \\
f_2(t, \lambda) &= t^{2\star} -_\star \lambda \cdot_\star t +_\star \lambda^{2\star} \\
&= e^{(\log t)^2} -_\star e^{\log \lambda \log t} +_\star e^{(\log \lambda)^2} \\
&= e^{(\log t)^2 - \log \lambda \log t + (\log \lambda)^2}, \quad (t, \lambda) \in \mathbb{R}_\star^2.
\end{aligned}
$$

Hence,

$$
f_{1t}(t, \lambda) = \left( \frac{2\log t}{t} + \frac{\log \lambda}{t} \right) e^{(\log t)^2 + \log \lambda \log t + (\log \lambda)^2},
$$

$$
f_{2t}(t, \lambda) = \left( \frac{2\log t}{t} - \frac{\log \lambda}{t} \right) e^{(\log t)^2 - \log \lambda \log t + (\log \lambda)^2},
$$

$$
f_{1\lambda}(t, \lambda) = \left( \frac{2\log \lambda}{\lambda} + \frac{\log t}{\lambda} \right) e^{(\log t)^2 + \log \lambda \log t + (\log \lambda)^2},
$$

$$
f_{2\lambda}(t, \lambda) = \left( \frac{2\log \lambda}{\lambda} - \frac{\log t}{\lambda} \right) e^{(\log t)^2 - \log \lambda \log t + (\log \lambda)^2},
$$

$(t, \lambda) \in \mathbb{R}_\star^2$, and

$$f_t(t, \lambda) = (f_{1t}(t, \lambda), f_{2t}(t, \lambda))$$

$$= \left( \left( \frac{2\log t}{t} + \frac{\log \lambda}{t} \right) e^{(\log t)^2 + \log \lambda \log t + (\log \lambda)^2}, \right.$$

$$\left( \frac{2\log t}{t} - \frac{\log \lambda}{t} \right) e^{(\log t)^2 - \frac{\log \lambda}{t} \log t + (\log \lambda)^2} \right),$$

$$f_\lambda(t, \lambda) = (f_{1\lambda}(t, \lambda), f_{2\lambda}(t, \lambda))$$

$$= \left( \left( \frac{2\log \lambda}{\lambda} + \frac{\log t}{\lambda} \right) e^{(\log t)^2 + \log \lambda \log t + (\log \lambda)^2}, \right.$$

$$\left( \frac{2\log \lambda}{\lambda} - \frac{\log t}{\lambda} \right) e^{(\log t)^2 - \log \lambda \log t + (\log \lambda)^2} \right),$$

$(t, \lambda) \in \mathbb{R}_\star^2$, and the equation of the multiplicative envelope is as follows:

$$0 = \left( \frac{2\log t}{t} + \frac{\lambda}{t} \right) \left( \frac{2\log \lambda}{\lambda} - \frac{\log t}{\lambda} \right) e^{2(\log t)^2 + 2(\log \lambda)^2}$$

$$- \left( \frac{2\log \lambda}{\lambda} + \frac{\log t}{\lambda} \right) \left( \frac{2\log t}{t} - \frac{\log \lambda}{t} \right) e^{2(\log t)^2 + 2(\log \lambda)^2}, \quad (t, \lambda) \in \mathbb{R}_\star^2,$$

whereupon

$$0 = \left( \frac{2\log t}{t} + \frac{\log \lambda}{t} \right) \left( \frac{2\log \lambda}{\lambda} - \frac{\log t}{\lambda} \right) - \left( \frac{2\log \lambda}{\lambda} + \frac{\log t}{\lambda} \right) \left( \frac{2\log t}{t} - \frac{\log \lambda}{t} \right),$$

$$0 = (2\log t + \log \lambda)(2\log \lambda - \log t) - (2\log \lambda + \log t)(2\log t + \log \lambda)$$

$$= 4\log t \log \lambda + 2(\log \lambda)^2 - 2(\log t)^2 - \log \lambda \log t$$

$$\quad - 4\log \lambda \log t + 2(\log \lambda)^2 - 2(\log t)^2 + \log \lambda \log t$$

$$= 4(\log \lambda)^2 - 4(\log t)^2, \quad (t, \lambda) \in \mathbb{R}_\star^2,$$

or

$$\log \lambda = \pm \log t, \quad (t, \lambda) \in \mathbb{R}_\star^2,$$

or

$$\lambda = t, \quad \lambda t = 1, \quad (t, \lambda) \in \mathbb{R}_\star^2.$$

**Example 3.3:** Let $f\colon \mathbb{R}_\star^2 \to \mathbb{R}_\star^2$ be given by

$$f(t, \lambda) = (t +_\star \lambda, t -_\star \lambda), \quad (t, \lambda) \in \mathbb{R}_\star^2.$$

We have

$$
\begin{aligned}
f_1(t, \lambda) &= t +_\star \lambda \\
&= t\lambda, \\
f_2(t, \lambda) &= t -_\star \lambda \\
&= \tfrac{t}{\lambda}, \quad (t, \lambda) \in \mathbb{R}_\star^2.
\end{aligned}
$$

Hence,

$$
\begin{aligned}
f_{1t}(t, \lambda) &= \lambda, \\
f_{2t}(t, \lambda) &= \tfrac{1}{\lambda}, \\
f_{1\lambda}(t, \lambda) &= t, \\
f_{2\lambda}(t, \lambda) &= -\tfrac{t}{\lambda^2}, \quad (t, \lambda) \in \mathbb{R}_\star^2,
\end{aligned}
$$

and

$$
\begin{aligned}
f_t(t, \lambda) &= (f_{1t}(t, \lambda), f_{2t}(t, \lambda)) \\
&= \left(\lambda, \tfrac{1}{\lambda}\right), \\
f_\lambda(t, \lambda) &= (f_{1\lambda}(t, \lambda), f_{2\lambda}(t, \lambda)) \\
&= \left(t, -\tfrac{t}{\lambda^2}\right), \quad (t, \lambda) \in \mathbb{R}_\star^2,
\end{aligned}
$$

and the equation of the multiplicative envelope is

$$0 = -\frac{t}{\lambda} - \frac{t}{\lambda}, \quad (t, \lambda) \in \mathbb{R}_\star^2,$$

which has no any solutions in $\mathbb{R}_\star^2$. Thus, the considered family has no any multiplicative envelopes.

**Exercise 3.1:** Let $f\colon \mathbb{R}_\star^2 \to \mathbb{R}_\star^2$ be given by

$$f(t, \lambda) = (t^2 +_\star \lambda^2, t^2 -_\star \lambda^2), \quad (t, \lambda) \in \mathbb{R}_\star^2.$$

Find the equation of the multiplicative envelope of the considered family.

**Answer 3.1:** *The considered family has no any multiplicative envelopes.*

**Exercise 3.2:** Let $f \colon \mathbb{R}_\star^2 \to \mathbb{R}_\star^2$ be given by

$$f(t, \lambda) = (t^2 + \lambda^2, t^2 - \lambda^2), \quad (t, \lambda) \in \mathbb{R}_\star^2.$$

Find the equation of the multiplicative envelope of the considered family.

**Answer 3.2:** *The considered family has no any multiplicative envelopes.*

Now, suppose that the family of multiplicative curves is given by the equation

$$F(x, y, \lambda) = 0_\star, \tag{3.2}$$

where $F$ is $\mathscr{C}_\star^1$-function with respect to its arguments.

**Theorem 3.2:** *The multiplicative envelope for the family (3.2) satisfies the system*

$$
\begin{aligned}
F(x, y, \lambda) &= 0_\star \\
F_\lambda^\star(x, y, \lambda) &= 0_\star.
\end{aligned}
\tag{3.3}
$$

*Proof.* Locally, around each point a multiplicative curve of the family can be represented in the form

$$
\begin{aligned}
x &= x(t, \lambda) \\
y &= y(t, \lambda)
\end{aligned}
$$

and the equation (3.2) can be written as follows:

$$F(x(t, \lambda), y(t, \lambda), \lambda) = 0_\star. \tag{3.4}$$

Let

$$f(t, \lambda) = (x(t, \lambda), y(t, \lambda)).$$

We have

$$
\begin{aligned}
f_t^\star(t, \lambda) &= (x_t^\star(t, \lambda), y_t^\star(t, \lambda)), \\
f_\lambda^\star(t, \lambda) &= (x_\lambda^\star(t, \lambda), y_\lambda^\star(t, \lambda)).
\end{aligned}
$$

By the equation

$$f_t \times_\star f_\lambda = 0_\star,$$

we get

$$x_t^\star(t, \lambda) \cdot_\star y_\lambda^\star(t, \lambda) -_\star x_\lambda^\star(t, \lambda) \cdot_\star y_t^\star(t, \lambda) = 0_\star$$

and then there is a constant $K \in \mathbb{R}_\star$ so that

$$x_t^\star(t, \lambda) = K \cdot_\star x_\lambda^\star(t, \lambda)$$
$$y_t^\star(t, \lambda) = K \cdot_\star y_\lambda(t,'\lambda).$$

Now, we multiplicative differentiate equation (3.4) with respect to $t$ and $\lambda$ and find

$$F_x^\star \cdot_\star x_t +_\star F_y^\star \cdot_\star y_t = 0_\star$$
$$F_x^\star \cdot_\star x_\lambda^\star +_\star F_y^\star \cdot_\star y_\lambda^\star +_\star F_\lambda^\star = 0_\star,$$

whereupon

$$
\begin{aligned}
0_\star &= F_x^\star \cdot_\star K \cdot_\star x_t^\star +_\star F_y^\star \cdot_\star K \cdot_\star y_t^\star +_\star F_\lambda^\star \\
&= K \cdot_\star (F_x^\star \cdot_\star x_t^\star +_\star F_y^\star \cdot_\star y_t^\star) +_\star F_\lambda^\star \\
&= F_\lambda^\star.
\end{aligned}
$$

This completes the proof.

The system (3.3) can be written as follows:

$$F(x, y, \lambda) = 1$$
$$F_\lambda(x, y, \lambda) = 0.$$

**Example 3.4:** Let us consider the family

$$x^2 - y^2 - \lambda^2 + \lambda = 1.$$

Then

$$F(x, y, \lambda) = x^2 - y^2 - \lambda^2 + \lambda,$$
$$F_\lambda(x, y, \lambda) = -2\lambda + 1.$$

Thus, we get the system

$$x^2 - y^2 - \lambda^2 - \lambda = 1$$
$$- 2\lambda + 1 = 0.$$

Therefore $\lambda = \frac{1}{2}$ and the equation of the multiplicative envelope to the considered family is

$$x^2 - y^2 = \frac{7}{4}.$$

**Example 3.5:** Let us consider the family

$$x^{2_\star} -_\star y^{2_\star} +_\star \lambda^{2_\star} = 0_\star.$$

Then

$$e^{(\log x)^2} -_\star e^{(\log y)^2} +_\star e^{(\log \lambda)^2} = 1$$

or

$$e^{(\log x)^2 - (\log y)^2 + (\log \lambda)^2} = 1.$$

Therefore we get the system

$$e^{(\log x)^2 - (\log y)^2 + (\log \lambda)^2} = e^0$$
$$\frac{2 \log \lambda}{\lambda} e^{(\log x)^2 - (\log y)^2 + (\log \lambda)^2} = 0.$$

Hence, $\lambda = 0_\star$ and the equation of the multiplicative envelope to the considered family is

$$x^{2_\star} -_\star y^{2_\star} = 0_\star.$$

**Example 3.6:** Let us consider the family

$$x^{2_\star} -_\star y^{2_\star} +_\star \lambda = 0_\star.$$

We have

$$e^{(\log x)^2 - (\log y)^2 + \log \lambda} = 1.$$

We obtain the system

$$e^{(\log x)^2 - (\log y)^2 + \log \lambda} = 1$$
$$\frac{1}{\lambda} e^{(\log x)^2 - (\log y)^2 + \log \lambda} = 0.$$

The last system has no any solution for $\lambda \in \mathbb{R}_\star$ and thus, the considered family has no any multiplicative envelope.

**Exercise 3.3:** Find the equation of the multiplicative envelope of the family

$$x^{2\star} +_\star y^{2\star} +_\star \lambda^{2\star} -_\star \lambda = 0_\star.$$

**Answer 3.3:**

$$x^{2\star} +_\star y^{2\star} +_\star e^{-\frac{1}{4}} = 0_\star.$$

---

## 3.2 The Multiplicative Evolute

Suppose that $I \subseteq \mathbb{R}_\star$.

**Definition 3.2:** Let $(I, f = f(t))$ be a multiplicative parameterized curve. The multiplicative evolute of $f$ is said to be the multiplicative envelope of the family of the multiplicative normals to $f$.
Let

$$f(t) = (f_1(t), f_2(t)), \quad t \in I.$$

Suppose that $f \in \mathscr{C}^2_\star(I)$. The equation of the multiplicative normals to $f$ is as follows:

$$(X -_\star f_1(t)) \cdot_\star f_1^\star(t) +_\star (Y -_\star f_2(t)) \cdot_\star f_2^\star(t) = 0_\star, \quad t \in I,$$

or

$$f_1^\star(t) \cdot_\star X +_\star f_2^\star(t) \cdot_\star Y = f_1(t) \cdot_\star f_1^\star(t) +_\star f_2(t) \cdot_\star f_2^\star(t), \quad t \in I.$$

We multiplicative differentiate it with respect to $t$ and we find

$$0_\star = (X \rightarrow_\star f_1(t)) \cdot_\star f_1^{\star\star}(t) \rightarrow_\star (f_1^\star(t))^{2_\star} +_\star (Y \rightarrow_\star f_2(t)) \cdot_\star f_2^{\star\star}(t)$$
$$\rightarrow_\star (f_2^\star(t))^{2_\star}, \ t \in I,$$

whereupon

$$f_1^{\star\star}(t) \cdot_\star X +_\star f_2^{\star\star}(t) \cdot_\star Y = f_1(t) \cdot_\star f_1^{\star\star}(t) +_\star f_2(t) \cdot_\star f_2^{\star\star}(t)$$
$$+_\star (f_1^\star(t))^{2_\star} +_\star (f_2^\star(t))^{2_\star}.$$

Therefore the multiplicative envelope of $f$ satisfies the system

$$f_1^\star(t) \cdot_\star X +_\star f_2^\star(t) \cdot_\star Y = f_1(t) \cdot_\star f_1^\star(t) +_\star f_2(t) \cdot_\star f_2^\star(t)$$
$$f_1^{\star\star}(t) \cdot_\star X +_\star f_2^{\star\star}(t) \cdot_\star Y = f_1(t) \cdot_\star f_1^{\star\star}(t) +_\star f_2(t) \cdot_\star f_2^{\star\star}(t)$$
$$+_\star (f_1^\star(t))^{2_\star} +_\star (f_2^\star(t))^{2_\star}, \quad t \in I.$$

**Example 3.7:** Let $f: \mathbb{R}_\star \to \mathbb{R}_\star^2$ be given by

$$f(t) = \left(e^{t^2}, e^{3t^2}\right), \ t \in I.$$

Here

$$f_1(t) = e^{t^2},$$
$$f_2(t) = e^{3t^2}, \ t \in I.$$

Then

$$f_1^\star(t) = e^{2t^2},$$
$$f_2^\star(t) = e^{6t^2},$$
$$f_1^{\star\star}(t) = e^{4t^2},$$
$$f_2^{\star\star}(t) = e^{12t^2}, \ t \in I.$$

Then we get the system

$$e^{2t^2} \cdot_\star X +_\star e^{6t^2} \cdot_\star Y = e^{t^2} \cdot_\star e^{2t^2} +_\star e^{3t^2} \cdot_\star e^{6t^2}$$
$$e^{4t^2} \cdot_\star X +_\star e^{12t^2} \cdot_\star Y = e^{t^2} \cdot_\star e^{4t^2} +_\star e^{3t^2} \cdot_\star e^{12t^2}, \ t \in I,$$

or

$$e^{2t^2\log X} +_\star e^{6t^2\log Y} = e^{2t^4} +_\star e^{18t^4}$$
$$e^{4t^2\log X} +_\star e^{12t^2\log Y} = e^{4t^4} +_\star e^{36t^4}, \quad t \in I,$$

or

$$e^{2t^2(\log X + 3\log Y)} = e^{20t^4}$$
$$e^{4t^2(\log X + 3\log Y)} = e^{40t^4}, \quad t \in I,$$

or

$$\log X + \log Y^3 = 10t^2, \quad t \in I,$$

or

$$\log(XY^3) = 10t^2, \quad t \in I,$$

or

$$XY^3 = e^{10t^2}, \quad t \in I.$$

**Exercise 3.4:** Let $f: \mathbb{R}_\star \to \mathbb{R}_\star^2$ be given by

$$f(t) = \left(e^{t^3}, e^{2t^3}\right), \quad t \in \mathbb{R}_\star.$$

Find the equation of its multiplicative evolute.

**Answer 3.4:**

$$XY^2 = e^{6t^3}, \quad t \in \mathbb{R}_\star.$$

## 3.3 The Multiplicative Complex Structure on $\mathbb{R}_\star^2$

**Definition 3.3:** The multiplicative complex structure on $\mathbb{R}_\star^2$ is the map $J_\star: \mathbb{R}_\star^2 \to \mathbb{R}_\star^2$ defined by

$$J_\star u = \left(\frac{1}{u_2}, u_1\right), \quad u = (u_1, u_2) \in \mathbb{R}_\star^2.$$

Below, we will deduct some of the properties of the multiplicative complex structure. Suppose that $\{e_1, e_2, e_3\}$ is a multiplicative orthonormal basis in $\mathbb{R}^3_\star$ and

$$u = (u_1, u_2),$$
$$v = (v_1, v_2) \in \mathbb{R}^2_\star.$$

1.

$$\langle J_\star u, J_\star v \rangle_\star = \langle u, v \rangle_\star.$$

*Proof.* We have

$$J_\star u = \left(\frac{1}{u_2}, u_1\right),$$
$$J_\star v = \left(\frac{1}{v_2}, v_1\right).$$

Then

$$\langle J_\star u, J_\star v \rangle = e^{\log \frac{1}{u_2} \log \frac{1}{v_2} + \log u_1 \log v_1}$$
$$= e^{\log u_2 \log v_2 + \log u_1 \log v_1}$$
$$= \langle u, v \rangle_\star.$$

This completes the proof.

2.

$$\langle J_\star u, u \rangle_\star = 0_\star.$$

*Proof.* We have

$$J_\star u = \left(\frac{1}{u_2}, u_1\right)$$

and

$$\langle J_\star u, u \rangle_\star = e^{\log \frac{1}{u_2} \log u_1 + \log u_1 \log u_2}$$
$$= e^{-\log u_1 \log u_2 + \log u_1 \log u_2}$$
$$= e^0$$
$$= 0_\star.$$

This completes the proof.

3.

$$\langle v, J_\star u \rangle_\star = \langle u \times_\star v, e_3 \rangle_\star.$$

*Proof.* We have

$$u \times_\star v = \begin{vmatrix} e_1 & e_2 & e_3 \\ u_1 & u_2 & 0_\star \\ v_1 & v_2 & 0_\star \end{vmatrix}_\star$$

$$= e_3 \cdot_\star \begin{vmatrix} u_1 & u_2 \\ v_1 & v_2 \end{vmatrix}_\star$$

$$= e_3 \cdot_\star (u_1 \cdot_\star v_2 -_\star u_2 \cdot_\star v_1)$$

and

$$J_\star u = \left( \frac{1}{u_2}, u_1 \right),$$

$$\langle u \times_\star v, e_3 \rangle_\star = u_1 \cdot_\star v_2 -_\star u_2 \cdot_\star v_1,$$

and

$$\langle v, J_\star u \rangle_\star = e^{\log v_1 \log \frac{1}{u_2} + \log v_2 \log u_1}$$

$$= e^{-\log v_1 \log u_2 + \log v_2 \log u_1}$$

$$= (u_1 \cdot_\star v_2 -_\star u_2 \cdot_\star v_1).$$

Thus, we get the desired result. This completes the proof.

4.

$$J_\star (J_\star u) = -_\star u.$$

*Proof.* We have

$$J_\star u = \left( \frac{1}{u_2}, u_1 \right)$$

and

$$J_\star (J_\star u) = \left( \tfrac{1}{u_1}, \tfrac{1}{u_2} \right)$$
$$= (-_\star u_1, -_\star u_2)$$
$$= -_\star (u_1, u_2)$$
$$= -_\star u.$$

This completes the proof.

---

## 3.4 Multiplicative Curvature of Multiplicative Plane Curves

Suppose that $(I, f = f(t))$ is a multiplicative parameterized plane curve.

**Definition 3.4:** The multiplicative curvature of $f$ is defined by

$$k_{\pm_\star} = (\langle f^{\star\star}, J_\star f^\star \rangle_\star) /_\star |f^\star|_\star^{3_\star}.$$

Let

$$f = (f_1, f_2).$$

Then

$$f^\star = \left( f_1^\star, f_2^\star \right),$$
$$f^{\star\star} = \left( f_1^{\star\star}, f_2^{\star\star} \right),$$
$$J_\star f^\star = \left( \tfrac{1}{f_2^\star}, f_1^\star \right)$$

and

$$\langle f^{\star\star}, J_\star f^\star \rangle_\star = e^{\log f_1^{\star\star} \log \frac{1}{f_2^\star} + \log f_2^{\star\star} \log f_1^\star}$$
$$= e^{\log f_1^\star \log f_2^{\star\star} - \log f_1^{\star\star} \log f_2^\star}$$

and

$$|f^\star|_\star = e^{\left(\left(\log f_1^\star\right)^2 + \left(\log f_2^\star\right)^2\right)^{\frac{1}{2}}},$$

$$|f^\star|_\star^{3_\star} = e^{\left(\left(\log f_1^\star\right)^2 + \left(\log f_2^\star\right)^2\right)^{\frac{3}{2}}}.$$

Thus,

$$k_{\pm_\star} = e^{\dfrac{\log f_1^\star \, \log f_2^{\star\star} - \log f_1^{\star\star} \, \log f_2^\star}{\left(\left(\log f_1^\star\right)^2 + \left(\log f_2^\star\right)^2\right)^{\frac{3}{2}}}}. \qquad (3.5)$$

**Example 3.8:** Let $f \colon \mathbb{R}_\star \to \mathbb{R}_\star^2$ be defined by

$$f(t) = (t^2, t + t^2), \quad t \in \mathbb{R}_\star.$$

Here

$$f_1(t) = t^2,$$
$$f_2(t) = t + t^2, \quad t \in \mathbb{R}_\star.$$

Then

$$f_1'(t) = 2t,$$
$$f_2'(t) = 1 + 2t, \quad t \in \mathbb{R}_\star,$$

and

$$f_1^\star(t) = e^{t \frac{f_1'(t)}{f_1(t)}}$$
$$= e^{t \frac{2t^2}{t^2}}$$
$$= e^2,$$
$$f_2^\star(t) = e^{t \frac{f_2'(t)}{f_2(t)}}$$
$$= e^{t \frac{1+2t}{t+t^2}}$$
$$= e^{\frac{1+2t}{1+t}}, \quad t \in \mathbb{R}_\star,$$

and

$$(f_1^\star)'(t) = 0,$$

$$(f_2^\star)'(t) = \frac{2(1+t) - (1+2t)}{(1+t)^2} e^{\frac{1+2t}{1+t}}$$

$$= \frac{1}{(1+t)^2} e^{\frac{1+2t}{1+t}}, \quad t \in \mathbb{R}_\star,$$

and

$$f_1^{\star\star}(t) = e^{t\frac{\left(f_1^\star\right)'(t)}{f_1^\star(t)}}$$

$$= 1,$$

$$f_2^{\star\star}(t) = e^{t\frac{\left(f_2^\star\right)'(t)}{f_2^\star(t)}}$$

$$= e^{t\frac{\frac{1}{(1+t)^2}e^{\frac{1+2t}{1+t}}}{e^{\frac{1+2t}{1+t}}}}$$

$$= e^{\frac{t}{(1+t)^2}}, \quad t \in \mathbb{R}_\star.$$

Therefore

$$f^\star(t) = \left(f_1^\star(t), f_2^\star(t)\right)$$

$$= \left(e^2, e^{\frac{1+2t}{1+t}}\right),$$

$$f^{\star\star}(t) = \left(f_1^{\star\star}(t), f_2^{\star\star}(t)\right)$$

$$= \left(1, e^{\frac{t}{(1+t)^2}}\right), \quad t \in \mathbb{R}_\star.$$

Hence,

$$k_{\pm_\star}(t) = e^{\dfrac{\log e^2 \log e(1+t)^2 \overline{\dfrac{t}{(1+t)^2}} - \log 1 \log e \overline{\dfrac{1+2t}{1+t}}}{\left(\left(\log e^2\right)^2 + \left(\log e \dfrac{1+2t}{1+t}\right)^2\right)^{\frac{1}{2}}}}$$

$$= e^{\dfrac{\frac{2t}{(1+t)^2}}{\left(4 + \frac{(1+2t)^2}{(1+t)^2}\right)^{\frac{3}{2}}}}$$

$$= e^{\dfrac{\frac{2t}{(1+t)^2}}{\frac{(4+8t+4t^2+1+4t+4t^2)^{\frac{3}{2}}}{(1+t)^3}}}$$

$$= e^{\dfrac{2t(1+t)}{(5+12t+8t^2)^{\frac{3}{2}}}}, \quad t \in \mathbb{R}_\star.$$

**Example 3.9:** Let $f \colon \mathbb{R}_\star \to \mathbb{R}_\star^2$ be defined by

$$f(t) = \left(e^{t^2}, e^{t^3}\right), \quad t \in \mathbb{R}_\star.$$

Here

$$f_1(t) = e^{t^2},$$
$$f_2(t) = e^{t^3}, \quad t \in \mathbb{R}_\star.$$

Then

$$f_1'(t) = 2t e^{t^2},$$
$$f_2'(t) = 3t^2 e^{t^3}, \quad t \in \mathbb{R}_\star,$$

and

$$f_1^\star(t) = e^{t \frac{f_1'(t)}{f_1(t)}}$$
$$= e^{2t^2},$$
$$f_2^\star(t) = e^{t \frac{f_2'(t)}{f_2(t)}}$$
$$= e^{3t^3}, \quad t \in \mathbb{R}_\star,$$

and

$$(f_1^\star)'(t) = 4te^{2t^2},$$
$$(f_2^\star)^\star(t) = 9t^2 e^{3t^3}, \ t \in \mathbb{R}_\star,$$

and

$$f_1^{\star\star}(t) = e^{t \frac{\left(f_1^\star\right)'(t)}{f_1^\star(t)}}$$
$$= e^{4t^2},$$

$$f_2^{\star\star}(t) = e^{t \frac{\left(f_2^\star\right)'(t)}{f_2^\star(t)}}$$
$$= e^{9t^3}, \ t \in \mathbb{R}_\star.$$

Consequently

$$k_{\pm_\star}(t) = e^{\frac{\log_e 2t^2 \log_e 9t^3 - \log_e 4t^2 \log_e 3t^3}{\left(\left(\log_e 2t^2\right)^2 + \left(\log_e 3t^3\right)^2\right)^{\frac{3}{2}}}}$$

$$= e^{\frac{18t^5 - 12t^5}{(4t^4 + 9t^6)^{\frac{3}{2}}}}$$

$$= e^{t^6 \frac{6t^5}{(4 + 9t^2)^{\frac{3}{2}}}}$$

$$= e^{t \frac{6}{(4 + 9t^2)^{\frac{3}{2}}}}, \ t \in \mathbb{R}_\star.$$

**Example 3.10:** Let $f: \mathbb{R}_\star \to \mathbb{R}_\star^2$ be given by

$$f(t) = (t, g(t)), \ t \in \mathbb{R}_\star,$$

where $g \in \mathscr{C}_\star^2(\mathbb{R}_\star)$. Then

$$f_1^\star(t) = e,$$
$$f_2^\star(t) = g^\star(t), \ t \in \mathbb{R}_\star,$$

and

$$f_1^{\star\star}(t) = 1,$$
$$f_2^{\star\star}(t) = g^{\star\star}(t), \ t \in \mathbb{R}_\star.$$

Hence,

$$k_{\pm_\star}(t) = e^{\dfrac{g^{\star\star}(t)}{\left(1+(g^\star(t))^2\right)^{\frac{3}{2}}}}, \ t \in \mathbb{R}_\star.$$

**Exercise 3.5:** Let $f \colon \mathbb{R}_)\star \to \mathbb{R}_\star^2$ be given by

$$f(t) = \left(e^{t^3}, e^{3t^2}\right), \ t \in \mathbb{R}_\star.$$

Find $k_{\pm_\star}(t), \ t \in \mathbb{R}_\star$.

**Answer 3.5:**

$$k_{\pm_\star}(t) = e^{-\dfrac{2}{3t(t^2+4)^{\frac{3}{2}}}}, \ t \in \mathbb{R}_\star.$$

By the properties of the multiplicative complex structure, it follows that

$$k_{\pm_\star} = (f^\star \times_\star f^{\star\star})/_\star |f^\star|_\star^{3_\star}.$$

Thus,

$$|k_{\pm_\star}|_\star = k.$$

**Theorem 3.3:** *Let* $(I, f = f(t))$ *be a multiplicative parameterized curve and* $(J, g = g(s(t)))$ *be a multiplicative naturally parameterized curve equivalent to f. Then*

$$k_{\pm_\star}(g)(s(t)) = k_{\pm_\star}(f)(t), \ t \in I.$$

*Proof.* We have

$$f(t) = g(s(t)), \ t \in I,$$

and

$$f^\star(t) = g^\star(s(t))\cdot_\star s^\star(t),$$
$$f^{\star\star}(t) = g^{\star\star}(s(t))\cdot_\star(s^\star(t))^{2_\star} +_\star g^\star(s(t))\cdot_\star s^{\star\star}(t), \quad t \in I.$$

Hence,

$$\langle f^{\star\star}(t), J_\star f^\star(t)\rangle_\star = \langle g^{\star\star}(s(t))\cdot_\star(s^\star(t))^{2_\star} +_\star g^\star(s(t))\cdot_\star s^{\star\star}(t), g^\star(s(t))$$
$$\cdot_\star s^\star(t)\rangle_\star = \langle g^{\star\star}(s(t))\cdot_\star(s^\star(t))^{2_\star}, s^\star(t)\cdot_\star J_\star(g^\star(s(t)))\rangle_\star$$
$$+_\star \langle g^\star(s(t))\cdot_\star s^{\star\star}(t), s^\star(t)\cdot_\star J_\star(g^\star(s(t)))\rangle_\star = (s^\star(t))^{3_\star}$$
$$\cdot_\star \langle g^{\star\star}(s(t))\cdot_\star J_\star(g^\star(s(t)))\rangle_\star +_\star (s^{\star\star}(t)\cdot_\star s^\star(t))$$
$$\cdot_\star \langle g^\star(s(t))\cdot_\star J_\star(g^\star(s(t)))\rangle_\star = (s^\star(t))^{3_\star}$$
$$\cdot_\star \langle g^{\star\star}(s(t)), J_\star(g^\star(s(t)))\rangle_\star = (s^\star(t))^{3_\star}\cdot_\star k_{\pm_\star}(g)(s(t))$$
$$= |f^\star(t)|_\star^{3_\star}\cdot_\star k_{\pm_\star}(g)(s(t)), \quad t \in I,$$

whereupon

$$k_{\pm_\star}(g)(s(t)) = \langle f^{\star\star}(t)\cdot_\star J_\star f^\star(t)\rangle_\star /_\star |f^\star(t)|_\star^{3_\star}$$
$$= k_{\pm_\star}(f)(t), \quad t \in I.$$

This completes the proof.

**Theorem 3.4:** *Let* $(I, f = f(s))$ *be a multiplicative naturally parameterized curve. Then*

$$f^{\star\star}(s) = k_{\pm_\star}(s)\cdot_\star f^\star(s), \quad s \in I.$$

*Proof.* We have

$$\langle f^\star(s), f^\star(s)\rangle_\star = 1_\star, \quad s \in I.$$

Then

$$\langle f^\star(s), f^{\star\star}(s)\rangle_\star = 0_\star, \quad s \in I,$$

and

$$f^\star \perp_\star f^{\star\star} \text{ on } I.$$

Since

$$f^\star \perp_\star J_\star f^\star \text{ on } I,$$

we get

$$f^{\star\star} \|_\star J_\star f^\star \text{ on } I,$$

and

$$f^{\star\star}(s) = \alpha(s) \cdot_\star J_\star f^\star(s), \ s \in I.$$

Hence,

$$
\begin{aligned}
k_{\pm_\star}(s) &= \langle f^{\star\star}(s), J_\star f^\star(s) \rangle_\star \\
&= \alpha(s) \cdot_\star \langle J_\star f^\star(s), J_\star f^\star(s) \rangle_\star \\
&= \alpha(s) \cdot_\star \langle f^\star(s), f^\star(s) \rangle_\star \\
&= \alpha(s), \ s \in I.
\end{aligned}
$$

Consequently

$$f^{\star\star}(s) = k_{\pm_\star}(s) \cdot_\star J_\star f^\star(s), \ s \in I.$$

This completes the proof.

---

## 3.5 Multiplicative Rotation Angle of Multiplicative Plane Curves

Let $(I, f = f(t))$ be a multiplicative parameterized plane curve.

**Definition 3.5:** The multiplicative rotation angle of $f$ is the function $\theta \colon I \to \mathbb{R}_\star$, defined by

$$\tau(t) = (\cos_\star \theta(t), \sin_\star \theta(t)), \ t \in I.$$

Here $\tau$ is the multiplicative unit tangent vector to $f$.

**Theorem 3.5:** *We have*

$$\theta^\star(t) = k_{\pm_\star}(t) \cdot_\star |f^\star(t)|_\star, \ t \in I.$$

*Proof.* We have

$$\tau(t) = f^{\star}(t)/_{\star} |f^{\star}(t)|_{\star} , \ t \in I.$$

Then

$$\tau^{\star}(t) = f^{\star\star}(t) \cdot_{\star} (1_{\star}/_{\star} |f^{\star}(t)|_{\star})$$
$$+_{\star} f^{\star}(t) \cdot_{\star} (1_{\star}/_{\star} |f^{\star}(t)|_{\star})^{\star}, \ t \in I,$$

and

$$\tau^{\star}(t) = (-_{\star} \sin_{\star} \theta(t) \cdot_{\star} \theta^{\star}(t), \ \cos_{\star} \theta(t) \cdot_{\star} \theta^{\star}(t))$$
$$= \theta^{\star}(t) \cdot_{\star} (-_{\star} \sin_{\star} \theta(t), \ \cos_{\star} \theta(t))$$
$$= \theta^{\star}(t) \cdot_{\star} J_{\star} \tau(t)$$
$$= \theta^{\star}(t) \cdot_{\star} J_{\star}((f^{\star}(t))/_{\star} |f^{\star}(t)|_{\star})$$
$$= \theta^{\star}(t) \cdot_{\star} J_{\star}((f^{\star}(t))/_{\star} |J_{\star} f^{\star}(t)|_{\star}), \ t \in I.$$

Thus,

$$f^{\star\star}(t) \cdot_{\star} (1_{\star}/_{\star} |f^{\star}(t)|_{\star}) +_{\star} f^{\star}(t) \cdot_{\star} (1_{\star}/_{\star} |f^{\star}(t)|_{\star})^{\star}$$
$$= \theta^{\star}(t) \cdot_{\star} J_{\star}((f^{\star}(t))/_{\star} |J_{\star} f^{\star}(t)|_{\star}), \ t \in I.$$

Hence,

$$\langle f^{\star\star}(t) \cdot_{\star} (1_{\star}/_{\star} |f^{\star}(t)|_{\star}), J_{\star} f^{\star}(t) \rangle_{\star} +_{\star} \langle f^{\star}(t) \cdot_{\star} (1_{\star}/_{\star} |f^{\star}(t)|_{\star})^{\star} \cdot J_{\star} f^{\star}(t) \rangle_{\star}$$
$$= \theta^{\star}(t) \cdot_{\star} (1_{\star}/_{\star} |J_{\star} f^{\star}(t)|_{\star}) \cdot_{\star} \langle J_{\star} f^{\star}(t), J_{\star} f^{\star}(t) \rangle_{\star}, \quad t \in I,$$

or

$$(1_{\star}/_{\star} |f^{\star}(t)|_{\star}) \cdot_{\star} \langle f^{\star\star}(t), J_{\star} f^{\star}(t) \rangle_{\star} = \theta^{\star}(t) \cdot_{\star} |J_{\star} f^{\star}(t)|_{\star}$$
$$= \theta^{\star}(t) \cdot_{\star} |f^{\star}(t)|_{\star}, \ t \in I,$$

whereupon

$$\theta^{\star}(t) = (\langle f^{\star\star}(t), J_{\star} f^{\star}(t) \rangle_{\star}/_{\star})(|f^{\star}(t)|_{\star}^{2_{\star}})$$
$$= k_{\pm_{\star}}(t) \cdot_{\star} |f^{\star}(t)|_{\star}, \ t \in I.$$

This completes the proof

## 3.6 Advanced Practical Problems

**Problem 3.1:** Let $f: \mathbb{R}_\star^2 \to \mathbb{R}_\star^2$ be given by

$$f(t, \lambda) = (t^2 +_\star \lambda^2, t^2 + \lambda^2), \quad (t, \lambda) \in \mathbb{R}_\star^2.$$

Find the equation of the multiplicative envelope of the considered family.

**Answer 3.6:**

$$t = \lambda, \quad (t, \lambda) \in \mathbb{R}_\star^2.$$

**Problem 3.2:** Find the equation of the multiplicative envelope of the family

$$x^{2_\star} -_\star y^{2_\star} +_\star \lambda^{2_\star} -_\star e^2 \cdot_\star \lambda = 0_\star.$$

**Answer 3.7:**

$$x^{2_\star} -_\star y^{2_\star} -_\star 1_\star = 0_\star.$$

**Problem 3.3:** Let $f: \mathbb{R}_\star \to \mathbb{R}_\star^2$ be given by

$$f(t) = (e^{4t}, e^t), \quad t \in \mathbb{R}_\star.$$

Find the equation of its multiplicative evolute.

**Answer 3.8:**

$$X^4 Y = e^{5t}, \quad t \in \mathbb{R}_\star.$$

**Problem 3.4:** Let $f: \mathbb{R}_\star \to \mathbb{R}_\star^2$ be defined by

$$f(t) = \left(e^{2t^2}, e^{t^4}\right), \quad t \in \mathbb{R}_\star.$$

Find $k_{\pm_\star}(t)$, $t \in \mathbb{R}_\star$.

**Answer 3.9:**

$$k_{\pm_\star}(t) = e^{\frac{1}{2(1+t^4)^{\frac{3}{2}}}}, \quad t \in \mathbb{R}_\star.$$

# 4

# General Theory of Multiplicative Surfaces

## 4.1 Multiplicative Parameterized Surfaces

Suppose that $U \subseteq \mathbb{R}_\star^2$.

**Definition 4.1:** A multiplicative regular parameterized surface in $\mathbb{R}_\star^3$ is a multiplicative smooth map $f: U \to \mathbb{R}_\star^3$, $(t_1, t_2) \to f(t_1, t_2)$, and

$$f_{t_1}^\star \times_\star f_{t_2}^\star \neq 0_\star \quad \text{on} \quad U. \tag{4.1}$$

**Definition 4.2:** The condition (4.1) is called the multiplicative regularity condition.

**Example 4.1:** Let $f: \mathbb{R}_\star^2 \to \mathbb{R}_\star^3$ be given by

$$f(t_1, t_2) = \left(t_1 + t_2, e^{t_1^2 + t_2}, e^{t_1 + t_2^2}\right), \quad (t_1, t_2) \in \mathbb{R}_\star^2.$$

Here

$$f_1(t_1, t_2) = t_1 + t_2,$$
$$f_2(t_1, t_2) = e^{t_1^2 + t_2},$$
$$f_3(t_1, t_2) = e^{t_1 + t_2^2}, \quad (t_1, t_2) \in \mathbb{R}_\star^2.$$

Then

DOI: 10.1201/9781003299844-4

$$f_{1t_1}(t_1, t_2) = 1,$$

$$f_{1t_2}(t_1, t_2) = 1,$$

$$f_{2t_1}(t_1, t_2) = 2t_1 e^{t_1^2 + t_2},$$

$$f_{2t_2}(t_1, t_2) = e^{t_1^2 + t_2},$$

$$f_{3t_1}(t_1, t_2) = e^{t_1 + t_2^2},$$

$$f_{3t_2}(t_1, t_2) = 2t_2 e^{t_1 + t_2^2}, \quad (t_1, t_2) \in \mathbb{R}_\star^2.$$

Hence,

$$f_{1t_1}^\star(t_1, t_2) = e^{t_1 \frac{f_{1t_1}(t_1,t_2)}{f_1(t_1,t_2)}}$$

$$= e^{\frac{t_1}{t_1 + t_2}},$$

$$f_{2t_1}^\star(t_1, t_2) = e^{t_1 \frac{f_{2t_1}(t_1,t_2)}{f_2(t_1,t_2)}}$$

$$= e^{2t_1^2},$$

$$f_{3t_1}^\star(t_1, t_2) = e^{t_1 \frac{f_{3t_1}(t_1,t_2)}{f_3(t_1,t_2)}}$$

$$= e^{t_1},$$

$$f_{1t_2}^\star(t_1, t_2) = e^{t_2 \frac{f_{1t_2}(t_1,t_2)}{f_1(t_1,t_2)}}$$

$$= e^{\frac{t_2}{t_1 + t_2}},$$

$$f_{2t_2}^\star(t_1, t_2) = e^{t_2 \frac{f_{2t_2}(t_1,t_2)}{f_2(t_1,t_2)}}$$

$$= e^{t_2},$$

$$f_{3t_2}^\star(t_1, t_2) = e^{t_2 \frac{f_{3t_2}(t_1,t_2)}{f_3(t_1,t_2)}}$$

$$= e^{2t_2^2}, \quad (t_1, t_2) \in \mathbb{R}_\star^2.$$

Therefore

$$f^\star_{t_1}(t_1, t_2) = \left(f^\star_{1t_1}(t_1, t_2), f^\star_{2t_1}(t_1, t_2), f^\star_{3t_1}(t_1, t_2)\right)$$

$$= \left(e^{\frac{t_1}{t_1+t_2}}, e^{2t_1^2}, e^{t_1}\right),$$

$$f^\star_{t_2}(t_1, t_2) = \left(f^\star_{1t_2}(t_1, t_2), f^\star_{2t_2}(t_1, t_2), f^\star_{3t_2}(t_1, t_2)\right)$$

$$= \left(e^{\frac{t_2}{t_1+t_2}}, e^{t_2}, e^{2t_2^2}\right), \qquad (t_1, t_2) \in \mathbb{R}^2_\star,$$

and

$$f^\star_{t_1}(t_1, t_2) \times_\star f^\star_{t_2}(t_1, t_2) = \left(e^{4t_1^2 t_2^2 - t_1 t_2}, e^{\frac{t_1 t_2}{t_1+t_2} - \frac{2t_1 t_2^2}{t_1+t_2}}, e^{\frac{t_1 t_2}{t_1+t_2} - \frac{2t_1^2 t_2}{t_1+t_2}}\right)$$

$$= \left(e^{t_1 t_2(4t_1 t_2 - 1)}, e^{\frac{t_1 t_2(1-2t_2)}{t_1+t_2}}, e^{\frac{t_1 t_2(1-2t_1)}{t_1+t_2}}\right)$$

$$\neq (0_\star, 0_\star, 0_\star), \qquad (t_1, t_2) \in \mathbb{R}^2_\star.$$

Thus, $f: \mathbb{R}^2_\star \to \mathbb{R}^3_\star$ is a multiplicative regular surface.

**Example 4.2:** Let $f: \mathbb{R}^2_\star \to \mathbb{R}^3_\star$ be given by

$$f(t_1, t_2) = \left(e^{t_1+t_2}, e^{3(t_1+t_2)}, e^{2(t_1+t_2)}\right), \qquad (t_1, t_2) \in \mathbb{R}^2_\star.$$

Here

$$f_1(t_1, t_2) = e^{t_1+t_2},$$
$$f_2(t_1, t_2) = e^{3(t_1+t_2)},$$
$$f_3(t_1, t_2) = e^{2(t_1+t_2)}, \qquad (t_1, t_2) \in \mathbb{R}^2_\star.$$

Hence,

$$f_{1t_1}(t_1, t_2) = e^{t_1 + t_2},$$

$$f_{1t_2}(t_1, t_2) = e^{t_1 + t_2},$$

$$f_{2t_1}(t_1, t_2) = 3e^{3(t_1 + t_2)},$$

$$f_{2t_2}(t_1, t_2) = 3e^{3(t_1 + t_2)},$$

$$f_{3t_1}(t_1, t_2) = 2e^{2(t_1 + t_2)},$$

$$f_{3t_2}(t_1, t_2) = 2e^{2(t_1 + t_2)}, \quad (t_1, t_2) \in \mathbb{R}_\star^2,$$

and

$$f_{1t_1}^\star(t_1, t_2) = e^{t_1 \frac{f_{1t_1}(t_1, t_2)}{f_1(t_1, t_2)}}$$

$$= e^{t_1},$$

$$f_{2t_1}^\star(t_1, t_2) = e^{t_1 \frac{f_{2t_1}(t_1, t_2)}{f_2(t_1, t_2)}}$$

$$= e^{3t_1},$$

$$f_{3t_1}^\star(t_1, t_2) = e^{t_1 \frac{f_{3t_1}(t_1, t_2)}{f_3(t_1, t_2)}}$$

$$= e^{2t_1},$$

$$f_{1t_2}^\star(t_1, t_2) = e^{t_2 \frac{f_{1t_2}(t_1, t_2)}{f_1(t_1, t_2)}}$$

$$= e^{t_2},$$

$$f_{2t_2}^\star(t_1, t_2) = e^{t_2 \frac{f_{2t_2}(t_1, t_2)}{f_2(t_1, t_2)}}$$

$$= e^{3t_2},$$

$$f_{3t_2}^\star(t_1, t_2) = e^{t_2 \frac{f_{3t_2}(t_1, t_2)}{f_3(t_1, t_2)}}$$

$$= e^{2t_2}, \quad (t_1, t_2) \in \mathbb{R}_\star^2.$$

Therefore

$$f_{t_1}^\star(t_1, t_2) = \left( f_{1t_1}^\star(t_1, t_2), f_{2t_1}^\star(t_1, t_2), f_{3t_1}^\star(t_1, t_2) \right)$$

$$= (e^{t_1}, e^{3t_1}, e^{2t_1}),$$

$$f_{t_2}^\star(t_1, t_2) = \left( f_{1t_2}^\star(t_1, t_2), f_{2t_2}^\star(t_1, t_2), f_{3t_2}^\star(t_1, t_2) \right)$$

$$= (e^{t_2}, e^{3t_2}, e^{2t_2}), \quad (t_1, t_2) \in \mathbb{R}_\star^2,$$

and

$$f^\star_{t_1}(t_1, t_2) \times_\star f^\star_{t_2}(t_1, t_2) = (e^{6t_1t_2 - 6t_1t_2}, e^{2t_1t_2 - 2t_1t_2}, e^{3t_1t_2 - 3t_1t_2})$$

$$= (0_\star, 0_\star, 0_\star), \quad (t_1, t_2) \in \mathbb{R}^2_\star.$$

Thus, $f: \mathbb{R}^2_\star \to \mathbb{R}^3_\star$ is not a multiplicative regular surface.

**Example 4.3:** Let $g \in \mathscr{C}^2_\star(U)$, $g > 0$ on $U$. Consider $f: U \to \mathbb{R}^3_\star$ be given by

$$f(t_1, t_2) = (t_1, t_2, g(t_1, t_2)), \quad (t_1, t_2) \in U.$$

Then

$$f^\star_{t_1}(t_1, t_2) = (1_\star, 0_\star, g^\star_{t_1}(t_1, t_2)),$$
$$f^\star_{t_2}(t_1, t_2) = (0_\star, 1_\star, g^\star_{t_2}(t_1, t_2)), \quad (t_1, t_2) \in U.$$

Hence,

$$f^\star_{t_1}(t_1, t_2) \times_\star f^\star_{t_2}(t_1, t_2) = (-_\star g^\star_{t_1}(t_1, t_2), -_\star g^\star_{t_2}(t_1, t_2), 1_\star)$$
$$\neq (0_\star, 0_\star, 0_\star), \quad (t_1, t_2) \in U.$$

Thus, $f: U \to \mathbb{R}^3_\star$ is a multiplicative regular surface.

**Example 4.4:** Let

$$g(t_1, t_2) = e^{t_1^2 + t_2^3},$$
$$f(t_1, t_2) = (t_1, t_2, g(t_1, t_2)), \quad (t_1, t_2) \in \mathbb{R}^2_\star.$$

Then

$$g_{t_1}(t_1, t_2) = 2t_1 e^{t_1^2 + t_2^3},$$
$$g_{t_2}(t_1, t_2) = 3t_2^2 e^{t_1^2 + t_2^3}, \quad (t_1, t_2) \in \mathbb{R}^2_\star,$$

and

$$g^\star_{t_1}(t_1, t_2) = e^{t_1 \frac{g_{t_1}(t_1, t_2)}{g(t_1, t_2)}}$$
$$= e^{2t_1^2},$$
$$g^\star_{t_2}(t_1, t_2) = e^{t_2 \frac{g_{t_2}(t_1, t_2)}{g(t_1, t_2)}}$$
$$= e^{3t_2^3}, \quad (t_1, t_2) \in \mathbb{R}^2_\star.$$

Hence,

$$f_{t_1}^\star (t_1, t_2) \times_\star f_{t_2}^\star (t_1, t_2) = \left( e^{-2t_1^2}, e^{-3t_2^3}, 1_\star \right), \quad (t_1, t_2) \in \mathbb{R}_\star^2,$$

and $f: \mathbb{R}_\star^2 \to \mathbb{R}_\star^3$ is a multiplicative regular surface.

**Exercise 4.1:** Let $f: \mathbb{R}_\star^2 \to \mathbb{R}_\star^3$ is given by

$$f(t_1, t_2) = \left( \frac{1}{t_1 + t - 2}, e^{t_1^2 + t_2}, t_1 t_2 \right), \quad (t_1, t_2) \in \mathbb{R}_\star^2.$$

Prove that $(\mathbb{R}_\star^2, f)$ is a multiplicative regular surface.
In Fig. 4.1 is shown the multiplicative regular surface

$$f(t_1, t_2) = (t_1, t_2, t_1 + t_2), \quad (t_1, t_2) \in [1, 10] \times [1, 5].$$

In Fig. 4.2 is shown the multiplicative regular surface

$$f(t_1, t_2) = (t_1, t_2, 2 + \sin(t_1 + t_2)), \quad (t_1, t_2) \in [1, 10] \times [1, 5].$$

In Fig. 4.3 is shown the multiplicative regular surface

$$f(t_1, t_2) = (t_1, t_2, 4 + \cos(t_1^2 + t_2^3)), \quad (t_1, t_2) \in [1, 10] \times [1, 5].$$

**Definition 4.3:** The set $f(U) \subseteq \mathbb{R}_\star^3$ is said to be the multiplicative support of the multiplicative parameterized surface $(U, f)$.

**Definition 4.4:** Let $m, n \in \mathbb{N}$, $M \subseteq \mathbb{R}_\star^m$, $N \subseteq \mathbb{R}_\star^n$. A multiplicative differentiable map $f: M \to N$ is called a multiplicative homeomorphism if it is bijection, differentiable. If $f \in \mathscr{C}_\star(M)$ and its inverse $f^{-1}: N \to M$ is multiplicative continuous.

**Definition 4.5:** Let $m, n \in \mathbb{N}$, $M \subseteq \mathbb{R}_\star^m$, $N \subseteq \mathbb{R}_\star^n$. A multiplicative differentiable map $f: M \to N$ is called a multiplicative diffeomorphism if it is bijection and its inverse $f^{-1}: N \to M$ is multiplicative differentiable. If $f \in \mathscr{C}_\star^r(M)$ and $f^{-1} \in \mathscr{C}_\star^r(N)$, then $f$ is said to be $\mathscr{C}_\star^r$-multiplicative diffeomorphism.

**Definition 4.6:** Two multiplicative parameterized surfaces $(U, f)$ and $(V, g)$ are said to be multiplicative equivalent if there is a multiplicative diffeomorphism $\phi: U \to V$ do that

$$f = g(\phi).$$

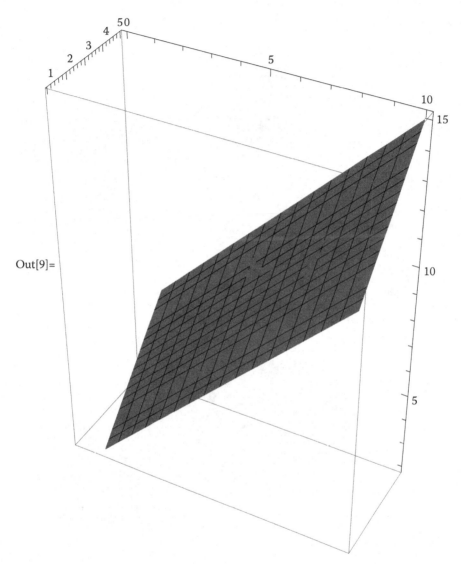

**FIGURE 4.1**
A multiplicative regular surface.

**Definition 4.7:** A subset $S$ of $\mathbb{R}^3_\star$ is called a multiplicative regular surface if for each points $a \in S$ there is a neighbourhood $W$ in $S$ and a homeomorphism $f : U \to W$ so that $f(U) = W$ and $(U, f)$ is a multiplicative parameterized surface. The pair $(U, f)$ is said to be a multiplicative local parameterization of the multiplicative surface $S$. The multiplicative support $f(U)$ is called the multiplicative domain of the multiplicative representation. If $f(U) = S$, then the multiplicative surface $S$ is said to be a multiplicative simple surface. We have the following representations of the multiplicative surfaces.

Out[16]=

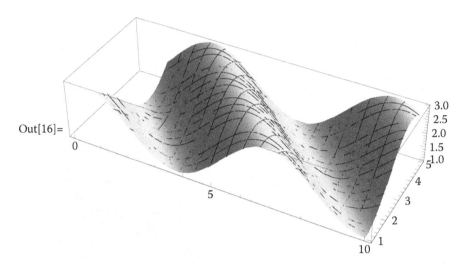

**FIGURE 4.2**
A multiplicative regular surface.

Out[17]=

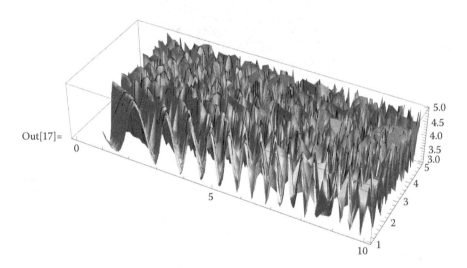

**FIGURE 4.3**
A multiplicative regular surface.

1. If $(U, f)$ is a multiplicative parameterized surface and

$$f(t_1, t_2) = (f_1(t_1, t_2), f_2(t_1, t_2), f_3(t_1, t_2)), \quad (t_1, t_2) \in U,$$

then the equations

$$f_1 = f_1(t_1, t_2)$$
$$f_2 = f_2(t_1, t_2)$$
$$f_3 = f_3(t_1, t_2), \quad (t_1, t_2) \in U,$$

are said to be parametric equations of the multiplicative parameterized surface $(U, f)$.

2. If $g: U \to \mathbb{R}_\star$, $g \in \mathscr{C}^1_\star(U)$ and

$$f(t_1, t_2) = (t_1, t_2, g(t_1, t_2)),$$

then the representation of the multiplicative parameterized surface $(U, f)$ is said to be an explicit representation.

**Definition 4.8:** The set

$$S = \{(t_1, t_2, t_3) \in \mathbb{R}^3_\star : F(t_1, t_2, t_3) = 0_\star\}$$

is said to be $0_\star$-level set of $F$.

**Definition 4.9:** The multiplicative vector

$$grad_\star F(t_1, t_2, t_3) = (F^\star_{t_1}(t_1, t_2, t_3), F^\star_{t_2}(t_1, t_2, t_3), F^\star_{t_3}(t_1, t_2, t_3)), \quad (t_1, t_2, t_3) \in W,$$

is said to be multiplicative gradient vector of $F$.

**Theorem 4.1:** *If*

$$grad_\star F(t_1, t_2, t_3) \neq (0_\star, 0_\star, 0_\star), \quad (t_1, t_2, t_3) \in W,$$

*then $S$ is a multiplicative regular surface.*

*Proof.* Let $(t^0_1, t^0_2, t^0_3) \in W$. Without loss of generality, suppose that

$$F^\star_{t_3}(t^0_1, t^0_2, t^0_3) \neq 0_\star.$$

Then, by the multiplicative implicit theorem, it follows that there is a neighbourhood $M$ of $(t^0_1, t^0_2, t^0_3)$ and $f \in \mathscr{C}^1_\star$ so that

$$t_3 = f(t_1, t_2), \quad (t_1, t_2, t_3) \in M.$$

This completes the proof.

3. Let $W \subseteq \mathbb{R}^3_*$ and $F \in \mathscr{C}^1_*(\mathbb{R}^3_*)$.

---

## 4.2 The Multiplicative Equivalence of Multiplicative Local Representations

**Definition 4.10:** Let $S$ be a multiplicative surface, $(U, f)$ be its multiplicative local parameterization and $W = f(U)$. Then the map $f^{-1} \colon W \to U$ is a bijection and will be called a multiplicative curvilinear coordinate system on $S$ or a multiplicative chart on $S$.

**Lemma 4.1:** *Let $(U, f)$ be a multiplicative local parameterization of the multiplicative surface $S$, $f(U) = W$ and $f^{-1} \colon W \to U$. Then for each point $a \in W$, there is a multiplicative open set $B$ in the topology of $\mathbb{R}^3_*$ and a multiplicative smooth map $G \colon B \to V$ so that $a \in B$ and*

$$f^{-1}_{|W \cap B} = G_{|W \cap B}.$$

*Proof.* Let

$$f(t_1, t_2) = (f_1(t_1, t_2), f_2(t_1, t_2), f_3(t_1, t_2)), \quad (t_1, t_2) \in U,$$

and

$$a = f(t_1^0, t_2^0).$$

Note that

$$rank_\star \begin{pmatrix} f^\star_{1t_1} & f^\star_{1t_2} \\ f^\star_{2t_1} & f^\star_{2t_2} \\ f^\star_{3t_1} & f^\star_{3t_2} \end{pmatrix} = 2.$$

Without loss of generality, suppose that

$$\begin{vmatrix} f^\star_{1t_1} & f^\star_{1t_2} \\ f^\star_{2t_1} & f^\star_{2t_2} \end{vmatrix} \neq 0_\star.$$

Then, by the multiplicative inverse function theorem, it follows that there is a multiplicative open neighbourhood $V$ of the point $(t_1^0, t_2^0)$ in $U$ and a multiplicative open neighbourhood $V_1$ of the point $(f_1(t_1^0, t_2^0), f_2(t_1^0, t_2^0))$ so that $f: V \to V_1$ is a multiplicative diffeomorphism. Because $f: U \to W$ is a multiplicative homeomorphism, we have that $f(V)$ is a multiplicative open neighbourhood in $S$ of the point $a$. Therefore there is a multiplicative open neighbourhood of the point $a$ so that

$$f(V) = B \cap S = B \cap W.$$

Now, define the map $\phi: \mathbb{R}_\star^3 \to \mathbb{R}_\star^2$ as follows:

$$\phi(t_1, t_2, t_3) = (t_1, t_2), \quad (t_1, t_2, t_3) \in \mathbb{R}_\star^3.$$

Let

$$G = (f^{-1}(\phi))\big|_B : B \to U.$$

Note that $G$ is a multiplicative smooth map. Next, to each point $(s_1, s_2, s_3) \in B \cap W$ corresponds a single point

$$(t_1, t_2) = f^{-1}(s_1, s_2, s_3) \in V.$$

Also, to each point $(s_1, s_2) \in V_1$ corresponds the point

$$(t_1, t_2) = f^{-1}(s_1, s_2) \in V.$$

Thus, if $(s_1, s_2, s_3) \in B \cap W$, then

$$\begin{aligned}
f^{-1}(s_1, s_2, s_3) &= (t_1, t_2) \\
&= f^{-1}(s_1, s_2) \\
&= f^{-1}(\phi(s_1, s_2, s_3)) \\
&= G(s_1, s_2, s_3).
\end{aligned}$$

This completes the proof.

**Theorem 4.2:** *Let $(U, f)$ and $(U_1, f_1)$ be two multiplicative local parameterizations of a multiplicative surface $S$ and $f(U) = f_1(U_1)$. Then there is a multiplicative diffeomorphism $\phi: U \to U_1$ so that*

$$f = f_1(\phi).$$

*Proof.* Let $W = f(U) = f_1(U_1)$ and

$$\phi = f_1^{-1} \cdot f.$$

Then $\phi\colon U \to U_1$. Since $f_1\colon U_1 \to W$ is a multiplicative homeomorphism, we have that $f_1^{-1}\colon W \to U_1$ is a multiplicative homeomorphism. Therefore $\phi\colon U \to U_1$ is a multiplicative homeomorphism. Now, we will prove that each point $(t_1^0, t_2^0) \in U$ has a multiplicative open neighbourhood $V \subset U$ so that the map $\phi_{|V}$ is a multiplicative smooth map. Let

$$a = f_1(\phi(t_1^0, t_2^0)).$$

By Lemma 4.1, it follows that there is a multiplicative open set $B$ of $\mathbb{R}_\star^3$ so that $G\colon B \to U_1$ is a multiplicative smooth map so that

$$f_1^{-1}\Big|_{B\cap W} = G\big|_{B\cap W}$$

and

$$V = f^{-1}(B \cap W).$$

Then

$$\begin{aligned}
\phi|_V &= f_1^{-1} \cdot f_{|V} \\
&= G \cdot f_{|V}
\end{aligned}$$

and $\phi|_V$ is a multiplicative smooth map. As above, $\phi^{-1}$ is a multiplicative smooth map. This completes the proof.

**Theorem 4.3:** *Let $(U, f)$ be a multiplicative regular parameterized surface. Then each point $(t_1^0, t_2^0) \in U$ has a multiplicative open neighbourhood $V' \subset U$ such that $f(V)$ is a multiplicative simple surface in $\mathbb{R}_\star^3$ for which $(V, f_{|V})$ is a multiplicative global representation.*

*Proof.* Let

$$f(t_1, t_2) = (f_1(t_1, t_2), f_2(t_1, t_2), f_3(t_1, t_2)), \qquad (t_1, t_2) \in U,$$

and

$$\begin{vmatrix} f^{\star}_{1t_1}(t^0_1, t^0_2) & f^{v}_{1t_2}(t^0_1, t^0_2) \\ f^{\star}_{2t_1}(t^0_1, t^0_2) & f^{\star}_{2t_2}(t^0_1, t^0_2) \end{vmatrix}_{\star} \neq 0_{\star}.$$

Then, by the multiplicative inverse function theorem, there is a multiplicative open neighbourhood $V \subset U$ of the point $(t^0_1, t^0_2)$ and a multiplicative open neighbourhood $V_1$ of the point $(s_{10}, s_{20}, s_{30}) = f(t^0_1, t^0_2)$ so that $f: V \to V_1$ is a multiplicative diffeomorphism. Let now, $(t_1, t_2), (y_1, y_2) \in V$ be such that

$$f_1(t_1, t_2) = f_1(y_1, y_2),$$

$$f_2(t_1, t_2) = f_2(y_1, y_2).$$

Since $f: V \to V_1$ is a multiplicative diffeomorphism, it is an injective map and then

$$(t_1, t_2) = (y_1, y_2).$$

Since $f: U \to \mathbb{R}^3_{\star}$ is multiplicative continuous, we have that $f_{|V}: V \to f(V)$ is multiplicative continuous. Note that

$$(x, y, z) \in f(V) \to (u, v, w) \in V_1 \to (t_1, t_2) = f^{-1}(u, v, w).$$

Thus, the inverse map of $f_{|V}: V \to f(V)$ is multiplicative continuous. This completes the proof.

---

## 4.3 Multiplicative Curves on Multiplicative Surfaces

**Theorem 4.4:** *Let $(U, f)$ be a multiplicative parameterization of the multiplicative surface $S$ and $(I, g = g(t))$ is a multiplicative smooth parameterized curve whose support is included in $f(U)$. Then there is a unique multiplicative smooth parameterized curve $(I, g_1)$ on $U$ so that*

$$g(t) = f(g_1(t)), \quad t \in I. \tag{4.2}$$

*Conversely, any multiplicative smooth parameterized curve $g_1$ on $U$ defines, through (4.2), a multiplicative smooth curve on $f(U)$. The multiplicative regularity of $g$ at $t$ is equivalent to the multiplicative regularity of $g_1$ at $t$.*

*Proof.* Since $f: U \to f(U)$ is a multiplicative homeomorphism and $g(I) \subset f(U)$, we get

$$g_1 = f^{-1} \circ g.$$

We have that $g_1$ is multiplicative continuous because it is a composition of two multiplicative continuous maps. Let $t \in I$. Then $g(t) \in f(U)$. By Lemma 4.1, it follows that there is a multiplicative open neighbourhood $B$ of the point $g(t)$ and a multiplicative smooth map $G: B \to U$ so that

$$f^{-1}_{|B \cap f(U)} = G_{|B \cap f(U)}.$$

Therefore, the map $g_1$ can be represented in the multiplicative neighbourhood of the point $t$ as a composition of $Rg$ and $G$. Since $G$ and $g$ are multiplicative smooth, we have that $g_1$ is multiplicative smooth. The converse assertion follows because by (4.2) and the multiplicative smoothness of $f$ and $g_1$, it follows that $g$ is multiplicative smooth. Let

$$g_1(t) = (u(t), v(t)), \quad t \in I.$$

Then, by (4.2), we find

$$g(t) = f(u(t), v(t)), \quad t \in I.$$

Now, we multiplicative differentiate the last equation with respect to $t$ and we find

$$g^\star(t) = f^\star_{t_1}(u(t), v(t)) \cdot_\star u^\star(t) +_\star f^\star_{t_2}(u(t), v(t)) \cdot_\star v^\star(t), \quad t \in I.$$

Since

$$f^\star_{t_1} \times_\star f^\star_{t_2} \neq 0_\star \quad \text{on} \quad U,$$

we have that $f^\star_{t_1}$ and $f^\star_{t_2}$ are not multiplicative collinear. Therefore

$$g^\star(t) = 0_\star, \quad t \in I,$$

if and only if

$$g^\star_1(t) = 0_\star, \quad t \in I.$$

This completes the proof.

**Definition 4.11:** Let $U$, $f$, $g$, $g_1$ and $I$ be as in Theorem 4.4. Then the multiplicative parameterized curve $g_1$ on $U$ is called multiplicative local parameterization of $g$ in the local parameterization $(U, f)$. The equations

$$u = u(t)$$
$$v = v(t), \quad t \in I,$$

are called multiplicative local equations of $g$.

---

## 4.4 The Multiplicative Tangent Vector Space, the Multiplicative Tangnet Plane, the Multiplicative Normal to a Multiplicative Surface

Let $a \in \mathbb{R}^3_\star$. With $\mathbb{R}^3_{\star a}$ we will denote the space of all multiplicative vectors with the origin $a$.

**Definition 4.12:** A multiplicative vector $p \in \mathbb{R}^3_\star$ is called a multiplicative tangent vector to the multiplicative surface $S$ if there is a multiplicative parameterized curve $(I, g = g(t))$ on $S$ and a $t_0 \in I$ so that $g(t_0) = a$ and

$$p = g^\Delta(t_0).$$

Thus, a multiplicative tangent vector to a multiplicative surface is a multiplicative tangent vector to a multiplicative parameterized curve on the multiplicative surface $S$.

With $T_{\star a}S$ we will denote the set of all multiplicative tangent vectors at $a$ to $S$. If

$$g(t) = f(u(t), v(t)), \quad t \in I,$$

then

$$g^\star(t) = f^\star_u(u(t), v(t)) \cdot_\star v^\star(t) +_\star f^\star_v(u(t), v(t)) \cdot_\star v^\star(t), \quad t \in I.$$

**Theorem 4.5:** *The set $T_{\star a}S$ is a two-dimensional multiplicative vector subspace of $\mathbb{R}^3_\star$. If $(U, f)$ is a multiplicative local parameterization of $S$, $a = f(u_0, v_0)$, then $f^\star_u(u_0, v_0)$ and $f^\star_v(u_0, v_0)$ make up a multiplicative basis to $T_{\star a}S$.*

*Proof.* Let $(I, g = g(t))$ be a multiplicative parameterized curve on $S$ and $g(t_0) = a$ for some $t_0 \in I$. Assume that $g(I) \subset f(U)$ and

$$g(t) = f(u(t), v(t)), \quad t \in I.$$

Then

$$g^\star(t) = f_u^\star(u(t), v(t)) \cdot_\star v^\star(t) +_\star f_v^\star(u(t), v(t)) \cdot_\star v^\star(t), \quad t \in I.$$

Note that any vector in the form

$$h = \alpha \cdot_\star f_u^\star(u_0, v_0) +_\star \beta \cdot_\star f_v^\star(u_0, v_0),$$

for some $\alpha, \beta \in \mathbb{R}_\star$, is a multiplicative tangent vector to the multiplicative curve with equations

$$u = u_0 +_\star \alpha \cdot_\star t$$
$$v = v_0 +_\star \beta \cdot_\star t,$$

which is a multiplicative curve on $S$ passing through the point $a$ for $t = t_0$. Thus, $h \in T_{\star a} S$. This completes the proof.

**Definition 4.13:** The multiplicative vector space $T_{\star a} S$ is called the multiplicative tangent space to $S$ at $a$. The multiplicative plane passing through $a$ and having $T_{\star a} S$ as a multiplicative directing plane is called the multiplicative tangent plane of $S$ at $a$.
    If

$$a = (f_1(t_1^0, t_2^0), f_2(t_1^0, t_2^0), f_3(t_1^0, t_2^0)),$$

then the equation of the multiplicative tangent plane is given by

$$0_\star = \begin{vmatrix} X -_\star f_1(t_1^0, t_2^0) & Y -_\star f_2(t_1^0, t_2^0) & Z -_\star f_3(t_1^0, t_2^0) \\ f_{1t_1}^\star(t_1^0, t_2^0) & f_{2t_1}^\star(t_1^0, t_2^0) & f_{3t_1}^\star(t_1^0, t_2^0) \\ f_{1t_2}^\star(t_1^0, t_2^0) & f_{2t_2}^\star(t_1^0, t_2^0) & f_{3t_2}^\star(t_1^0, t_2^0) \end{vmatrix}_\star .$$

**Example 4.5:** Let $f: \mathbb{R}_\star^2 \to \mathbb{R}_\star^3$ be as in Example 4.1. We will find the equation of its multiplicative tangent plane at the point $f(1, 1)$. We have

$$f(t_1, t_2) = (t_1 + t_2, e^{t_1^2 + t_2}, e^{t_1 + t_2^2}),$$

$$f_{t_1}^\star(t_1, t_2) = \left( e^{\frac{t_1}{t_1 + t_2}}, e^{2t_1^2}, e^{t_1} \right),$$

$$f_{t_2}^\star(t_1, t_2) = \left( e^{\frac{t_2}{t_1 + t_2}}, e^{t_2}, e^{2t_2^2} \right), \quad (t_1, t_2) \in \mathbb{R}_\star^2.$$

Then

$$f(1, 1) = (2, e^2, e^2),$$

$$f^{\star}_{t_1}(1, 1) = \left(e^{\frac{1}{2}}, e^2, e\right),$$

$$f^{\star}_{t_2}(1, 1) = \left(e^{\frac{1}{2}}, e, e^2\right).$$

Therefore

$$0_{\star} = \begin{vmatrix} X \rightarrow_{\star} 2 & Y \rightarrow_{\star} e^2 & Z \rightarrow_{\star} e^2 \\ e^{\frac{1}{2}} & e^2 & e \\ e^{\frac{1}{2}} & e & e^2 \end{vmatrix}_{\star}$$

$$= \begin{vmatrix} \dfrac{X}{2} & \dfrac{Y}{e^2} & \dfrac{Z}{e^2} \\ e^{\frac{1}{2}} & e^2 & e \\ e^{\frac{1}{2}} & e & e^2 \end{vmatrix}_{\star}$$

$$= \dfrac{X}{2} \cdot_{\star} \begin{vmatrix} e^2 & e \\ e & e^2 \end{vmatrix}_{\star} \rightarrow_{\star} \dfrac{Y}{e^2} \cdot_{\star} \begin{vmatrix} e^{\frac{1}{2}} & e \\ e^{\frac{1}{2}} & e^2 \end{vmatrix}_{\star}$$

$$+_{\star} \dfrac{Z}{e^2} \cdot_{\star} \begin{vmatrix} e^{\frac{1}{2}} & e^2 \\ e^{\frac{1}{2}} & e \end{vmatrix}_{\star}$$

$$= \dfrac{X}{2} \cdot_{\star} e^{4-1} \rightarrow_{\star} \dfrac{Y}{e^2} \cdot_{\star} e^{1-\frac{1}{2}} +_{\star} \dfrac{Z}{e^2} \cdot_{\star} e^{\frac{1}{2}-1}$$

$$= \dfrac{X}{2} \cdot_{\star} e^3 \rightarrow_{\star} \dfrac{Y}{e^2} \cdot_{\star} e^{\frac{1}{2}} +_{\star} \dfrac{Z}{e^2} \cdot_{\star} e^{-\frac{1}{2}}$$

$$= e^{3\log \frac{X}{2} - \frac{1}{2}\log \frac{Y}{e^2} - \frac{1}{2}\log \frac{Z}{e^2}}$$

$$= e^{3\log X - 3\log 2 - \frac{1}{2}\log Y + 1 - \frac{1}{2}\log Z + 1}$$

$$= e^{\log X^3 - \log \sqrt{Y} - \log \sqrt{Z} + 2 - 3\log 2}$$

$$= e^{\log \frac{X^3}{\sqrt{YZ}} + \log e^2 - \log 8}$$

$$= e^{\log \frac{X^3}{\sqrt{YZ}} + \log \frac{e^2}{8}}$$

$$= e^{\log \frac{X^3 e^2}{8\sqrt{YZ}}}$$

$$= e^0,$$

whereupon

$$e^2 X^3 = 8\sqrt{YZ}.$$

**Example 4.6:** Let $f\colon \mathbb{R}^2_\star \to \mathbb{R}^3_\star$ be given as in Example 4.4. We will find its multiplicative tangent plane at $f(1, 1)$. We have

$$f(t_1, t_2) = (t_1, t_2, e^{t_1^2 + t_2^3}),$$
$$f^\star_{t_1}(t_1, t_2) = (1_\star, 0_\star e^{2t_1^2}),$$
$$f^\star_{t_2}(t_1, t_2) = (0_\star, 1_\star, e^{3t_2^3}), \quad (t_1, t_2) \in \mathbb{R}^2_\star.$$

Then

$$f(1, 1) = (1, 1, e^2),$$
$$f^\star_{t_1}(1, 1) = (e, 1, e^2),$$
$$f^\star_{t_2}(1, 1) = (1, e, e^3).$$

Hence,

$$0_\star = e^0$$

$$= \begin{vmatrix} X \to_\star 1 & Y \to_\star 1 & Z \to_\star e^2 \\ e & 1 & e^2 \\ 1 & e & e^3 \end{vmatrix}_\star$$

$$= \begin{vmatrix} X & Y & \dfrac{Z}{e^2} \\ e & 1 & e^2 \\ 1 & e & e^3 \end{vmatrix}_\star$$

$$= X \cdot_\star \begin{vmatrix} 1 & e^2 \\ e & e^3 \end{vmatrix}_\star -_\star Y \cdot_\star \begin{vmatrix} e & e^2 \\ 1 & e^3 \end{vmatrix}_\star$$

$$+_\star \dfrac{Z}{e^2} \cdot_\star \begin{vmatrix} e & 1 \\ 1 & e \end{vmatrix}_\star$$

$$= X \cdot_\star e^{-2} -_\star Y \cdot_\star e^3 +_\star \dfrac{Z}{e^2} \cdot_\star e$$

$$= e^{\log X \log e^{-2} - \log Y \log e^3 + \log e \log \frac{Z}{e^2}}$$

$$= e^{-2\log X - 3\log Y + \log Z - 2}$$

$$= e^{\log \frac{Z}{X^2 Y^3} - 2}$$

$$= e^{\log \frac{Z}{e^2 X^2 Y^3}},$$

whereupon

$$Z = e^2 X^2 Y^3.$$

**Exercise 4.2:** Let $f: \mathbb{R}_\star^2 \to \mathbb{R}_\star^3$ be given by

$$f(t_1, t_2) = \left(t_1^2, t_2^3, e^{t_1 + t_2^2}\right), \quad (t_1, t_2) \in \mathbb{R}_\star^2.$$

Find the equation of the multiplicative tangent plane through the point $f(1, 1)$.

Let now, $(U, f = f(u, v))$ be a multiplicative parameterized surface $S$ and $(u_0, v_0) \in U$. Then

$$f(u_0 +_\star \alpha \cdot_\star h, v_0 +_\star \beta \cdot_\star h) = f(u_0, v_0) +_\star h \cdot_\star \left( \alpha \cdot_\star f_{t_1}^\star(u_0, v_0) \right.$$

$$\left. +_\star \beta \cdot_\star f_{t_2}^\star(u_0, v_0) \right) +_\star h \cdot_\star \varepsilon,$$

where

$$\lim_{h \to 0_\star} \varepsilon = 0_\star.$$

Let $\Pi$ be the multiplicative plane in $\mathbb{R}_\star^3$ passing through the point $a_0 = f(u_0, v_0)$ and $d$ be the multiplicative distance from the point

$$a = f(u_0 +_\star \alpha \cdot_\star h, v_0 +_\star \beta \cdot_\star h)$$

to the multiplicative plane $\Pi$ and $\delta$ be the multiplicative distance between $a_0$ and $a$.

**Theorem 4.6:** *The multiplicative plane $\Pi$ is the multiplicative tangent plane to $S$ at $a_0$ if and only if for any $\alpha, \beta \in \mathbb{R}_\star$, $\alpha^{2\star} +_\star \beta^{2\star} \neq 0_\star$, we have*

$$\lim_{h \to 0_\star} d /_\star \delta = 0_\star. \tag{4.3}$$

*Proof.* Let $n$ be the multiplicative versor of the multiplicative normal to $\Pi$. Then

$$d = \langle f(u_0 +_\star \alpha \cdot_\star h, v_0 +_\star \beta \cdot_\star h) - f(u_0, v_0), n \rangle_\star,$$
$$\delta = |f(u_0 +_\star \alpha \cdot_\star h, v_0 +_\star \beta \cdot_\star h) - f(u_0, v_0)|_\star.$$

Then

$$\lim_{h \to 0_\star} d/_\star \delta = \lim_{h \to 0_\star} \left( \langle f(u_0 +_\star \alpha \cdot_\star h, v_0 +_\star \beta \cdot_\star h) - f(u_0, v_0), n \rangle_\star \right.$$

$$\left. \Big/_\star |f(u_0 +_\star \alpha \cdot_\star h, v_0 +_\star \beta \cdot_\star h) - f(u_0, v_0)|_\star \right)$$

$$= \lim_{h \to 0_\star} \left( \langle h \cdot_\star (\alpha \cdot_\star f^\star_{t_1}(u_0, v_0) +_\star \beta \cdot_\star f^\star_{t_2}(u_0, v_0)) +_\star h \cdot_\star \epsilon, n \rangle_\star \Big/_\star |h$$

$$\cdot_\star (\alpha \cdot_\star f^\star_{t_1}(u_0, v_0) +_\star \beta \cdot_\star f^\star_{t_2}(u_0, v_0)) +_\star h \cdot_\star \epsilon|_\star \right)$$

$$= \pm_\star \left( \langle \alpha \cdot_\star f^\star_{t_1}(u_0, v_0) +_\star \beta \cdot_\star f^\star_{t_2}(u_0, v_0), n \rangle_\star \Big/_\star |\alpha \cdot_\star f^\star_{t_1}(u_0, v_0) \right.$$

$$\left. +_\star \beta \cdot_\star f^\star_{t_2}(u_0, v_0)|_\star \right).$$

Thus, (4.3) holds if and only if

$$\left\langle \alpha \cdot_\star f^\star_{t_1}(u_0, v_0) +_\star \beta \cdot_\star f^\star_{t_2}(u_0, v_0), n \right\rangle_\star = 0_\star. \tag{4.4}$$

1. Let $\Pi$ be the multiplicative tangent plane to $S$ at $f(u_0, v_0)$. Then

$$n \perp_\star f^\star_{t_1}(u_0, v_0), \quad n \perp_\star f^\star_{t_2}(u_0, v_0)$$

and (4.4) holds. Hence, (4.3) holds.

2. Let (4.3) holds. Then (4.4) holds. For $\alpha = 1_\star$, $\beta = 0_\star$, we get

$$\left\langle f^\star_{t_1}(u_0, v_0), n \right\rangle_\star = 0_\star.$$

For $\alpha = 0_\star$, $\beta = 1_\star$, we find

$$\left\langle f^\star_{t_2}(u_0, v_0), n \right\rangle_\star = 0_\star.$$

Thus,

$$n \perp_\star f^\star_{t_1}(u_0, v_0), \quad n \perp_\star f^\star_{t_2}(u_0, v_0)$$

and $\Pi$ is the multiplicative tangent plane to $S$ at $f(u_0, v_0)$. This completes the proof.

**Definition 4.14:** The multiplicative straight line passing through a point of the multiplicative surface $S$ that is multiplicative perpendicular to the multiplicative tangent plane of the multiplicative surface at that point, is called the multiplicative normal to the multiplicative surface $S$ at the considered point.

Let $(U, f = f(u, v))$ be a multiplicative surface $S$, $a = f(u_0, v_0) \in S$,

$$f(u, v) = (f_1(u, v), f_2(u, v), f_3(u, v)), \quad (u, v) \in U.$$

Then the equations of the multiplicative normal to $S$ at $a$ are given by the following equations

$$(X \to_\star f_1(u_0, v_0))/_\star \begin{vmatrix} f_{2u}^\star(u_0, v_0) & f_{3u}^\star(u_0, v_0) \\ f_{2v}^\star(u_0, v_0) & f_{3v}^\star(u_0, v_0) \end{vmatrix}_\star$$

$$= (Y \to_\star f_2(u_0, v_0))/_\star \begin{vmatrix} f_{3u}^\star(u_0, v_0) & f_{1u}^\star(u_0, v_0) \\ f_{3v}^\star(u_0, v_0) & f_{1v}^\star(u_0, v_0) \end{vmatrix}_\star$$

$$= (Z \to_\star f_3(u_0, v_0))/_\star \begin{vmatrix} f_{1u}^\star(u_0, v_0) & f_{2u}^\star(u_0, v_0) \\ f_{1v}^\star(u_0, v_0) & f_{2v}^\star(u_0, v_0) \end{vmatrix}_\star.$$

**Example 4.7:** Let $f: \mathbb{R}_\star^2 \to \mathbb{R}_\star^3$ be as in Example 4.1. Then, by the computations in Example 4.5, we get the following equations of the multiplicative normal to $S$ at $f(1, 1)$

$$\left(\frac{X}{2}\right)/_\star e^3 = \left(\frac{Y}{e^2}\right)/_\star e^{-\frac{1}{2}} = \left(\frac{Z}{e^2}\right)/_\star e^{-\frac{1}{2}}$$

or

$$e^{\frac{\log \frac{X}{2}}{3}} = e^{\frac{\log \frac{Y}{e^2}}{-\frac{1}{2}}} = e^{\frac{\log \frac{Z}{e^2}}{-\frac{1}{2}}},$$

or

$$e^{\log \frac{\sqrt[3]{X}}{\sqrt[3]{2}}} = e^{\log \frac{e^4}{Y^2}} = e^{\log \frac{e^4}{Z^2}},$$

or

$$\frac{\sqrt[3]{X}}{\sqrt[3]{2}} = \frac{e^4}{Y^2} = \frac{e^4}{Z^2}.$$

**Example 4.8:** Let $f\colon \mathbb{R}^2_\star \to \mathbb{R}^3_\star$ be as in Example 4.4. Then, by the computations of Example 4.6, we find

$$X/_\star\, e^{-2} = Y/_\star\, e^{-3} = \left(\frac{Z}{e^2}\right)/_\star\, e$$

or

$$e^{\frac{\log X}{-2}} = e^{\log \frac{\log Y}{e^{-3}}} = e^{\log \frac{Z}{e^2}},$$

or

$$e^{\log \frac{1}{\sqrt{X}}} = e^{\log \frac{1}{\sqrt[3]{Y}}} = e^{\log \frac{Z}{e^2}},$$

or

$$\frac{1}{\sqrt{X}} = \frac{1}{\sqrt[3]{Y}} = \frac{Z}{e^2}.$$

**Exercise 4.3:** Let $f\colon \mathbb{R}^2_\star \to \mathbb{R}^3_\star$ be given by

$$f(t_1, t_2) = \left(e^{t_1 - t_2},\, e^{t_1 - 2t_2^2},\, t_1^3 - t_2\right), \quad (t_1, t_2) \in \mathbb{R}^2_\star.$$

Find the equations of the multiplicative normal line at $f(1, 1)$.

**Theorem 4.7:** *Let $(x_0, y_0, z_0)$ be a given point of the multiplicative surface given by the equation*

$$F(x, y, z) = 0_\star.$$

*Then the multiplicative gradient vector*

$$grad_\star F(x_0, y_0, z_0) = (F^\star_x(x_0, y_0, z_0),\, F^\star_y(x_0, y_0, z_0),\, F^\star_z(x_0, y_0, z_0))$$

*is multiplicative perpendicular to the multiplicative tangent plane of the multiplicative surface at this point.*

*Proof.* Let

$$f = (x(u, v), y(u, v), z(u, v))$$

be a multiplicative local parameterization of the considered multiplicative surface. Then

$$f_u = (x_u, y_u, z_u),$$
$$f_v = (x_v, y_v, z_v)$$

and

$$0_\star = F_x^\star \cdot x_u +_\star F_y^\star \cdot y_u +_\star F_z^\star \cdot_\star z_u$$
$$= \langle grad_\star F, f_u \rangle_\star$$

and

$$0_\star = F_x^\star \cdot x_v +_\star F_y^\star \cdot y_v +_\star F_z^\star \cdot_\star z_v$$
$$= \langle grad_\star F, f_v \rangle_\star.$$

Thus,

$$grad_\star F \perp_\star f_u, \quad grad_\star F \perp_\star f_v.$$

This completes the proof.

Suppose that all conditions of Theorem 4.7 hold. Then the equation of the multiplicative tangent plane of $S$ at $(x_0, y_0, z_0)$ is

$$(X -_\star x_0) \cdot_\star F_x^\star(x_0, y_0, z_0) +_\star (Y -_\star y_0) \cdot_\star F_y^\star(x_0, y_0, z_0)$$
$$+_\star (Z -_\star z_0) \cdot_\star F_z^\star(x_0, y_0, z_0) = 0_\star.$$

The equations of the multiplicative normal line to $S$ at $(x_0, y_0, z_0)$ is

$$(X -_\star x_0) /_\star F_x^\star(x_0, y_0, z_0) = (Y -_\star y_0) /_\star F_y^\star(x_0, y_0, z_0)$$
$$= (Z -_\star z_0) /_\star F_z^\star(x_0, y_0, z_0).$$

**Example 4.9:** Let $S$ be a multiplicative surface given by the equation

$$F(x, y, z) = e^{x^2 - 2y^3 + 4z^5} = 0_\star, \quad (x, y, z) \in \mathbb{R}_\star^3.$$

Let also, $(x_0, y_0, z_0) = (1, 1, 1)$. We have

$$F_x(x, y, z) = 2xe^{x^2-2y^3+4z^5},$$
$$F_y(x, y, z) = -6y^2e^{x^2-2y^3+4z^5},$$
$$F_z(x, y, z) = 20z^4e^{x^2-2y^3+4z^5}, \quad (x, y, z) \in \mathbb{R}^3_\star,$$

and

$$F_x^\star(x, y, z) = e^{x \cdot \frac{F_x(x,y,z)}{F(x,y,z)}}$$
$$= e^{2x^2},$$
$$F_y^\star(x, y, z) = e^{y \cdot \frac{F_y(x,y,z)}{F(x,y,z)}}$$
$$= e^{-6y^3},$$
$$F_z^\star(x, y, z) = e^{z \cdot \frac{F_z(x,y,z)}{F(x,y,z)}}$$
$$= e^{20z^5}, \quad (x, y, z) \in \mathbb{R}^3_\star,$$

and

$$F_x^\star(1, 1, 1) = e^2,$$
$$F_y^\star(x, y, z) = e^{-6},$$
$$F_z^\star(x, y, z) = e^{20}.$$

Then the equation of the multiplicative tangent plane to $S$ at $(1, 1, 1)$ is

$$(X \to_\star 1) \cdot_\star e^2 +_\star (Y \to_\star 1) \cdot_\star e^{-6} +_\star (Z \to_\star 1) \cdot_\star e^{20} = 0_\star$$

or

$$X \cdot_\star e^2 +_\star Y \cdot_\star e^{-6} +_\star Z \cdot_\star e^{20} = 0_{\star\prime}$$

or

$$e^{2 \log X - 6 \log Y + 20 \log Z} = 0_{\star\prime}$$

or

$$e^{\log X^2 - \log Y^6 + \log Z^{20}} = e^0,$$

or

$$e^{\log \frac{X^2 Z^{20}}{Y^6}} = e^0,$$

or

$$X^2 Z^{20} = Y^6.$$

The equations of the multiplicative normal line to $S$ at $(1, 1, 1)$ are given by

$$(X \to_\star 1)/_\star e^2 = (Y \to_\star 1)/_\star e^{-6} = (Z \to_\star 1)/_\star e^{20}$$

or

$$X/_\star e^2 = Y/_\star e^{-6} = Z/_\star e^{20},$$

or

$$e^{\frac{\log X}{2}} = e^{\frac{\log Y}{-6}} = e^{\frac{\log Z}{20}},$$

or

$$\frac{\log X}{2} = \frac{\log Y}{-6} = \frac{\log Z}{20},$$

or

$$\log X = \frac{\log Y}{-3} = \frac{\log Z}{10},$$

or

$$\log X = \log \frac{1}{\sqrt[3]{Y}} = \log \sqrt[10]{Z},$$

or

$$X = \frac{1}{\sqrt[3]{Y}} = \sqrt[10]{Z}.$$

**Exercise 4.4:** Let $S$ be a multiplicative surface given by the equation

$$F(x, y, z) = e^{x^6 - y^2 + xyz^2} = 0_\star, \quad (x, y, z) \in \mathbb{R}_\star^3,$$

and $(x_0, y_0, z_0) = (1, 1, 1)$. Find the equation of the multiplicative tangent plane and the equations of the multiplicative normal line to $S$ at $(x_0, y_0, z_0)$.

**Definition 4.15:** An orientation of a multiplicative surface $S$ is a choice of a multiplicative normal vector to $T_{\star a} S$, $n$.

**Definition 4.16:** Let $S$ be an oriented multiplicative surface with orientation $n(a)$. A multiplicative local parameterization $(U, f)$ of $S$ is said to be multiplicative compatible if

$$n = (f_u^\star \times_\star f_v^\star)/_\star \left| f_u^\star \times_\star f_v^\star \right|_\star .$$

## 4.5 Multiplicative Differentiable Maps on a Multiplicative Surface

Let $U \subseteq \mathbb{R}_\star^2, V \subseteq \mathbb{R}_\star^3$.

**Definition 4.17:** Suppose that $S$ is a multiplicative surface on $\mathbb{R}_\star^3$. A map $g: S \to V$ is said to be a multiplicative differentiable if for any multiplicative local parameterization $(U, f)$ of $S$ the map $g \circ f: U \to V$ is multiplicative smooth. The map $g \circ f$ is said to be multiplicative local representation of $g$ with respect to the multiplicative local parameterization $(U, f)$ of the multiplicative surface $S$.

**Example 4.10:** Let $S \subseteq V$ be a multiplicative surface. The inclusion $i: S \to V$ is defined by

$$i(a) = a, \quad a \in S.$$

Note that the inclusion $i$ is a multiplicative smooth map for any multiplicative local parameterization $(U, f)$ of $S$. The multiplicative parameterization of $i$ is given by

$$i_f = i \circ f$$
$$= f.$$

Now, suppose that $S$ is a multiplicative surface with a multiplicative local parameterization $(U, f)$. Let $g: S \to V$ be a multiplicative smooth map. We have

$$f(t_1, t_2) = (f_1(t_1, t_2), f_2(t_1, t_2), f_3(t_1, t_2)), \quad (t_1, t_2) \in U,$$

and

$$g(f)(t_1, t_2) = (g_1(f)(t_1, t_2), g_2(f)(t_1, t_2), g_3(f)(t_1, t_2)), \quad (t_1, t_2) \in U.$$

Moreover,

$$g_{lt_j}^\star(f)(t_1, t_2) = g_{lf_1}^\star(f)(t_1, t_2) \cdot_\star f_{1t_j}^\star(t_1, t_2) +_\star g_{lf_2}^\star(f)(t_1, t_2) \cdot_\star f_{2t_j}^\star(t_1, t_2)$$
$$+_\star g_{lf_3}^\star(f)(t_1, t_2) \cdot_\star f_{3t_j}^\star(t_1, t_2), \quad (t_1, t_2) \in U,$$

$l \in \{1, 2, 3\}, j \in \{1, 2\}$.

**Definition 4.18:** Let $S_1, S_2 \subseteq \mathbb{R}_\star^3$ be two multiplicative surfaces. A map $F: S_1 \to S_2$ is called multiplicative smooth if the map

$$F_1 = i \circ F: S_1 \to \mathbb{R}_\star^3$$

is multiplicative smooth.

**Definition 4.19:** Let $S_1, S_2 \subseteq \mathbb{R}_\star^3$ be two multiplicative surfaces. A map $F: S_1 \to S_2$ is said to be multiplicative diffeomorphism if $F$ is bijective and $F$, $F^{-1}$ are multiplicative smooth maps.

## 4.6 The Multiplicative Differential of a Multiplicative Smooth Map between Two Multiplicative Surfaces

Let $I \subseteq \mathbb{R}_\star$, $V \subseteq \mathbb{R}_\star^3$ and $G: V \to \mathbb{R}_\star^3$ be a multiplicative smooth map,

$$G(x, y, z) = (g_1(x, y, z), g_2(x, y, z), g_3(x, y, z)), \quad (x, y, z) \in V.$$

**Definition 4.20:** For any $a = (x_0, y_0, z_0) \in V$ the multiplicative differential of $G$ at $a$,

$$d_{\star a}G: \mathbb{R}^3_{\star a} \rightarrow \mathbb{R}^3_{\star G(a)},$$

is a linear map with the matrix

$$\mathscr{G}(x_0, y_0, z_0) = \begin{pmatrix} g^\star_{1x}(x_0, y_0, z_0) & g^\star_{2x}(x_0, y_0, z_0) & g^\star_{3x}(x_0, y_0, z_0) \\ g^\star_{1y}(x_0, y_0, z_0) & g^\star_{2y}(x_0, y_0, z_0) & g^\star_{3y}(x_0, y_0, z_0) \\ g^\star_{1z}(x_0, y_0, z_0) & g^\star_{2x}(x_0, y_0, z_0) & g^\star_{3z}(x_0, y_0, z_0) \end{pmatrix}.$$

Let now,

$$f(t) = (f_1(t), f_2(t), f_3(t)), \quad t \in I,$$

be a multiplicative parameterized curve. Then

$$G \circ f(t) = (g_1(f_1(t), f_2(t), f_3(t)), g_2(f_1(t), f_2(t), f_3(t)), g_3(f_1(t), f_2(t), f_3(t))), \quad t \in I.$$

Hence,

$$\begin{aligned}(G\circ f)^\star(t) = \Big( &g^\star_{1f_1}(f_1(t), f_2(t), f_3(t)) \cdot_\star f_1^\star(t) +_\star g^\star_{1f_2}(f_1(t), f_2(t), f_3(t)) \cdot_\star f_2^\star(t) \\ &+_\star g^\star_{1f_3}(f_1(t), f_2(t), f_3(t)) \cdot_\star f_3^\star(t), g^\star_{2f_1}(f_1(t), f_2(t), f_3(t)) \cdot_\star f_1^\star(t) \\ &+_\star g^\star_{2f_2}(f_1(t), f_2(t), f_3(t)) \cdot_\star f_2^\star(t) +_\star g^\star_{2f_3}(f_1(t), f_2(t), f_3(t)) \cdot_\star f_3^\star(t), \\ &g^\star_{3f_1}(f_1(t), f_2(t), f_3(t)) \cdot_\star f_1^\star(t) +_\star g^\star_{3f_2}(f_1(t), f_2(t), f_3(t)) \cdot_\star f_2^\star(t) \\ &+_\star g^\star_{3f_3}(f_1(t), f_2(t), f_3(t)) \cdot_\star f_3^\star(t) \Big), \quad t \in I.\end{aligned}$$

Thus, $d_{\star a}G$ assigns to any multiplicative tangent vector to $f(t)$ at $t = t_0$ the multiplicative tangent vector to $G(f)(t)$ at $t = t_0$.

**Definition 4.21:** Let $S_1$ and $S_2$ be two multiplicative surfaces, $F: S_1 \rightarrow S_2$ be a multiplicative smooth map between $S_1$ and $S_2$, $a \in S_1$. Then to any multiplicative smooth curve $(I, f)$ on $S_1$ corresponds a multiplicative smooth curve $(I, F\circ f)$ on $S_2$. If $a = f(t_0)$, $t_0 \in I$, then $F\circ f(t)$ passes through $F(a)$ at $t = t_0$. The map $T_{\star a}S_1 \rightarrow T_{\star F(a)}S_2$ assigning to each multiplicative tangent vector $f^\star(t_0)$ to a multiplicative parameterized curve $f(t)$ on $S$, with $f(t_0) = a$, the multiplicative tangent vector $(F\circ f)^\star(t_0)$ to the multiplicative parameterized curve $F\circ f$ at $t = t_0$, is said to be the multiplicative differential of the multiplicative smooth map $F: S_1 \rightarrow S_2$ at the point $a$.

## 4.7 The Multiplicative Spherical Map. The Multiplicative Shape Operator

Let $S \subseteq \mathbb{R}^3_\star$ be a multiplicative oriented surface and $S^2_\star$ be the multiplicative unit sphere in $\mathbb{R}^3_\star$ centred at the multiplicative origin $(0_\star, 0_\star, 0_\star)$. Let $a \in S$ and the multiplicative orientation of $S$ is the multiplicative unit normal to $S$.

**Definition 4.22:** The map $\Gamma \colon S \to S^2_\star$,

$$\Gamma(a) = n(a),$$

is said to be the multiplicative spherical map of the surface $S$.

**Theorem 4.8:** *The multiplicative spherical map* $\Gamma \colon S \to S^2_\star$ *is a multiplicative smooth map.*

*Proof.* Let $(U, f)$ be a multiplicative local parameterization of $S$ that is multiplicative compatible with the multiplicative orientation of $S$. Then

$$n(a) = \left( f^\star_u \times_\star f^\star_v \right) /_\star \left| f^\star_u \times_\star f^\star_v \right|_\star .$$

Hence,

$$\Gamma_{\circ} f(u, v) = \Gamma(f(u, v))$$
$$= n(u, v), \quad (u, v) \in U.$$

Therefore $\Gamma_{\circ} f$ is a multiplicative smooth map. This completes the proof.

**Definition 4.23:** The linear operator

$$d_{\star a} \Gamma \colon T_{\star a} S \to T_{\star a} S$$

will be called the multiplicative shape operator of $S$ at $a$. It will be denoted by $A$ or $A_a$.

Below, suppose that $(U, f)$ is a multiplicative local parameterization of $S$ that is multiplicative compatible with the multiplicative orientation of $S$, and the multiplicative orientation of $S$ is given by $n(u, v)$. Then

$$n(u, v) = \left( f^\star_u \times_\star f^\star_v \right) /_\star \left| f^\star_u \times_\star f^\star_v \right|_\star .$$

Let $h \in T_{\left(f_u^\star \times_\star f_v^\star\right)/_\star \left|f_u^\star \times_\star f_v^\star\right|_\star f(u,v)} S$, $h = (h_1, h_2)$ with respect to the multi-
plicative orthonormal basis $\{f_u^\star, f_v^\star\}$. Then

$$A(h) = h_1 \cdot_\star n_u^\star +_\star h_2 \cdot_\star n_v^\star.$$

In particular, we have

$$A(f_u^\star) = n_u^\star,$$
$$A(f_v^\star) = n_v^\star.$$

**Theorem 4.9:** *The multiplicative shape operator A is a multiplicative self-adjoint operator, i.e.,*

$$\langle A(h), p \rangle_\star = \langle h, A(p) \rangle_\star$$

*for any $h, p \in T_{\star a} S$.*

*Proof.* Consider the equations

$$\langle n, f_u^\star \rangle_\star = 0_\star,$$
$$\langle n, f_v^\star \rangle_\star = 0_\star,$$

which we multiplicative differentiate with respect to $v$ and $u$, respectively, and we get

$$0_\star = \langle n_v^\star, f_u^\star \rangle_\star +_\star \langle n, f_{uv}^\star \rangle_\star,$$
$$0_\star = \langle n_u^\star, f_v^\star \rangle_\star +_\star \langle n, f_{uv}^\star \rangle_\star.$$

By the last two equations, we conclude that

$$\langle n_v^\star, f_u^\star \rangle_\star = \langle n_u^\star, f_v^\star \rangle_\star.$$

Let $h = (h_1, h_2)$, $p = (p_1, p_2) \in T_{\star a} S$ with respect to the multiplicative basis $\{f_u^\star, f_v^\star\}$. Then

$$h = h_1 \cdot_\star f_u^\star +_\star h_2 \cdot_\star f_v^\star,$$
$$p = p_1 \cdot_\star f_u^\star +_\star p_2 \cdot_\star f_v^\star,$$
$$A(h) = h_1 \cdot_\star n_u^\star +_\star h_2 \cdot_\star n_v^\star,$$
$$A(p) = p_1 \cdot_\star n_u^\star +_\star p_2 \cdot_\star n_v^\star.$$

Hence,

$$<A(h), p>_\star = <n_u^*\cdot_\star h_1 +_\star n_v^*\cdot_\star h_2, f_u^*\cdot_\star p_1 +_\star f_v^*\cdot_\star p_2>_\star$$

$$= <n_u^*\cdot_\star h_1, f_u^*\cdot_\star p_1>_\star$$

$$+_\star <n_u^*\cdot_\star h_1, f_v^*\cdot_\star p_2>_\star$$

$$+_\star <n_v^*\cdot_\star h_2, f_u^*\cdot_\star p_1>_\star$$

$$+_\star <n_v^*\cdot_\star h_2, f_u^*\cdot_\star p_2>_\star$$

$$= (h_{1}\cdot_\star p_1)\cdot_\star <n_u^*, f_u^*>_\star +_\star (h_{1}\cdot_\star p_2)\cdot_\star <n_u^*, f_v^*>_\star$$

$$+_\star (h_{2}\cdot_\star p_1)\cdot_\star <n_v^*, f_u^*>_\star +_\star (h_{2}\cdot_\star p_2)\cdot_\star <n_v^*, f_v^*>_\star$$

$$= (h_{1}\cdot_\star p_1)\cdot_\star <n_u^*, f_u^*>_\star +_\star (h_{1}\cdot_\star p_2)\cdot_\star <n_v^*, f_u^*>_\star$$

$$+_\star (h_{2}\cdot_\star p_1)\cdot_\star <n_u^*, f_v^*>_\star +_\star (h_{2}\cdot_\star p_2)\cdot_\star <n_u^*, f_u^*>_\star$$

$$= <f_u^*\cdot_\star h_1, n_u^*\cdot_\star p_1>_\star$$

$$+_\star <f_u^*\cdot_\star h_1, n_v^*\cdot_\star p_2>_\star$$

$$+_\star <f_v^*\cdot_\star h_2, n_u^*\cdot_\star p_1>_\star$$

$$+_\star <f_v^*\cdot_\star h_2, n_v^*\cdot_\star p_2>_\star$$

$$= <f_u^*\cdot_\star h_1, A(p)>_\star$$

$$+_\star <f_v^*\cdot_\star h_2, A(p)>_\star$$

$$= <h, A(p)>_\star .$$

This completes the proof.

**Corollary 4.1:** *In each multiplicative tangent space $T_{\star a}S$ there is a multiplicative orthonormal basis made up of the multiplicative eigenvectors of the multiplicative shape operator A.*

*Proof.* Let $\lambda_1, \lambda_2 \in \mathbb{R}_\star, \lambda_1 \neq \lambda_2$, be multiplicative eigenvalues of the operator $A$ that correspond to multiplicative eigenvectors $p$ and $q$, respectively. Then

$$A(p) = \lambda_1\cdot_\star p,$$
$$A(q) = \lambda_2\cdot_\star q.$$

Hence,

$$\langle A(p), q\rangle_\star = \langle \lambda_1\cdot_\star p, q\rangle_\star$$
$$= \lambda_1\cdot_\star \langle p, q\rangle_\star$$

and

$$\langle p, A(q) \rangle_\star = \langle p, \lambda_{2}\cdot_\star q \rangle_\star$$
$$= \lambda_{2}\cdot_\star \langle p, q \rangle_\star.$$

Now, by Theorem 4.9, we have that

$$\langle A(p), q \rangle_\star = \langle p, A(q) \rangle_\star.$$

From here, we find

$$\lambda_{1}\cdot_\star \langle p, q \rangle_\star = \lambda_{2}\cdot_\star \langle p, q \rangle_\star,$$

from where

$$(\lambda_{1} -_\star \lambda_{2})\cdot_\star \langle p, q \rangle_\star = 0_\star,$$

and

$$\langle p, q \rangle_\star = 0_\star.$$

Let

$$p_{1} = p/_\star |p|_\star,$$
$$q_{1} = q/_\star |q|_\star.$$

Then

$$|p_{1}|_\star = 1_\star, \qquad |q_{1}|_\star = 1_\star,$$

and

$$\langle p_{1}, q_{1} \rangle_\star = 0_\star.$$

This completes the proof.

## 4.8 The First Multiplicative Fundamental Form of a Multiplicative Surface

Suppose that $S$ is a multiplicative oriented surface.

**Definition 4.24:** For any $a \in S$ and $p, q \in T_{\star a}S$, the first multiplicative fundamental form of $S$ at $a$ is defined by

$$\phi_1(p, q) = \langle p, q \rangle_\star.$$

Let $(U, f)$ be a multiplicative local parameterization of $S$ that is multiplicative compatible with the multiplicative orientation of $S$. Let also, $p = (p_1, p_2)$, $q = (q_1, q_2) \in T_{\star a}S$ with respect to the multiplicative basis $\{f_u^\star, f_v^\star\}$. Then

$$p = p_1 \cdot_\star f_u^\star +_\star p_2 \cdot_\star f_v^\star,$$
$$q = q_1 \cdot_\star f_u^\star +_\star q_2 \cdot_\star f_v^\star.$$

Then

$$\begin{aligned}
\phi_1(p, q) &= \langle p, q \rangle_\star \\
&= \langle p_1 \cdot_\star f_u^\star +_\star p_2 \cdot_\star f_v^\star, q_1 \cdot_\star f_u^\star +_\star q_2 \cdot_\star f_v^\star \rangle_\star \\
&= (p_1 \cdot_\star q_1) \cdot_\star \langle f_u^\star, f_u^\star \rangle_\star +_\star (p_1 \cdot_\star q_2) \cdot_\star \langle f_u^\star, f_v^\star \rangle_\star \\
&\quad +_\star (p_2 \cdot_\star q_1) \cdot_\star \langle f_u^\star, f_v^\star \rangle_\star +_\star (p_2 \cdot_\star q_2) \cdot_\star \langle f_v^\star, f_v^\star \rangle_\star.
\end{aligned}$$

Define

$$\begin{aligned}
U(u, v) &= \langle f_u^\star, f_u^\star \rangle_\star, \\
F(u, v) &= \langle f_u^\star, f_v^\star \rangle_\star, \\
G(u, v) &= \langle f_v^\star, f_v^\star \rangle_\star, \quad (u, v) \in U,
\end{aligned}$$

and the matrix

$$\mathscr{G}(u, v) = \begin{pmatrix} E(u, v) & F(u, v) \\ F(u, v) & G(u, v) \end{pmatrix}, \quad (u, v) \in U.$$

Then

$$\phi_1(p, q) = (p_1 \cdot_\star q_1) \cdot_\star E(u, v) +_\star (p_1 \cdot_\star q_2) \cdot_\star F(u, v)$$
$$+_\star (p_2 \cdot_\star q_1) \cdot_\star F(u, v) +_\star (p_2 \cdot_\star q_2) \cdot_\star G(u, v)$$

$$= p_1 \cdot_\star \left( E(u, v) \cdot_\star q_1 +_\star F(u, v) \cdot_\star q_2 \right)$$

$$+_\star p_2 \cdot_\star \left( F(u, v) \cdot_\star q_1 +_\star G(u, v) \cdot_\star q_2 \right)$$

$$= p \cdot_\star G(u, v) \cdot_\star q.$$

**Example 4.11:** Let $f: \mathbb{R}^2_\star \to \mathbb{R}^3_\star$ be given as in Example 4.1. Then

$$f^\star_u(u, v) = \left( e^{\frac{u}{u+v}}, e^{2u^2}, e^u \right),$$
$$f^\star_v(u, v) = \left( e^{\frac{v}{u+v}}, e^v, e^{2v^2} \right), \quad (u, v) \in \mathbb{R}^2_\star.$$

Hence,

$$E(u, v) = \langle f^\star_u(u, v), f^\star_v(u, v) \rangle_\star$$
$$= e^{\frac{u^2}{(u+v)^2} + 4u^4 + u^2},$$
$$F(u, v) = \langle f^\star_u(u, v), f^\star_v(u, v) \rangle_\star$$
$$= e^{\frac{uv}{(u+v)^2} + 2u^2v + 2uv^2},$$
$$G(u, v) = \langle f^\star_v(u, v), f^\star_v(u, v) \rangle_\star$$
$$= e^{\frac{v^2}{(u+v)^2} + v^2 + 4v^4}, \quad (u, v) \in U,$$

and

$$\mathscr{G}(u, v) = \begin{pmatrix} E(u, v) & F(u, v) \\ F(u, v) & G(u, v) \end{pmatrix}$$

$$= \begin{pmatrix} e^{\frac{u^2}{(u+v)^2} + 4u^4 + u^2} & e^{\frac{uv}{(u+v)^2} + 2uv(u+v)} \\ e^{\frac{uv}{(u+v)^2} + 2uv(u+v)} & e^{\frac{v^2}{(u+v)^2} + v^2 + 4v^4} \end{pmatrix}, \quad (u, v) \in U.$$

**Example 4.12:** Let $f: \mathbb{R}^2_\star \to \mathbb{R}^3_\star$ be given as in Example 4.4. Then

$$f_u^\star(u, v) = (1_\star, 0_\star, e^{2u^2}),$$
$$f_v^\star(u, v) = (0_\star, 1_\star, e^{3v^3}), \quad (u, v) \in \mathbb{R}_\star^2,$$

and

$$E(u, v) = \langle f_u^\star, f_u^\star \rangle_\star$$
$$= e^{4u^4},$$
$$F(u, v) = \langle f_u^\star, f_v^\star \rangle_\star$$
$$= e^{6u^2v^3},$$
$$G(u, v) = \langle f_v^\star, f_v^\star \rangle_\star$$
$$= e^{9v^6}, \quad (u, v) \in \mathbb{R}_\star^2,$$

and

$$\mathscr{G}(u, v) = \begin{pmatrix} E(u, v) & F(u, v) \\ F(u, v) & G(u, v) \end{pmatrix}$$
$$= \begin{pmatrix} e^{4u^4} & e^{6u^2v^3} \\ e^{6u^2v^3} & e^{9v^6} \end{pmatrix}, \quad (u, v) \in \mathbb{R}_\star^2.$$

**Exercise 4.5:** Let $f: \mathbb{R}_\star^2 \to \mathbb{R}_\star^3$ be given by

$$f(u, v) = \left( \frac{u + v}{u^2 + v^2}, e^{u - 3v^3}, e^{u^2 + v} \right), \quad (u, v) \in \mathbb{R}_\star^2.$$

Find the matrix of the first multiplicative fundamental form.

## 4.9 Applications of the First Multiplicative Fundamental Form

Let $S$ be a multiplicative oriented surface with a multiplicative local parameterization $(U, f)$ that is multiplicative compatible with the multiplicative orientation of $S$.

### 4.9.1 *The Multiplicative Length of a Multiplicative Segment of a Multiplicative Curve on a Multiplicative Surface*

Let $(I, g = g(t))$ be a multiplicative parameterized curve on $S$, $g(I) \subset f(U)$. Let also

$$g(t) = (u(t), v(t)), \quad t \in I.$$

Then

$$g^\star(t) = (u^\star(t), v^\star(t)), \quad t \in I,$$

and

$$\langle g^\star(t), g^\star(t) \rangle_\star = g^\star(t) \cdot_\star G \cdot_\star g^\star(t)$$

$$= (u^\star(t), v^\star(t)) \cdot_\star \begin{pmatrix} E(u(t), v(t)) & F(u(t), v(t)) \\ F(u(t), v(t)) & G(u(t), v(t)) \end{pmatrix}^{\cdot_\star} \begin{pmatrix} u^\star(t) \\ v^\star(t) \end{pmatrix}$$

$$= (u^\star(t), v^\star(t)) \cdot_\star \begin{pmatrix} E(u(t), v(t)) \cdot_\star u^\star(t) +_\star F(u(t), v(t)) \cdot_\star v^\star(t) \\ F(u(t), v(t)) \cdot_\star u^\star(t) +_\star G(u(t), v(t)) \cdot_\star v^\star(t) \end{pmatrix}$$

$$= (u^\star(t))^{2\star} \cdot_\star E(u(t), v(t)) +_\star e^{2} \cdot_\star F(u(t), v(t)) \cdot_\star u^\star(t) \cdot_\star v^\star(t)$$

$$+_\star (v^\star(t))^{2\star} \cdot_\star G(u(t), v(t)), \quad t \in I.$$

**Definition 4.25:** The multiplicative length of the multiplicative segment of the multiplicative curve between $t_1$ and $t_2$ on the multiplicative surface $S$ is defined by

$$l(t_1, t_2) = \int_{\star t_1}^{t_2} |g^\star(t)|_\star^{\frac{1}{2}\star} \cdot_\star d_\star t.$$

We have

$$l(t_1, t_2) = \int_{\star t_1}^{t_2} ((u^\star(t))^{2\star} \cdot_\star E(u(t), v(t)) +_\star e^{2} \cdot_\star F(u(t), v(t)) \cdot_\star u^\star(t) \cdot_\star v^\star(t)$$

$$+_\star (v^\star(t))^{2\star} \cdot_\star G(u(t), v(t)))^{\frac{1}{2}\star} \cdot_\star d_\star t.$$

### 4.9.2 The Multiplicative Angle between Two Multiplicative Curves on a Multiplicative Surface

Let $(I, g_1 = g_1(t))$ and $(J, g_2 = g_2(s))$ be two multiplicative parameterized curves on $S$ such that

$$g_1(t_0) = g_2(s_0) = f(u_0, v_0).$$

Let also,

$$g_1(t) = (u_1(t), v_1(t)), \quad t \in I,$$
$$g_2(s) = (u_2(s), v_2(s)), \quad s \in J.$$

Then

$$g_1^\star(t) = (u_1^\star(t), v_1^\star(t)), \quad t \in I,$$
$$g_2^\star(s) = (u_2^\star(s), v_2^\star(s)), \quad s \in J.$$

**Definition 4.26:** The multiplicative angle $\theta$ between the multiplicative curves $g_1$ and $g_2$ at the point $f(u_0, v_0)$ we define as follows:

$$\cos{}_\star \theta = \langle g_1^\star(t_0), g_2^\star(s_0) \rangle_\star /_\star \left( \left| g_1^\star(t_0) \right|_\star \cdot_\star \left| g_2^\star(s_0) \right|_\star \right).$$

We have

$$\langle g_1^\star(t_0), g_2^\star(s_0) \rangle_\star = (u_1^\star(t_0), v_1^\star(t_0)) \cdot_\star \begin{pmatrix} E(u_0, v_0) & F(u_0, v_0) \\ F(u_0, v_0) & G(u_0, v_0) \end{pmatrix} \cdot_\star \begin{pmatrix} u_2^\star(s_0) \\ v_2^\star(s_0) \end{pmatrix}$$

$$= (u_1^\star(t_0), v_1^\star(t_0)) \cdot_\star \begin{pmatrix} E(u_0, v_0) \cdot_\star u_2^\star(s_0) +_\star F(u_0, v_0) \cdot_\star v_2^\star(s_0) \\ F(u_0, v_0) \cdot_\star u_2^\star(s_0) +_\star G(u_0, v_0) \cdot_\star v_2^\star(s_0) \end{pmatrix}$$

$$= (u_1^\star(t_0) \cdot_\star u_2^\star(s_0)) \cdot_\star E(u_0, v_0)$$
$$+_\star (u_1^\star(t_0) \cdot_\star v_2^\star(s_0) +_\star v_1^\star(t_0) \cdot_\star u_2^\star(s_0)) \cdot_\star F(u_0, v_0)$$
$$+_\star (v_1^\star(t_0) \cdot_\star v_2^\star(s_0)) \cdot_\star G(u_0, v_0)$$

and

$$\langle g_1^\star(t_0), g_1^\star(t_0) \rangle_\star = (u_1^\star(t_0), v_1^\star(t_0)) \cdot_\star \begin{pmatrix} E(u_0, v_0) & F(u_0, v_0) \\ F(u_0, v_0) & G(u_0, v_0) \end{pmatrix} \cdot_\star \begin{pmatrix} u_1^\star(t_0) \\ v_1^\star(t_0) \end{pmatrix}$$

$$= (u_1^\star(t_0), v_1^\star(t_0)) \cdot_\star \begin{pmatrix} E(u_0, v_0) \cdot_\star u_1^\star(t_0) +_\star F(u_0, v_0) \cdot_\star v_1^\star(t_0) \\ F(u_0, v_0) \cdot_\star u_1^\star(t_0) +_\star G(u_0, v_0) \cdot_\star v_1^\star(t_0) \end{pmatrix}$$

$$= (u_1^\star(t_0))^{2_\star} \cdot_\star E(u_0, v_0)$$
$$+_\star e^{2_\star} \cdot_\star (u_1^\star(t_0) \cdot_\star v_1^\star(t_0)) \cdot_\star F(u_0, v_0)$$
$$+_\star (v_1^\star(t_0))^{2_\star} \cdot_\star G(u_0, v_0),$$

and

$$\langle g_2^\star(s_0), g_2^\star(s_0) \rangle_\star = (u_2^\star(s_0), v_2^\star(s_0)) \cdot_\star \begin{pmatrix} E(u_0, v_0) & F(u_0, v_0) \\ F(u_0, v_0) & G(u_0, v_0) \end{pmatrix} \cdot_\star \begin{pmatrix} u_2^\star(s_0) \\ v_2^\star(s_0) \end{pmatrix}$$

$$= (u_2^\star(s_0), v_2^\star(s_0)) \cdot_\star \begin{pmatrix} E(u_0, v_0) \cdot_\star u_2^\star(s_0) +_\star F(u_0, v_0) \cdot_\star v_2^\star(s_0) \\ F(u_0, v_0) \cdot_\star u_2^\star(s_0) +_\star G(u_0, v_0) \cdot_\star v_2^\star(s_0) \end{pmatrix}$$

$$= (u_2^\star(s_0))^{2_\star} \cdot_\star E(u_0, v_0)$$
$$+_\star e^{2_\star} \cdot_\star (u_2^\star(s_0) \cdot_\star v_2^\star(s_0)) \cdot_\star F(u_0, v_0)$$
$$+_\star (v_2^\star(s_0))^{2_\star} \cdot_\star G(u_0, v_0).$$

Consequently

$$\cos_\star \theta = ((u_1^\star(t_0) \cdot_\star u_2^\star(s_0)) \cdot_\star E(u_0, v_0)$$
$$+_\star (u_1^\star(t_0) \cdot_\star v_2^\star(s_0) +_\star v_1^\star(t_0) \cdot_\star u_2^\star(s_0)) \cdot_\star F(u_0, v_0)$$
$$+_\star (v_1^\star(t_0) \cdot_\star v_2^\star(s_0)) \cdot_\star G(u_0, v_0))$$
$$/_\star (((u_1^\star(t_0))^{2_\star} \cdot_\star E(u_0, v_0)$$
$$+_\star e^{2_\star} \cdot_\star (u_1^\star(t_0) \cdot_\star v_1^\star(t_0)) \cdot_\star F(u_0, v_0)$$
$$+_\star (v_1^\star(t_0))^{2_\star} \cdot_\star G(u_0, v_0))^{\frac{1}{2}_\star}$$
$$\star_\star ((u_2^\star(s_0))^{2_\star} \cdot_\star E(u_0, v_0)$$
$$+_\star e^{2_\star} \cdot_\star (u_2^\star(s_0) \cdot_\star v_2^\star(s_0)) \cdot_\star F(u_0, v_0)$$
$$+_\star (v_2^\star(s_0))^{2_\star} \cdot_\star G(u_0, v_0))^{\frac{1}{2}_\star}).$$

### 4.9.3 *The Multiplicative Area of a Multiplicative Parameterized Surface*

**Definition 4.27:** The multiplicative area of the multiplicative surface $S$ is defined as follows:

$$A = \iint_{\star U} (E(u, v) \cdot_\star G(u, v) +_\star (F(u, v))^{2_\star})^{\frac{1}{2}_\star} \cdot_\star d_\star u \cdot_\star d_\star v.$$

## 4.10 The Multiplicative Matrix of the Multiplicative Shape Operator

Let $(U, f)$ be a multiplicative local parameterization of $S$. Let also, $\{f_u^\star, f_v^\star\}$ be the multiplicative basis. With $\mathscr{A}$ we will denote the multiplicative matrix of the multiplicative shape operator $A$. We have

$$A(f_u^\star) = n_u^\star,$$
$$A(f_v^\star) = n_v^\star.$$

Then

$$(n_u^\star n_v^\star) = (f_u^\star f_v^\star) \cdot_\star \mathscr{A}. \tag{4.5}$$

Note that

$$\begin{pmatrix} f_u^\star \\ f_v^\star \end{pmatrix} \cdot_\star (n_u^\star n_v^\star) = \begin{pmatrix} \langle f_u^\star, n_u^\star \rangle_\star & \langle f_u^\star, n_v^\star \rangle_\star \\ \langle f_v^\star, n_u^\star \rangle & \langle f_v^\star, n_v^\star \rangle_\star \end{pmatrix}.$$

Let

$$L = \langle f_u^\star, f_u^\star \rangle_\star,$$
$$M = \langle f_u^\star, n_v^\star \rangle_\star$$
$$= \langle f_v^\star, n_u^\star \rangle_\star,$$
$$N = \langle f_v^\star, n_v^\star \rangle_\star.$$

Set

$$\mathscr{H} = \begin{pmatrix} L & M \\ M & N \end{pmatrix}.$$

Therefore

$$\begin{pmatrix} f_u^\star \\ f_v^\star \end{pmatrix} \cdot_\star (n_u^\star n_v^\star) = \mathscr{H}.$$

Hence and (4.5), we get

$$\mathcal{H} = \begin{pmatrix} L & M \\ M & N \end{pmatrix}$$

$$= \begin{pmatrix} \langle f_u^\star, n_u^\star \rangle_\star & \langle f_u^\star, n_v^\star \rangle_\star \\ \langle f_v^\star, n_u^\star \rangle_\star & \langle f_v^\star, n_v^\star \rangle_\star \end{pmatrix}$$

$$= \begin{pmatrix} f_u^\star \\ f_v^\star \end{pmatrix} \cdot_\star (n_u^\star n_v^\star)$$

$$= \begin{pmatrix} f_u^\star \\ f_v^\star \end{pmatrix} \cdot_\star (f_u^\star f_v^\star) \cdot_\star \mathcal{A}$$

$$= \begin{pmatrix} \langle f_u^\star, f_u^\star \rangle_\star & \langle f_u^\star, f_v^\star \rangle_\star \\ \langle f_u^\star, f_v^\star \rangle_\star & \langle f_v^\star, f_v^\star \rangle_\star \end{pmatrix} \cdot_\star \mathcal{A}$$

$$= \begin{pmatrix} E & F \\ F & G \end{pmatrix} \cdot_\star \mathcal{A}$$

$$= G \cdot_\star A,$$

whereupon

$$\mathcal{A} = \mathcal{G}^{-1\star} \cdot_\star \mathcal{H}.$$

Note that

$$\mathcal{G}^{-1\star} = \left( 1_\star /_\star (E \cdot_\star G -_\star F^{2\star}) \right) \cdot_\star \begin{pmatrix} G & -_\star F \\ -_\star F & E \end{pmatrix}.$$

Thus,

$$\mathcal{A} = \left( 1_\star /_\star (E \cdot_\star G -_\star F^{2\star}) \right) \cdot_\star \begin{pmatrix} G & -_\star F \\ -_\star F & E \end{pmatrix} \cdot_\star \begin{pmatrix} L & M \\ M & N \end{pmatrix}$$

$$= \left( 1_\star /_\star (E \cdot_\star G -_\star F^{2\star}) \right) \cdot_\star \begin{pmatrix} G \cdot_\star L -_\star F \cdot_\star M & G \cdot_\star M -_\star F \cdot_\star N \\ E \cdot_\star M -_\star F \cdot_\star L & E \cdot_\star M -_\star F \cdot_\star N \end{pmatrix}.$$

Now, we multiplicative differentiate with respect to $u$ the equation

$$\langle f_u^\star, n \rangle_\star = 0_\star \qquad\qquad\qquad (4.6)$$

and we obtain

$$0_\star = \langle f_u^{\star\star}, n \rangle_\star +_\star \langle f_u^\star, n_u^\star \rangle_\star,$$

whereupon

$$L = -_\star \langle n, f_u^{\star\star} \rangle_\star.$$

We multiplicative differentiate with respect to $v$ the equation (4.6) and we arrive at

$$0_\star = \langle f_{uv}^{\star\star}, n \rangle_\star +_\star \langle f_u^\star, n_v^\star \rangle_\star,$$

from where

$$M = -_\star \langle f_{uv}^{\star\star}, n \rangle_\star.$$

We multiplicative differentiate with respect to $v$ the equation

$$\langle f_v^\star, n \rangle_\star = 0_\star$$

and we find

$$0_\star = \langle f_v^{\star\star}, n \rangle_\star +_\star \langle f_v^\star, n_v^\star \rangle_\star,$$

whereupon

$$N = -_\star \langle n, f_v^{\star\star} \rangle_\star.$$

Note that

$$\begin{aligned}
\langle f_u^\star \times_\star f_v^\star, f_u^\star \times_\star f_v^\star \rangle_\star &= \langle f_u^\star, f_u^\star \rangle_\star \cdot_\star \langle f_v^\star, f_v^\star \rangle_\star \\
&\quad -_\star \langle f_u^\star, f_v^\star \rangle_\star \cdot_\star \langle f_u^\star, f_v^\star \rangle_\star \\
&= E \cdot G -_\star F^2_\star
\end{aligned}$$

and

$$|f_u^\star \times_\star f_v^\star|_\star = \left( E \cdot G -_\star F^2_\star \right)^{\frac{1}{2}}_\star.$$

Let

$$H = \left( E \cdot G -_\star F^{2\star} \right)^{\frac{1}{2}\star}.$$

Then

$$f_u^\star \times_\star f_v^\star = H \cdot_\star n$$

and

$$L = -_\star (1_\star /_\star H) \cdot_\star \langle f_u^\star \times_\star f_v^\star, f_u^{\star\star} \rangle_\star,$$
$$M = -_\star (1_\star /_\star H) \cdot_\star \langle f_u^\star \times_\star f_v^\star, f_{uv}^{\star\star} \rangle_\star,$$
$$N = -_\star (1_\star /_\star H) \cdot_\star \langle f_u^\star \times_\star f_v^\star, f_v^{\star\star} \rangle_\star.$$

**Example 4.13:** Let $f: \mathbb{R}_\star^2 \to \mathbb{R}_\star^3$ be given by

$$f(u, v) = \left( u, v, e^{u^2 + v^3} \right), \quad (u, v) \in \mathbb{T}_\star^2.$$

Then

$$f_u^\star (u, v) = (e, 1, e^{2u^2}),$$
$$f_v^\star (u, v) = (1, e, e^{3v^3}),$$
$$f_u^{\star\star} (u, v) = (1, 1, e^{4u^2}),$$
$$f_v^{\star\star} (u, v) = (1, 1, e^{9v^3}),$$
$$f_{uv}^{\star\star} (u, v) = (1, 1, 1), \quad (u, v) \in \mathbb{R}_\star^2,$$

and

$$f_u^\star (u, v) \times_\star f_v^\star (u, v) = (e^{-2u^2}, e^{-3v^3}, e),$$

$$|f_u^\star (u, v) \times_\star f_v^\star (u, v)|_\star = e^{(4u^4 + 9v^6)^{\frac{1}{2}}}, \quad (u, v) \in \mathbb{R}_\star^2,$$

and

$$n(u, v) = \left[ f_u^\star(u, v) \times_\star f_v^\star(u, v) \right] \Big/ \star \, |f_u^\star(u, v) \times_\star f_v^\star(u, v)|_\star$$

$$= \left( e^{-\frac{2u^2}{(4u^4+9v^6)^{\frac{1}{2}}}}, e^{\frac{2u^2-3v^3}{(4u^4+9v^6)^{\frac{1}{2}}}}, e^{\frac{1}{(4u^4+9v^6)^{\frac{1}{2}}}} \right), \quad (u, v) \in \mathbb{R}_\star^2,$$

and

$$E(u, v) = \langle f_u^\star(u, v), f_u^\star(u, v) \rangle_\star$$
$$= e^{1+4u^4},$$
$$F(u, v) = \langle f_u^\star(u, v), f_v^v(u, v) \rangle_\star$$
$$= e^{6u^2v^3},$$
$$G(u, v) = \langle f_v^\star(u, v), f_v^\star(u, v) \rangle_\star$$
$$= e^{1+9v^6}, \quad (u, v) \in \mathbb{R}_\star^2,$$

and

$$L(u, v) = -_\star \langle n(u, v), f_{uu}^{\star\star}(u, v) \rangle_\star$$
$$= -_\star e^{\frac{4u^2}{(4u^4+9v^6)^{\frac{1}{2}}}}, M(u, v)$$
$$= -_\star \langle n(u, v), f_{uv}^{\star\star}(u, v) \rangle_\star$$
$$= 1, N(u, v)$$
$$= -_\star \langle n(u, v), f_{vv}^{\star\star}(u, v) \rangle_\star$$
$$= -_\star e^{\frac{9v^3}{(4u^4+9v^6)^{\frac{1}{2}}}}, \quad (u, v) \in \mathbb{R}_\star^2,$$

and

$$E(u, v) \cdot_\star G(u, v) -_\star (F(u, v))^2_\star = e^{(1+4u^4)(1+9v^6)-36u^4v^6}$$
$$= e^{1+4u^4+9v^6}, \quad (u, v) \in \mathbb{R}_\star^2,$$

and

$$G(u, v)\cdot_\star L(u, v) -_\star F(u, v)\cdot_\star M(u, v) = e^{-\frac{4u^2(1+9v^6)}{(4u^4+9v^6)^{\frac{1}{2}}}},$$

$$G(u, v)\cdot_\star M(u, v) -_\star F(u, v)\cdot_\star N(u, v) = e^{-\frac{54u^2v^6}{(4u^4+9v^6)^{\frac{1}{2}}}},$$

$$E(u, v)\cdot_\star M(u, v) -_\star F(u, v)\cdot_\star L(u, v) = e^{-\frac{24u^4v^3}{(4u^4+9v^6)^{\frac{1}{2}}}},$$

$$E(u, v)\cdot_\star N(u, v) -_\star F(u, v)\cdot_\star M(u, v) = e^{-\frac{9v^3(1+4u^4)}{(4u^4+9v^6)^{\frac{1}{2}}}},$$

$(v, v)\in'\mathbb{R}^2_\star$. Consequently

$$A = \begin{pmatrix} e^{-\frac{4u^2(1+9v^6)}{\left(4u^4+9v^6\right)^{\frac{1}{2}}}} & e^{-\frac{54u^2v^6}{\left(4u^4+9v^6\right)^{\frac{1}{2}}}} \\ e^{-\frac{24u^4v^3}{\left(4u^4+9v^6\right)^{\frac{1}{2}}}} & e^{-\frac{9v^3(1+4u^4)}{\left(4u^4+9v^6\right)^{\frac{1}{2}}}} \end{pmatrix},$$

$$(u, v) \in \mathbb{R}^2_\star.$$

**Example 4.14:** Let $f: \mathbb{R}^2_\star \to \mathbb{R}^3_\star$ be given by

$$f(u, v) = \left(u, v, e^{u^2+v}\right), \quad (u, v) \in \mathbb{R}^2_\star.$$

Then

$$f_u^\star(u, v) = (e, 1, e^{2u^2}),$$
$$f_v^\star(u, v) = (1, e, e^v),$$
$$f_u^{\star\star}(u, v) = (1, 1, e^{4u^2}),$$
$$f_v^{\star\star}(u, v) = (1, 1, e^v),$$
$$f_{uv}^{\star\star}(u, v) = (1, 1, 1), \quad (u, v) \in \mathbb{R}^2_\star,$$

and

$$f_u^\star(u, v) \times_\star f_v^\star(u, v) = (e^{-2u^2}, e^{-v}, e),$$
$$\left| f_u^\star(u, v) \times_\star f_v^\star(u, v) \right|_\star = e^{(4u^4+v^2+1)^{\frac{1}{2}}}, \quad (u, v) \in \mathbb{R}^2_\star,$$

and

$$n(u, v) = \left( f_u^\star (u, v) \times_\star f_v^\star (u, v) \right) \Big/ \left| f_u^\star (u, v) \times_\star f_v^\star (u, v) \right|_\star$$

$$= \left( e^{-\dfrac{2u^2}{\left(4u^4+v^2+1\right)^{\frac{1}{2}}}}, \; e^{-\dfrac{v}{\left(4u^4+v^2+1\right)^{\frac{1}{2}}}}, \; e^{\dfrac{1}{\left(4u^4+v^2+1\right)^{\frac{1}{2}}}} \right),$$

$(u, v) \in \mathbb{R}_\star^2$, and

$$E(u, v) = \langle f_u^\star (u, v), f_u^\star (u, v) \rangle_\star$$
$$= e^{1+4u^4},$$
$$F(u, v) = \langle f_u^\star (u, v), f_v^\star (u, v) \rangle_\star$$
$$= e^{2u^2v},$$
$$G(u, v) = \langle f_v^\star (u, v), f_v^\star (u, v) \rangle_\star$$
$$= e^{1+v^2}, \quad (u, v) \in \mathbb{R}_\star^2,$$

and

$$L(u, v) = {}_\star \langle n(u, v), f_{uu}^{\star\star} (u, v) \rangle_\star$$
$$= {}_\star e^{\left( \dfrac{4u^2}{\left(4u^4+v^2+1\right)^{\frac{1}{2}}} \right)}$$
$$= e^{-\dfrac{4u^2}{\left(4u^4+v^2+1\right)^{\frac{1}{2}}}},$$
$$M(u, v) = {}_\star \langle n(u, v), f_{uv}^{\star\star} (u, v) \rangle_\star$$
$$= 1,$$
$$N(u, v) = {}_\star \langle n(u, v), f_{vv}^{\star\star} (u, v) \rangle_\star$$
$$= {}_\star e^{\left( \dfrac{v}{\left(4u^4+v^2+1\right)^{\frac{1}{2}}} \right)}$$
$$= e^{-\dfrac{v}{\left(4u^4+v^2+1\right)^{\frac{1}{2}}}}, \quad (u, v) \in \mathbb{R}_\star,$$

and

$$E(u, v)\cdot_\star G(u, v) -_\star (F(u, v))^{2\star} = e^{(1+4u^4)(1+v^2)-4u^4v^2}$$
$$= e^{1+4u^4+v^2}, \quad (u, v) \in \mathbb{R}^2_\star,$$

and

$$G(u, v)\cdot_\star L(u, v) -_\star F(u, v)\cdot_\star M(u, v) = e^{-\frac{4u^2(1+v^2)}{\left(4u^4+v^2+1\right)^{\frac{1}{2}}}},$$

$$G(u, v)\cdot_\star M(u, v) -_\star F(u, v)\cdot_\star N(u, v) = e^{\frac{2u^2v^2}{\left(4u^4+v^2+1\right)^{\frac{1}{2}}}},$$

$$E(u, v)\cdot_\star M(u, v) -_\star F(u, v)\cdot_\star L(u, v) = e^{\frac{8u^4v}{\left(4u^4+v^2+1\right)^{\frac{1}{2}}}},$$

$$E(u, v)\cdot_\star N(u, v) -_\star F(u, v)\cdot_\star M(u, v) = e^{-\frac{(1+4u^4)v}{\left(4u^4+v^2+1\right)^{\frac{1}{2}}}},$$

$(u, v) \in \mathbb{R}^2_\star$. Consequently

$$\mathscr{A} = \begin{pmatrix} e^{-\frac{4u^2(1+v^2)}{\left(4u^4+v^2+1\right)^{\frac{1}{2}}}} & e^{\frac{2u^2v^2}{\left(4u^4+v^2+1\right)^{\frac{1}{2}}}} \\[4mm] e^{\frac{8u^4v}{\left(4u^4+v^2+1\right)^{\frac{1}{2}}}} & e^{-\frac{(1+4u^4)v}{\left(4u^4+v^2+1\right)^{\frac{1}{2}}}} \end{pmatrix},$$

$$(u, v) \in \mathbb{R}^2_\star.$$

**Exercise 4.6:** Let $f\colon \mathbb{R}^2_\star \to \mathbb{R}^3_\star$ be given by

$$f(u, v) = \left(u, v, e^{2u^2+3v^3}\right), \quad (u, v) \in \mathbb{R}^2_\star.$$

Find the matrix $\mathscr{A}$.

## 4.11 The Second Multiplicative Fundamental Form of a Multiplicative Surface

Let $S$ be a multiplicative oriented surface.

**Definition 4.28:** For any $a \in S$, the second multiplicative fundamental form is defined by

$$\phi_2(\xi, \eta) = -_\star \phi_1(A(\xi), \eta), \quad \xi, \eta \in T_{\star a}S.$$

**Theorem 4.10:** *For each $a \in S$, the second multiplicative fundamental form is a symmetric bilinear form.*

*Proof.* Let $\xi, \eta \in T_{\star a}S$. Then

$$
\begin{aligned}
\phi_2(\eta, \xi) &= -_\star \phi_1(A(\eta), \xi) \\
&= -_\star \phi_1(\eta, A(\xi)) \\
&= -_\star \phi_1(A(\xi), \eta) \\
&= \phi_2(\xi, \eta).
\end{aligned}
$$

Let now, $\xi_1, \xi_2, \eta_1, \eta_2 \in T_{\star a}S$, $\alpha_1, \alpha_2, \beta_1, \beta_2 \in \mathbb{R}_\star$. Then

$$
\begin{aligned}
\phi_2 \star \alpha_1 \cdot_\star \xi_1 +_\star \alpha_2 \cdot_\star \xi_2, \eta_1) &= -_\star \phi_1(A(\alpha_1 \cdot_\star \xi_1 +_\star \alpha_2 \cdot_\star \xi_2), \eta_1) \\
&= -_\star \phi_1(\alpha_1 \cdot_\star A(\xi_1) +_\star \alpha_2 \cdot_\star A(\xi_2), \eta_1) \\
&= -_\star \alpha_1 \cdot_\star \phi_1(A(\xi_1), \eta_1) -_\star \alpha_2 \cdot_\star \phi_1(A(\xi_2), \eta_1) \\
&= \alpha_1 \cdot_\star \phi_2(\xi_1, \eta_1) +_\star \alpha_2 \cdot_\star \phi_2(\xi_2, \eta_1)
\end{aligned}
$$

and

$$
\begin{aligned}
\phi_2(\xi_1, \beta_1 \cdot_\star \eta_1 +_\star \beta_2 \cdot_\star \eta_2) &= -_\star \phi_1(A(\xi_1), \beta_1 \cdot_\star \eta_1 +_\star \beta_2 \cdot_\star \eta_2) \\
&= -_\star \beta_1 \cdot_\star \phi_1(A(\xi_1), \eta_1) -_\star \beta_2 \cdot_\star \phi_1(A(\xi_1), \eta_2) \\
&= \beta_1 \cdot_\star \phi_2(\xi_1, \eta_1) +_\star \beta_2 \cdot_\star \phi_2(\xi_1, \eta_2).
\end{aligned}
$$

This completes the proof.

Let $(U, f)$ be a multiplicative local representation of $S$ that is multiplicative compatible with the multiplicative orientation of $S$. Then

$$\phi_2(f_u^\star, f_u^\star) = -_\star \phi_1(A(f_u^\star), f_u^\star)$$
$$= -_\star \phi_1(n_u^\star, f_u^\star)$$
$$= -_\star \langle n_u^\star, f_u^\star \rangle_\star$$
$$= -_\star L.$$

As above,

$$\phi_2(f_u^\star, f_v^\star) = -_\star \langle n_u^\star, f_v^\star \rangle_\star$$
$$= -_\star M,$$
$$\phi_2(f_v^\star, f_u^\star) = -_\star \langle n_u^\star, f_v^\star \rangle_\star$$
$$= -_\star M,$$
$$\phi_2(f_v^\star, f_v^\star) = -_\star \langle n_v^\star, f_v^\star \rangle_\star$$
$$= -_\star N.$$

Consequently

$$[\phi_2] = \begin{pmatrix} \phi_2(f_u^\star, f_u^\star) & \phi_2(f_u^\star, f_v^\star) \\ \phi_2(f_u^\star, f_v^\star) & \phi_2(f_v^\star, f_v^\star) \end{pmatrix}$$
$$= -_\star \begin{pmatrix} \langle n_u^\star, f_u^\star \rangle_\star & \langle n_u^\star, f_v^\star \rangle_\star \\ \langle n_u^\star, f_v^\star \rangle_\star & \langle n_v^\star, f_v^\star \rangle_\star \end{pmatrix}$$
$$= -_\star \begin{pmatrix} L & M \\ M & N \end{pmatrix}$$

is the multiplicative matrix of the second multiplicative fundamental form $\phi_2$. Let

$$D = n \cdot_\star f_u^{\star\star},$$
$$D^\star = n \cdot_\star f_{uv}^{\star\star},$$
$$D^{\star\star} = n \cdot_\star f_v^{\star\star}.$$

Therefore

$$[\phi_2] = \begin{pmatrix} D & D^\star \\ D^\star & D^{\star\star} \end{pmatrix}.$$

**Example 4.15:** Let $f\colon \mathbb{R}^2_\star \to \mathbb{R}^3_\star$ be as in Example 4.13. Then

$$[\phi_2](U, v) = {}_{\star}\!\!\begin{pmatrix} L & M \\ M & N \end{pmatrix}$$

$$= -{}_{\star}\!\!\begin{pmatrix} e^{-\dfrac{4u^2}{(4u^4+9v^6)^{\frac{1}{2}}}} & 1 \\ 1 & e^{-\dfrac{9v^3}{(4u^4+9v^6)^{\frac{1}{2}}}} \end{pmatrix}$$

$$= \begin{pmatrix} e^{\dfrac{4u^2}{(4u^4+9v^6)^{\frac{1}{2}}}} & 1 \\ 1 & e^{\dfrac{9v^3}{(4u^4+9v^6)^{\frac{1}{2}}}} \end{pmatrix}, \quad (u, v) \in \mathbb{R}^2_\star.$$

**Example 4.16:** Let $f\colon \mathbb{R}^2_\star \to \mathbb{R}^3_\star$ be as in Example 4.14. Then

$$[\phi_2](U, v) = {}_{\star}\!\!\begin{pmatrix} L & M \\ M & N \end{pmatrix}$$

$$= -{}_{\star}\!\!\begin{pmatrix} e^{-\dfrac{4u^2}{\left(4u^4+v^2+1\right)^{\frac{1}{2}}}} & 1 \\ 1 & e^{-\dfrac{v}{\left(4u^4+v^2+1\right)^{\frac{1}{2}}}} \end{pmatrix}$$

$$= \begin{pmatrix} e^{\dfrac{4u^2}{\left(4u^4+v^2+1\right)^{\frac{1}{2}}}} & 1 \\ 1 & e^{\dfrac{v}{\left(4u^4+v^2+1\right)^{\frac{1}{2}}}} \end{pmatrix}.$$

**Exercise 4.7:** Let $f\colon \mathbb{R}^2_\star \to \mathbb{R}^3_\star$ be given by

$$f(u, v) = \left(e^{u^2-5u^4+uv}, v, u\right), \quad (u, v) \in \mathbb{R}^2_\star.$$

Find the matrix $[\phi_2]$.

## 4.12 The Multiplicative Normal Curvature. The Multiplicative Meusnier Theorem

Let $S$ be a multiplicative surface and $(U, f)$ be a multiplicative local parameterization that is multiplicative compatible with the multiplicative orientation of $S$.

Suppose that $(I, g = g(t))$ be a multiplicative parameterized curve lying on $S$ and $n(g(t))$, $t \in I$, be the multiplicative normal vector of $S$. Set

$$\theta(t) = \angle_\star(g(t), n(g(t))), \quad t \in I.$$

**Definition 4.29:** The multiplicative normal curvature of $g$ is defined as follows:

$$k_n(t) = k(t) \cdot_\star \cos_\star \theta(t), \quad t \in I.$$

**Theorem 4.11:** *The Multiplicative Meusnier Theorem We have*

$$k_n(t) = \phi_2(g^\star(t), g^\star(t)) /_\star \phi_1(g^\star, g^\star(t)), \quad t \in I.$$

*Proof.* Let $g_1$ be the multiplicative naturally parameterized curve that is multiplicative equivalent to $g$ with multiplicative arc length parameter $s$. Let also,

$$g_1(s) = f(u(s), v(s)).$$

Then

$$g^{\star\star}(s) = f_u^{\star\star} \cdot_\star (u^\star)^{2\star} +_\star e^{2} \cdot_\star f_{uv}^{\star\star} \cdot_\star u^\star \cdot_\star v^\star$$
$$+_\star f_v^{\star\star} \cdot_\star (v^\star)^{2\star} +_\star f_u^\star \cdot_\star u^{\star\star} +_\star f_v^\star \cdot_\star v^{\star\star}.$$

Hence,

$$k_n(s) = \langle \mathbf{k}(s), n(g_1(s)) \rangle_\star$$
$$= \langle g_1^{\star\star}(s), n(g_1(s)) \rangle_\star$$
$$= \langle f_u^{\star\star} \cdot_\star (u^\star)^{2\star} +_\star e^2 \cdot_\star f_{uv}^{\star\star} \cdot_\star u^\star \cdot_\star v^\star$$
$$+_\star f_v^{\star\star} \cdot_\star (v^\star)^{2\star} +_\star f_u^\star \cdot_\star u^{\star\star} +_\star f_v^\star \cdot_\star v^{\star\star}, n(g_1(s)) \rangle_\star$$
$$= \langle f_u^{\star\star} \cdot_\star (u^\star)^{2\star}, n(g_1(s)) \rangle_\star +_\star e^2 \cdot_\star \langle f_{uv}^{\star\star}, n(g_1(s)) \rangle_\star \cdot_\star u^\star \cdot_\star v^\star$$
$$+_\star \langle f_v^{\star\star}, n(g_1(s)) \rangle_\star \cdot_\star (v^\star)^{2\star} +_\star \langle f_u^\star, n(g_1(s)) \rangle_\star \cdot_\star u^{\star\star}$$
$$+_\star \langle f_v^\star, n(g_1(s)) \rangle_\star \cdot_\star v^{\star\star}$$
$$= -_\star L \cdot_\star (u^\star)^{2\star} -_\star e^2 \cdot_\star M \cdot_\star u^\star \cdot_\star v^\star -_\star N \cdot_\star (v^\star)^{2\star}$$
$$= \phi_2(g_1^\star(s), g_1^\star(s)).$$

Now, using that

$$g_1^\star(s) = g^\star(t)/_\star |g^\star(t)|_\star,$$

we get

$$k_n(t) = \phi_2(g^\star(t)/_\star |g^\star(t)|_\star, g^\star(t)/_\star |g^\star(t)|_\star)$$
$$= \phi_2(g^\star(t), g^\star(t))/_\star (\langle g^\star(t), g^\star(t)\rangle_\star)$$
$$= \phi_2(g^\star(t), g^\star(t))/_\star \phi_1(g^\star(t), g^\star(t)), \quad t \in I.$$

This completes the proof.

**Corollary 4.2:** *If two multiplicative curves on S have a common multiplicative point and they have the same multiplicative tangent lines at the common point, then they have the same multiplicative normal curvature.*

*Proof.* Let $p$ and $q$ be the multiplicative tangent vectors to both curves. Then

$$q = \alpha \cdot_\star p.$$

Hence,

$$\phi_2(q, q)/_\star \phi_1(q, q) = \phi_2(\alpha \cdot_\star p, \alpha \cdot_\star p)/_\star \phi_1(\alpha \cdot_\star p, \alpha \cdot_\star p)$$
$$= (\alpha^{2\star} \cdot_\star \phi_2(p, p))/_\star (\alpha^{2\star} \cdot_\star \phi_1(p, p))$$
$$= \phi_2(p, p)/_\star \phi_1(p, p).$$

This completes the proof.

## 4.13 Multiplicative Asymptotic Directions. Multiplicative Asymptotic Lines

Let $S$ be a multiplicative oriented surface and $(U, f)$ be its multiplicative local representation that is multiplicative compatible with the multiplicative orientation of $S$.

**Definition 4.30:** For any $a \in S$, a multiplicative directions $h \in T_{\star a}S$ is said to be a multiplicative asymptotic direction if

$$\phi_2(h, h) = 0_\star.$$

**Theorem 4.12:** *For any $a \in S$, a multiplicative direction $h \in T_{\star a}S$ is multiplicative asymptotic direction if and only if*

$$D\cdot_\star D_2 -_\star (D_1)^{2\star} \leq 0_\star. \tag{4.7}$$

*Proof.* Let $h = (h_1, h_2) \neq 0_\star \in T_{\star a}S$. Without loss of generality, suppose that $h_2 \neq 0_\star$. We have

$$0_\star = \phi_2(h, h)$$

$$= (h_1 h_2)\cdot_\star \begin{pmatrix} D & D^\star \\ D^\star & D^{\star\star} \end{pmatrix} \cdot_\star \begin{pmatrix} h_1 \\ h_2 \end{pmatrix}$$

$$= (h_1 h_2)\cdot_\star \begin{pmatrix} D\cdot_\star h_1 +_\star D^\star\cdot_\star h_2 \\ D^\star\cdot_\star h_1 +_\star D^{\star\star}\cdot_\star (h_2)^{2\star} \end{pmatrix}$$

$$= D\cdot_\star h_1^{2\star} +_\star e^{2}\cdot_\star D^\star\cdot_\star h_1\cdot_\star h_2 +_\star D^{\star\star}\cdot_\star h_2^{2\star}$$

if and only if

$$D\cdot_\star (h_1/_\star h_2)^{2\star} +_\star e^{2}\cdot_\star D^\star\cdot_\star (h_1/_\star h_2) +_\star D^{\star\star} = 0_\star$$

if and only if (4.7) holds. This completes the proof.

**Definition 4.31:** A point $a \in S$ is said to be

1. Multiplicative elliptic if the second multiplicative fundamental form is multiplicative positive defined at $a$.
2. Multiplicative hyperbolic if the second multiplicative fundamental form is multiplicative negative defined at $a$.

3. Multiplicative parabolic if the second multiplicative fundamental form is $0_\star$ and at least one of its coefficients is different than $0_\star$ at $a$.

4. Multiplicative flat if the coefficients of the second multiplicative fundamental form are $0_\star$ at $a$.

---

## 4.14 Multiplicative Principal Directions, Multiplicative Principal Curvatures, Multiplicative Gauss Curvature and Multiplicative Mean Curvature

Suppose that $S$ is a multiplicative oriented surface and $(U, f)$ is a multiplicative local parameterization of $S$ that is multiplicative compatible with the multiplicative orientation of $S$.

**Definition 4.32:** The multiplicative directions on the multiplicative tangent plane to $S$ at a point $a \in S$ that are multiplicative eigenvectors of the multiplicative shape operator of $S$ are said to be multiplicative principal directions of $S$.

**Definition 4.33:** A multiplicative curve on $S$ is said to be a multiplicative principal line if its multiplicative tangent directions at each point are multiplicative principal directions.

**Definition 4.34:** A multiplicative principal curvature of $S$ is the multiplicative normal curvature of $S$ in a multiplicative principal direction.

**Theorem 4.13:** *The multiplicative normal curvatures of $S$ are the multiplicative eigenvalues of the multiplicative shape operator with multiplicative opposite sign.*

*Proof.* Let $e$ be a multiplicative eigenvector of $A$. Then there is a $\lambda \in \mathbb{R}_\star$ so that

$$A(e) = \lambda \cdot_\star e.$$

Hence,

$$
\begin{aligned}
k_n(e) &= \phi_2(e, e) /_\star \phi_1(e, e) \\
&= -_\star \langle A(e), e \rangle_\star /_\star \langle e, e \rangle_\star \\
&= -_\star \lambda \cdot_\star \langle e, e \rangle_\star /_\star \langle e, e \rangle_\star \\
&= -_\star \lambda.
\end{aligned}
$$

This completes the proof.

Below, suppose that $k_1$ and $k_2$ are the multiplicative principal curvatures of $S$ and $k_1 \geq k_2$.

**Definition 4.35:** A multiplicative orthonormal basis $\{e_1, e_2\}$ of the multiplicative tangent space at a point of $S$ is called a multiplicative basis of the multiplicative principal directions.

Let $\{e_1, e_2\}$ be a multiplicative basis of the multiplicative principal directions. Then

$$A(e_j) = -_\star k_j \cdot_\star e_j, \quad j \in \{1, 2\}.$$

Let $|e|_\star = 1_\star$. Then

$$e = e_1 \cdot_\star \cos_\star \theta +_\star e_2 \cdot_\star \sin_\star \theta,$$

and

$$
\begin{aligned}
\phi_2(e, e) &= -_\star \langle A(e_1 \cdot_\star \cos_\star \theta +_\star e_2 \cdot_\star \sin_\star \theta), e_1 \cdot_\star \cos_\star \theta +_\star e_2 \cdot_\star \sin_\star \theta \rangle_\star \\
&= -_\star \langle -_\star k_1 \cdot_\star e_1 \cdot_\star \cos_\star \theta -_\star k_2 \cdot_\star e_2 \cdot_\star \sin_\star \theta, e_1 \cdot_\star \cos_\star \theta +_\star e_2 \cdot_\star \sin_\star \theta \rangle_\star \\
&= k_1 \cdot_\star (\cos_\star \theta)^{2_\star} +_\star k_2 \cdot_\star (\sin_\star \theta)^{2_\star}
\end{aligned}
$$

and

$$\phi_1(e, e) = 1_\star.$$

Hence,

$$k_n(e) = k_1 \cdot_\star (\cos_\star \theta)^{2_\star} +_\star k_2 \cdot_\star (\sin_\star \theta)^{2_\star}. \tag{4.8}$$

**Definition 4.36:** The formula (4.8) is said to be the multiplicative Euler formula.

**Definition 4.37:** The quantity

$$K_t = k_1 \cdot_\star k_2$$

will be called the multiplicative Gaussian curvature of $S$.

**Definition 4.38:** The quantity

$$K_m = (1_\star /_\star e^2) \cdot_\star (k_1 +_\star k_2)$$

is said to be the multiplicative mean curvature.
We have that

$$K_t = \det{}_\star \mathscr{A}$$

and

$$K_m = -_\star (1_\star /_\star e^2) \cdot_\star \operatorname{Tr}_\star A.$$

**Theorem 4.14: (The Multiplicative Jochanstahl Theorem)** *Let $S_1$ and $S_2$ be two multiplicative oriented surfaces and $\gamma$ be a multiplicative parameterized curve that lies on the intersection of $S_1$ and $S_2$. Let also, $S_1$ and $S_2$ intersect under a multiplicative constant angle. Then $\gamma$ is a multiplicative curvature line on $S_1$ if and only if $\gamma_2$ is a multiplicative curvature line on $S_2$.*

*Proof.* Let $(I, f = f(t))$ be a multiplicative local parameterization of $\gamma$. Let also, $n_1$ and $n_2$ be multiplicative normal directions to $S_1$ and $S_2$, respectively. Then

$$\langle n_1, n_2 \rangle_\star = const.$$

Hence,

$$0_\star = \langle n_1^\star, n_2 \rangle_\star +_\star \langle n_1, n_2^\star \rangle_\star.$$

Suppose that $\gamma$ is a multiplicative curvature line on $S_1$. Then

$$n_1^\star = -_\star l \cdot_\star f^\star,$$

where $l$ is one of the multiplicative principal curvatures of $S_1$. Since $\gamma$ lies on $S_2$, we have that

$$\langle f^\star, n_2 \rangle_\star = 0_\star.$$

Thus,

$$0_\star = \langle -_\star l \cdot_\star f^\star, n_2 \rangle_\star +_\star \langle n_1, n_2^\star \rangle_\star$$
$$= \langle n_1, n_2^\star \rangle_\star.$$

Hence,

$$\langle n_1^\star, n_2 \rangle_\star = 0_\star.$$

Since $n_2^\star \perp_\star n_2$ and $f^\star \perp_\star n_2$, we conclude that $n_2^\star \|_\star f_2^\star$ and there is a $k_2 \in \mathbb{R}_\star$ so that

$$n_2^\star = -_\star k_2 \cdot_\star f^\star,$$

i.e., $\gamma$ is a multiplicative curvature line on $S_2$. This completes the proof.

## 4.15 The Computation of the Multiplicative Curvatures of a Multiplicative Surface

Let $S$ be a multiplicative oriented surface and $(U, f)$ be a multiplicative local representation of $S$ that is multiplicative compatible with the multiplicative orientation of $S$.

**Theorem 4.15:** *We have*

$$K_l = (D \cdot_\star D_2 -_\star D_1^{2\star})/_\star H^{2\star},$$
$$K_m = (D \cdot_\star G -_\star e^2 \cdot_\star D_1 \cdot_\star F +_\star D_2 \cdot_\star E)/_\star (e^2 \cdot_\star H^{2\star}.$$

*Proof.* We have

$$\mathscr{A} = (1_\star/_\star H^{2\star}) \cdot_\star \begin{pmatrix} G \cdot_\star L -_\star F \cdot_\star M & G \cdot_\star M -_\star F \cdot_\star N \\ E \cdot_\star M -_\star F \cdot_\star L & E' \cdot_\star N -_\star F \cdot_\star M \end{pmatrix}$$

$$= (1_\star/_\star H^{2\star} \cdot_\star \begin{pmatrix} -_\star G \cdot_\star D +_\star F \cdot_\star D_1 & -_\star G \cdot_\star D_1 +_\star F \cdot_\star D_2 \\ -_\star E \cdot_\star D_1 +_\star F \cdot_\star D & -_\star E \cdot_\star D_2 +_\star F \cdot_\star D_1 \end{pmatrix}.$$

Then

$$\det_\star \mathscr{A} = (1_\star /_\star H^{4_\star} \cdot_\star (G \cdot_\star E \cdot_\star D \cdot_\star D_2 \to_\star G \cdot_\star F \cdot_\star D \cdot_\star D_1$$
$$\to_\star F \cdot_\star E' \cdot_\star D_) 1 \cdot_\star D_2 +_\star F^{2_\star} \cdot_\star D_1^{2_\star} \to_\star E \cdot_\star G \cdot_\star D_1^{2_\star}$$
$$+_\star E \cdot_\star F \cdot_\star D_1 \cdot_\star D_2 \to_\star F^{2_\star} \cdot_\star D \cdot_\star D_2 +_\star F' \cdot_\star G \cdot_\star D \cdot_\star D_1)$$
$$= (1_\star /_\star H^{4_\star} \cdot_\star ((G \cdot_\star E \to_\star F^{2_\star}) \cdot_\star D \cdot_\star D_2$$
$$\to_\star (G \cdot_\star E \to_\star F^{2_\star}) \cdot_\star D_1^{2_\star})$$
$$= (1_\star /_\star H^{4_\star} \cdot_\star)((G \cdot_\star E \to_\star F^{2_\star})) \cdot_\star (D \cdot_\star D_2 \to_\star D_1^{2_\star}))$$
$$= 1_\star /_\star H^{2_\star} \cdot_\star (D \cdot_\star D_2 \to_\star D_1^{2_\star}).$$

Next,

$$K_m = Tr_\star A$$
$$= (D \cdot_\star G \to_\star e^2 \cdot_\star D_1 \cdot_\star F +_\star D_2 \cdot_\star E) /_\star (e^2 \cdot_\star H^{2_\star}).$$

This completes the proof.

---

## 4.16 Advanced Practical Problems

**Problem 4.1:** Let $f: \mathbb{R}_\star^2 \to \mathbb{R}_\star^3$ be given by

$$f(t_1, t_2) = \left( t_1^2 t_2, e^{t_1 + 3t_2^2}, e^{t_1^2 + 2t_2^2} \right), \quad (t_1, t_2) \in \mathbb{R}_\star^2.$$

Prove that $(\mathbb{R}_\star^2, f)$ is a multiplicative regular surface.

**Problem 4.2:** Let $f: \mathbb{R}_\star^2 \to \mathbb{R}_\star^3$ be given by

$$f(t_1, t_2) = \left( e^{t_1^2}, e^{t_1^2 t_2^2}, t_1^2 + t_2^2 \right), \quad (t_1, t_2) \in \mathbb{R}_\star^2.$$

Find the equation of the multiplicative tangent plane at $f(1, 1)$.

**Problem 4.3:** Let $f: \mathbb{R}_\star^2 \to \mathbb{R}_\star^3$ be given by

$$f(t_1, t_2) = \left( e^{t_1 t_2}, e^{t_1 - 3t_1^2 t_2^2}, t_1 + t_2 \right), \quad (t_1, t_2) \in \mathbb{R}_\star^2.$$

Find the equation of the multiplicative normal line at $f(1, 1)$.

**Problem 4.4:** Let $S$ be a multiplicative surface given by the equation

$$F(x, y, z) = e^{x^6 y^2 + z^8 + 3x^2 y^3 z^2} = 0_\star, \quad (x, y, z) \in \mathbb{R}^3_\star,$$

and $(x_0, y_0, z_0) = (1, 1, 1)$. Find the equation of the multiplicative tangent plane and the equations of the multiplicative normal line to $S$ at $(x_0, y_0, z_0)$.

**Problem 4.5:** Let $f \colon \mathbb{R}^2_\star \to \mathbb{R}^3_\star$ be given by

$$f(u, v) = \left( u^2 + v^3 + 3uv^4, \, e^{u^4 + v}, \, e^{u + v^3} \right), \quad (u, v) \in \mathbb{R}^2_\star.$$

Find the matrix of the first multiplicative fundamental form.

**Problem 4.6:** Let $f \colon \mathbb{R}^2_\star \to \mathbb{R}^3_\star$ be given by

$$f(u, v) = \left( u, \, v, \, e^{u + 4v^4} \right), \quad (u, v) \in \mathbb{R}^2_\star.$$

Find the matrix $\mathscr{A}$.

**Problem 4.7:** Let $f \colon \mathbb{R}^2_\star \to \mathbb{R}^3_\star$ be given by

$$f(u, v) = \left( v, \, e^{u^4 - v^2}, \, u \right), \quad (u, v) \in \mathbb{R}^2_\star.$$

Find the matrix $[\phi_2]$.

# 5

## Multiplicative Fundamental Equations of a Multiplicative Surface

### 5.1 Some Relations

Suppose that $(U, f)$ is a multiplicative regular parameterized surface. Then $\{f_u^\star, f_v^\star, n\}$ is a multiplicative basis in $\mathbb{R}^3_{\star f(u,v)}$. We have

$$E = \langle f_u^\star, f_u^\star \rangle_\star,$$
$$F = \langle f_u^\star, f_v^\star \rangle_\star,$$
$$G = \langle f_v^\star, f_v^\star \rangle_\star.$$

Then

$$E_u^\star = \langle f_u^{\star\star}, f_u^\star \rangle_\star +_\star \langle f_u^\star, f_u^{\star\star} \rangle_\star$$
$$= e^{2 \cdot}{}_\star \langle f_u^{\star\star}, f_u^\star \rangle_\star,$$

whereupon

$$\langle f_u^\star, f_u^\star \rangle_\star = (1_\star /_\star e^2) \cdot_\star E_u^\star.$$

Also,

$$E_v^\star = \langle f_{uv}^{\star\star}, f_u^\star \rangle_\star +_\star \langle f_u^\star, f_{uv}^{\star\star} \rangle_\star$$
$$= e^{2 \cdot}{}_\star \langle f_u^\star, f_{uv}^{\star\star} \rangle_\star$$

and

$$\langle f_u^\star, f_{uv}^{\star\star} \rangle_\star = (1_\star /_\star e^2) \cdot_\star E_v^\star.$$

DOI: 10.1201/9781003299844-5

Consider $G$. We have

$$G_u^\star = \langle f_{uv}^{\star\star}, f_v^\star \rangle_\star +_\star \langle f_v^\star, f_{uv}^{\star\star} \rangle_\star$$
$$= e^2 \cdot_\star \langle f_{uv}^{\star\star}, f_v^\star \rangle_\star,$$

whereupon

$$\langle f_{uv}^{\star\star}, f_v^\star \rangle_\star = (1_\star /_\star e^2) \cdot_\star G_u^\star,$$

and

$$G_v^\star = \langle f_v^{\star\star}, f_v^\star \rangle_\star +_\star \langle f_v^\star, f_v^{\star\star} \rangle_\star$$
$$= e^2 \cdot_\star \langle f_v^{\star\star}, f_v^\star \rangle_\star,$$

from where

$$\langle f_v^{\star\star}, f_v^\star \rangle_\star = (1_\star /_\star e^2) \cdot_\star G_v^\star.$$

Moreover,

$$F_u^\star = \langle f_u^{\star\star}, f_v^\star \rangle_\star +_\star \langle f_u^\star, f_{uv}^{\star\star} \rangle_\star$$
$$= \langle f_u^{\star\star}, f_v^\star \rangle_\star +_\star (1_\star /_\star e^2) \cdot_\star E_v^\star,$$

whereupon

$$\langle f_u^{\star\star}, f_v^\star \rangle_\star = F_u^\star -_\star (1_\star /_\star e^2) \cdot_\star E_v^\star,$$

and

$$F_v^\star = \langle f_{uv}^{\star\star}, f_v^\star \rangle_\star +_\star \langle f_u^\star, f_v^{\star\star} \rangle_\star$$
$$= (1_\star /_\star e^2) \cdot_\star G_u^\star +_\star \langle f_u^\star, f_v^{\star\star} \rangle_\star,$$

from where

$$\langle f_u^\star, f_v^{\star\star} \rangle_\star = F_v^\star -_\star (1_\star /_\star e^2) \cdot_\star G_u^\star.$$

## 5.2 The Multiplicative Christoffel Coefficients

In this section, we will represent $f_u^{\star\star}$, $f_{uv}^{\star\star}$ and $f_v^{\star\star}$ by the multiplicative basis $\{f_u^{\star}, f_v^{\star}, n\}$ in the following way

$$
\begin{aligned}
f_u^{\star\star} &= \Gamma_{11}^1 {}_{\cdot\star} f_u^{\star} +_{\star} \Gamma_{11}^2 {}_{\cdot\star} f_v^{\star} +_{\star} A_{\cdot\star} n \\
f_{uv}^{\star\star} &= \Gamma_{12}^1 {}_{\cdot\star} f_u^{\star} +_{\star} \Gamma_{12}^2 {}_{\cdot\star} f_v^{\star} +_{\star} B_{\cdot\star} n \\
f_v^{\star\star} &= \Gamma_{22}^1 {}_{\cdot\star} f_u^{\star} +_{\star} \Gamma_{22}^2 {}_{\cdot\star} f_v^{\star} +_{\star} C_{\cdot\star} n,
\end{aligned}
\tag{5.1}
$$

where $A$, $B$, $C$ satisfy the second multiplicative fundamental form.

**Definition 5.1:** The coefficients $\Gamma_{ij}^k$, $i, j, k \in \{1, 2\}$, are said to be the multiplicative Christoffel coefficients.

By the first equation of the system (5.1), we find

$$
\begin{aligned}
\langle f_u^{\star\star}, f_u^{\star} \rangle_{\star} &= \Gamma_{11}^1 {}_{\cdot\star} \langle f_u^{\star}, f_u^{\star} \rangle_{\star} +_{\star} \Gamma_{11}^2 {}_{\cdot\star} \langle f_u^{\star}, f_v^{\star} \rangle_{\star} \\
\langle f_u^{\star\star}, f_v^{\star} \rangle_{\star} &= \Gamma_{11}^1 {}_{\cdot\star} \langle f_u^{\star}, f_v^{\star} \rangle_{\star} +_{\star} \Gamma_{11}^2 {}_{\cdot\star} \langle f_u^{\star}, f_v^{\star} \rangle_{\star}
\end{aligned}
$$

or

$$
\begin{aligned}
\Gamma_{11}^1 {}_{\cdot\star} E +_{\star} \Gamma_{11}^2 {}_{\cdot\star} F &= (1_{\star}/_{\star} e^2)_{\cdot\star} E_u^{\star} \\
\Gamma_{11}^1 {}_{\cdot\star} F +_{\star} \Gamma_{11}^2 {}_{\cdot\star} G &= F_u^{\star} -_{\star} (1_{\star}/_{\star} e^2)_{\cdot\star} E_v^{\star}.
\end{aligned}
$$

Therefore

$$
\begin{aligned}
\Gamma_{11}^1 &= ((1_{\star}/_{\star} e^2)_{\cdot\star} E_u^{\star} {}_{\cdot\star} G -_{\star} F_u^{\star} {}_{\cdot\star} F +_{\star} (1_{\star}/_{\star} e^2)_{\cdot\star} E_v^{\star} {}_{\cdot\star} F) /_{\star} H^2{}_{\star} \\
\Gamma_{11}^2 &= (-_{\star} (1_{\star}/_{\star} e^2)_{\cdot\star} E_u^{\star} {}_{\cdot\star} F +_{\star} E_{\cdot\star} F_u^{\star} -_{\star} (1_{\star}/_{\star} e^2)_{\cdot\star} E_{\cdot\star} E_v^{\star}) /_{\star} H^2{}_{\star},
\end{aligned}
$$

By the second equation of the system (5.1), we obtain

$$
\begin{aligned}
\langle f_{uv}^{\star\star}, f_u^{\star} \rangle_{\star} &= \Gamma_{12}^1 {}_{\cdot\star} \langle f_u^{\star}, f_u^{\star} \rangle_{\star} +_{\star} \Gamma_{12}^2 {}_{\cdot\star} \langle f_u^{\star}, f_v^{\star} \rangle_{\star} \\
\langle f_{uv}^{\star\star}, f_v^{\star} \rangle_{\star} &= \Gamma_{12}^1 {}_{\cdot\star} \langle f_u^{\star}, f_v^{\star} \rangle_{\star} +_{\star} \Gamma_{12}^2 {}_{\cdot\star} \langle f_v^{\star}, f_v^{\star} \rangle_{\star}
\end{aligned}
$$

or

$$
\begin{aligned}
\Gamma_{12}^1 {}_{\cdot\star} E +_{\star} \Gamma_{12}^2 {}_{\cdot\star} F &= (1_{\star}/_{\star} e^2)_{\cdot\star} E_v^{\star} \\
\Gamma_{12}^1 {}_{\cdot\star} F +_{\star} \Gamma_{12}^2 {}_{\cdot\star} G &= (1_{\star}/_{\star} e^2)_{\cdot\star} G_u^{\star}.
\end{aligned}
$$

Consequently

$$\Gamma^1_{12} = (1_\star /_\star e^2)\cdot_\star (E^\star_\star \cdot_\star G \to_\star G^\star_u \cdot_\star F)/_\star H^{2_\star}$$
$$\Gamma^2_{12} = (1_\star /_\star e^2)\cdot_\star (\to_\star E^\star_v \cdot_\star F +_\star E_\star \cdot_\star G^\star_u)/_\star H^{2_\star}.$$

By the third equation of the system (5.1), we obtain

$$\langle f^{\star\star}_v, f^\star_u \rangle_\star = \Gamma^1_{22}\cdot_\star \langle f^\star_u, f^\star_u \rangle_\star +_\star \Gamma^2_{22}\cdot_\star \langle f^\star_u, f^\star_v \rangle_\star$$
$$\langle f^{\star\star}_v, f^\star_v \rangle_\star = \Gamma^1_{22}\cdot_\star \langle f^\star_u, f^\star_v \rangle_\star +_\star \Gamma^2_{22}\cdot_\star \langle f^\star_v, f^\star_v \rangle_\star$$

or

$$\Gamma^1_{22}\cdot_\star E +_\star \Gamma^2_{22}\cdot_\star F = F^\star_v \to_\star (1_\star /_\star e^2)\cdot_\star G^\star_u$$
$$\Gamma^1_{22}\cdot_\star F +_\star \Gamma^2_{22}\cdot_\star G = (1_\star /_\star e^2)\cdot_\star G^\star_v.$$

Therefore

$$\Gamma^1_{22} = (F^\star_v \cdot_\star G \to_\star (1_\star /_\star e^2)\cdot_\star G^\star_u \cdot_\star G \to_\star (1_\star /_\star e^2)\cdot_\star G^\star_v \cdot_\star F)/_\star H^{2_\star}$$
$$\Gamma^2_{22} = (\to_\star F_\star \cdot_\star F^\star_v +_\star (1_\star /_\star e^2)\cdot_\star F_\star \cdot_\star G^\star_u +_\star (1_\star /_\star e^2)\cdot_\star E_\star \cdot_\star G^\star_v)/_\star H^{2_\star}.$$

## 5.3 The Multiplicative Weingarten Coefficients

In this section, we will represent $n^\star_u$ and $n^\star_v$ in the following way

$$n^\star_u = a_{11}\cdot_\star f^\star_u +_\star a_{12}\cdot_\star f^\star_v$$
$$n^\star_v = a_{21}\cdot_\star f^\star_u +_\star a_{22}\cdot_\star f^\star_v. \tag{5.2}$$

**Definition 5.2:** The coefficients $a_{ij}$, $i, j \in \{1, 2\}$, are said to be the multiplicative Weingarten coefficients.

By the first equation of the system (5.2), we find

$$\langle n^\star_u, f^\star_u \rangle_\star = a_{11}\cdot_\star \langle f^\star_u, f^\star_u \rangle_\star +_\star a_{12}\cdot_\star \langle f^\star_u, f^\star_v \rangle_\star$$
$$\langle n^\star_u, f^\star_v \rangle_\star = a_{11}\cdot_\star \langle f^\star_u, f^\star_v \rangle_\star +_\star a_{12}\cdot_\star \langle f^\star_v, f^\star_v \rangle_\star$$

or

$$a_{11}\cdot_\star E +_\star a_{12}\cdot_\star F = -_\star D$$
$$a_{11}\cdot_\star F +_\star a_{12}\cdot_\star G = -_\star D^\star.$$

Therefore

$$a_{11} = (F\cdot_\star D^\star -_\star G\cdot_\star D)/_\star H^2{}_\star$$
$$a_{12} = (F\cdot_\star D -_\star E\cdot_\star D^\star))/_\star H^2{}_\star.$$

By the second equation of the system (5.2), we arrive at

$$\langle n_v^\star, f_u^\star \rangle_\star = a_{21}\cdot_\star \langle f_u^\star, f_u^\star \rangle_\star +_\star a_{22}\cdot_\star \langle f_u^\star, f_v^\star \rangle_\star$$
$$\langle n_v^\star, f_v^\star \rangle_\star = a_{21}\cdot_\star \langle f_u^\star, f_v^\star \rangle_\star +_\star a_{22}\cdot_\star \langle f_v^\star, f_v^\star \rangle_\star$$

or

$$a_{21}\cdot_\star E +_\star a_{22}\cdot_\star F = -_\star D^\star$$
$$a_{21}\cdot_\star F +_\star a_{22}\cdot_\star G = -_\star D^{\star\star}.$$

Consequently

$$a_{21} = (-_\star G\cdot_\star D^\star +_\star F\cdot_\star D^{\star\star})/_\star H^2{}_\star$$
$$a_{22} = (F\cdot_\star D^\star -_\star E\cdot_\star D^{\star\star})/_\star H^2{}_\star.$$

## 5.4 The Multiplicative Gauss and Godazzi-Mainardi Equations

In this section, we will deduct the multiplicative Gauss and Godazzi-Mainardi equations. Let $S$ be a multiplicative oriented surface and $(U, f)$ be a multiplicative local representation of $S$ that is multiplicative compatible with the multiplicative orientation of $S$.

**Theorem 5.1:** *We have the following systems of equations*

$$(\Gamma^1_{11})^\star_v \to_\star (\Gamma^1_{12})^\star_u +_\star \Gamma^2_{11}\cdot_\star \Gamma^1_{22} \to_\star \Gamma^2_{12}\cdot_\star \Gamma^1_{12}$$

$$= \to_\star a_{21}\cdot_\star D_1 +_\star a_{11}\cdot_\star D_1, (\Gamma^2_{11})^\star_v \to_\star (\Gamma^2_{12})^\star_u +_\star \Gamma^1_{11}\cdot_\star \Gamma^2_{12} +_\star \Gamma^2_{11}\cdot_\star \Gamma^2_{22} \to_\star \Gamma^1_{12}$$

$$\cdot_\star \Gamma^2_{11} \to_\star (\Gamma^2_{12})^{2\star}$$

$$= D_1\cdot_\star a_{12} \to_\star D\cdot_\star a_{22}, (\Gamma^1_{12})^\star_v \to_\star (\Gamma^1_{22})^\star_u +_\star (\Gamma^1_{12})^{2\star} +_\star \Gamma^2_{12}\cdot_\star \Gamma^1_{22} \to_\star \Gamma^2_{22}\cdot_\star \Gamma^1_{11}$$

$$\to_\star \Gamma^2_{22}\cdot_\star \Gamma^1_{12}$$

$$= a_{11}\cdot_\star D_2 \to_\star D_1\cdot_\star a_{21}, (\Gamma^2_{12})^\star_v \to_\star (\Gamma^2_{22})^\star_u +_\star \Gamma^1_{12}\cdot_\star \Gamma^2_{12} +_\star \Gamma^2_{12}\cdot_\star \Gamma^2_{22} \to_\star \Gamma^1_{22}$$

$$\cdot_\star \Gamma^2_{11} \to_\star \Gamma^2_{22}\cdot_\star \Gamma^2_{12}$$

$$= a_{12}\cdot_\star D_2 \to_\star a_{22}\cdot_\star D_1$$

$$(5.3)$$

*and*

$$
\begin{aligned}
D^\star_v - D^\star_{1v} &= \to_\star \Gamma^1_{11}\cdot_\star D_1 \to_\star \Gamma^2_{11}\cdot_\star D_2 +_\star \Gamma^1_{12}\cdot_\star D +_\star \Gamma^2_{12}\cdot_\star D_1 \\
D^\star_{1v} \to_\star D^\star_{2v} &= \to_\star \Gamma^1_{12}\cdot_\star D_1 \to_\star \Gamma^2_{12}\cdot_\star D_2 +_\star \Gamma^1_{22}\cdot_\star D +_\star \Gamma^2_{22}\cdot_\star D_1.
\end{aligned}
$$

$$(5.4)$$

**Definition 5.3:** The equations (5.3) will be called the multiplicative Gauss equations.

**Definition 5.4:** The equations (5.4) will be called multiplicative Godazzi-Mainardi equations.

*Proof.* To deduct the multiplicative Gauss equations and the multiplicative Godazzi-Mainardi equations, we will use the relations

$$f^{\star\star\star}_{u^2 v} = f^{\star\star\star}_{uvu} \tag{5.5}$$

and

$$f^{\star\star\star}_{uv^2} = f^{\star\star\star}_{v^2 u}. \tag{5.6}$$

We have

$$f^{\star\star}_{u^2} = \Gamma^1_{11}\cdot_\star f^\star_u +_\star \Gamma^2_{11}\cdot_\star f^\star_v +_\star D\cdot_\star n$$

and then

$$f_{u^2v}^{\star\star\star} = (\Gamma_{11}^1)_v^{\star}{}^{\cdot}\star f_u^{\star} +_{\star} \Gamma_{11}^1{}^{\cdot}\star f_{uv}^{\star\star}$$

$$+_{\star} (\Gamma_{11}^2)_v^{\star}{}^{\cdot}\star f_v^{\star} +_{\star} \Gamma_{11}^2{}^{\cdot}\star f_{vv}^{\star\star}$$

$$+_{\star} D_v^{\star}{}^{\cdot}\star n +_{\star} D{}^{\cdot}\star n_v^{\star}$$

$$= (\Gamma_{11}^1)_v^{\star}{}^{\cdot}\star f_u^{\star} +_{\star} (\Gamma_{11}^2)_v^{\star}{}^{\cdot}\star f_v^{\star} +_{\star} D_v^{\star}{}^{\cdot}\star n$$

$$+_{\star} \Gamma_{11}^1{}^{\cdot}\star(\Gamma_{12}^1{}^{\cdot}\star f_u^{\star} +_{\star} \Gamma_{12}^2{}^{\cdot}\star f_v^{\star} +_{\star} D_1{}\star n)$$

$$+_{\star} \Gamma_{11}^2{}^{\cdot}\star(\Gamma_{22}^1{}^{\cdot}\star f_u^{\star} +_{\star} \Gamma_{22}^2{}^{\cdot}\star f_v^{\star} +_{\star} D_2{}\star n)$$

$$+_{\star} D{}^{\cdot}\star(a_{21}{}^{\cdot}\star f_u^{\star} +_{\star} a_{22}{}^{\cdot}\star f_v^{\star})$$

$$= ((\Gamma_{11}^1)_v^{\star} +_{\star} \Gamma_{11}^1{}^{\cdot}\star\Gamma_{12}^1 +_{\star} \Gamma_{11}^2{}^{\cdot}\star\Gamma_{22}^1 +_{\star} D{}^{\cdot}\star a_{21}){}^{\cdot}\star f_u^{\star}$$

$$+_{\star} ((\Gamma_{11}^2)_v^{\star} +_{\star} \Gamma_{11}^1{}^{\cdot}\star\Gamma_{12}^2 +_{\star} \Gamma_{11}^2{}^{\cdot}\star\Gamma_{22}^2 +_{\star} D{}^{\cdot}\star a_{22}){}^{\cdot}\star f_v^{\star}$$

$$+_{\star} (D_v^{\star} +_{\star} \Gamma_{11}^1{}^{\cdot}\star D_1 +_{\star} \Gamma_{11}^2{}^{\cdot}{}_{+}\star D_2){}^{\cdot}\star n,$$

i.e.,

$$f_{u^2v}^{\star\star\star} = \left((\Gamma_{11}^1)_v^{\star} +_{\star} \Gamma_{11}^1{}^{\cdot}\star\Gamma_{12}^1 +_{\star} \Gamma_{11}^2{}^{\cdot}\star\Gamma_{22}^1 +_{\star} D{}^{\cdot}\star a_{21}\right){}^{\cdot}\star f_u^{\star}$$

$$+_{\star}\left((\Gamma_{11}^2)_v^{\star} +_{\star} \Gamma_{11}^1{}^{\cdot}\star\Gamma_{12}^2 +_{\star} \Gamma_{11}^2{}^{\cdot}\star\Gamma_{22}^2 +_{\star} D{}^{\cdot}\star a_{22}\right){}^{\cdot}\star f_v^{\star} \qquad (5.7)$$

$$+_{\star}(D_v^{\star} +_{\star} \Gamma_{11}^1{}^{\cdot}\star D_1 +_{\star} \Gamma_{11}^2{}^{\cdot}\star D_2){}^{\cdot}\star n.$$

Now, using that

$$f_{uv}^{\star\star} = \Gamma_{12}^1{}^{\cdot}\star f_u^{\star} +_{\star} \Gamma_{12}^2{}^{\cdot}\star f_v^{\star} +_{\star} D_1{}\star n,$$

we obtain

$$f_{uvu}^{\star\star\star} = (\Gamma_{12}^1)_u^{\star} {}_{\star} f_u^{\star} \; +_{\star} \; \Gamma_{12}^1 {}_{\star} f_{u^2}^{\star\star}$$

$$+_{\star} \; (\Gamma_{12}^2)_u^{\star} {}_{\star} f_v^{\star} \; +_{\star} \; \Gamma_{12}^2 {}_{\star} f_{uv}^{\star\star}$$

$$+_{\star} \; D_{1u}^{\star} {}_{\star} n \; +_{\star} \; D_1 {}_{\star} n_u$$

$$= (\Gamma_{12}^1)_u^{\star} {}_{\star} f_u^{\star} \; +_{\star} \; (\Gamma_{12}^2)_u^{\star} {}_{\star} f_v^{\star} \; +_{\star} \; D_{1u}^{\star} {}_{\star} n$$

$$+_{\star} \; \Gamma_{12}^1 {}_{\star} \left( \Gamma_{11}^1 {}_{\star} f_u^{\star} \; +_{\star} \; \Gamma_{11}^2 {}_{\star} f_v^{\star} \; +_{\star} \; D {}_{\star} n \right)$$

$$+_{\star} \; \Gamma_{12}^2 {}_{\star} \left( \Gamma_{12}^1 {}_{\star} f_u^{\star} \; +_{\star} \; \Gamma_{12}^2 {}_{\star} f_v^{\star} \; +_{\star} \; D_1 {}_{\star} n \right)$$

$$+_{\star} \; D_1 {}_{\star} \left( a_{11} {}_{\star} f_u^{\star} \; +_{\star} \; a_{12} {}_{\star} f_v^{\star} \right)$$

$$= \left( (\Gamma_{12}^1)_u^{\star} \; +_{\star} \; \Gamma_{12}^1 {}_{\star} \Gamma_{11}^1 \; +_{\star} \; \Gamma_{12}^2 {}_{\star} \Gamma_{12}^1 \; +_{\star} \; a_{11} {}_{\star} D_1 \right) {}_{\star} f_u^{\star}$$

$$+_{\star} \; \left( (\Gamma_{12}^2)_u^{\star} \; +_{\star} \; \Gamma_{12}^1 {}_{\star} \Gamma_{11}^2 \; +_{\star} \; (\Gamma_{12}^2)^{2\star} \; +_{\star} \; a_{12} {}_{\star} D_1 \right) {}_{\star} f_v^{\star}$$

$$+_{\star} \; (D_{1u}^{\star} \; +_{\star} \; \Gamma_{12}^1 {}_{\star} D \; +_{\star} \; \Gamma_{12}^2 {}_{\star} D_1) {}_{\star} n,$$

i.e.,

$$f_{uvu}^{\star\star\star} = \left( (\Gamma_{12}^1)_u^{\star} \; +_{\star} \; \Gamma_{12}^1 {}_{\star} \Gamma_{11}^1 \; +_{\star} \; \Gamma_{12}^2 {}_{\star} \Gamma_{12}^1 \; +_{\star} \; a_{11} {}_{\star} D_1 \right) {}_{\star} f_u^{\star}$$

$$+_{\star} \; \left( (\Gamma_{12}^2)_u^{\star} \; +_{\star} \; \Gamma_{12}^1 {}_{\star} \Gamma_{11}^2 \; +_{\star} \; (\Gamma_{12}^2)^{2\star} \; +_{\star} \; a_{12} {}_{\star} D_1 \right) {}_{\star} f_v^{\star}$$

$$+_{\star} \; (D_{1u}^{\star} \; +_{\star} \; \Gamma_{12}^1 {}_{\star} D \; +_{\star} \; \Gamma_{12}^2 {}_{\star} D_1) {}_{\star} n.$$

By the last equation and (5.7) and (5.5), we find the first two multiplicative Gauss equations and the first multiplicative Godazzi-Mainardi equation. Next,

$$
\begin{aligned}
f_{uv^2}^{\star\star\star} =\ & (\Gamma_{12}^1)_v^{\star}{}_{\star} f_u^{\star} +_{\star} \Gamma_{12}^1{}_{\star} f_{uv}^{\star\star} \\
& +_{\star} (\Gamma_{12}^2)_v^{\star}{}_{\star} f_v^{\star} +_{\star} \Gamma_{12}^2{}_{\star} f_v^{\star\star} \\
& +_{\star} D_{1v}^{\star}{}_{\star} n +_{\star} D_1{}_{\star} n_v \\
=\ & (\Gamma_{12}^1)_v^{\star}{}_{\star} f_u^{\star} +_{\star} (\Gamma_{12}^2)_v^{\star}{}_{\star} f_v^{\star} +_{\star} D_{1v}^{\star}{}_{\star} n \\
& +_{\star} \Gamma_{12}^1{}_{\star}\left( \Gamma_{12}^1{}_{\star} f_u^{\star} +_{\star} \Gamma_{12}^2{}_{\star} f_v^{\star} +_{\star} D_1{}_{\star} n \right) \\
& +_{\star} \Gamma_{12}^2{}_{\star}\left( \Gamma_{22}^1{}_{\star} f_u^{\star} +_{\star} \Gamma_{22}^2{}_{\star} f_v^{\star} +_{\star} D_2{}_{\star} n \right) \\
& +_{\star} D_1{}_{\star}\left( a_{21}{}_{\star} f_u^{\star} +_{\star} a_{22}{}_{\star} f_v^{\star} \right) \\
=\ & \left( (\Gamma_{12}^1)_{\star}^{\star} +_{\star} (\Gamma_{12}^1)^{2}{}_{\star} +_{\star} \Gamma_{12}^2{}_{\star}\Gamma_{22}^1 +_{\star} D_1{}_{\star} a_{21} \right){}_{\star} f_u^{\star} \\
& +_{\star} \left( (\Gamma_{12}^2)_v^{\star} +_{\star} \Gamma_{12}^1{}_{\star}\Gamma_{12}^2 +_{\star} \Gamma_{12}^2{}_{\star}\Gamma_{22}^2 +_{\star} D_1{}_{\star} a_{22} \right){}_{\star} f_v^{\star} \\
& +_{\star} (D_{1v}^{\star} +_{\star} \Gamma_{12}^1{}_{\star} D_1 +_{\star} \Gamma_{12}^2{}_{\star} D_2){}_{\star} n,
\end{aligned}
$$

i.e.,

$$
\begin{aligned}
f_{uv^2}^{\star\star\star} =\ & \left( (\Gamma_{12}^1)_{\star}^{\star} +_{\star} (\Gamma_{12}^1)^{2}{}_{\star} +_{\star} \Gamma_{12}^2{}_{\star}\Gamma_{22}^1 +_{\star} D_1{}_{\star} a_{21} \right){}_{\star} f_u^{\star} \\
& +_{\star}\left( (\Gamma_{12}^2)_v^{\star} +_{\star} \Gamma_{12}^1{}_{\star}\Gamma_{12}^2 +_{\star} \Gamma_{12}^2{}_{\star}\Gamma_{22}^2 +_{\star} D_1{}_{\star} a_{22} \right){}_{\star} f_v^{\star} \\
& +_{\star} (D_{1v}^{\star} +_{\star} \Gamma_{12}^1{}_{\star} D_1 +_{\star} \Gamma_{12}^2{}_{\star} D_2){}_{\star} n.
\end{aligned} \tag{5.8}
$$

Now, using that

$$
f_v^{\star\star} = \Gamma_{22}^1{}_{\star} f_u^{\star} +_{\star} \Gamma_{22}^2{}_{\star} f_v^{\star} +_{\star} D_2{}_{\star} n,
$$

we arrive at

$$
\begin{aligned}
f^{\star\star\star}_{v^2 u} = {} & (\Gamma^1_{22})^\star_u {\cdot}_\star f^\star_u +_\star \Gamma^1_{22}{\cdot}_\star f^{\star\star}_{u^2} \\
& +_\star (\Gamma^2_{22})^\star_u {\cdot}_\star f^\star_v +_\star \Gamma^2_{22}{\cdot}_\star f^{\star\star}_{uv} \\
& +_\star D^\star_{2v}{\cdot}_\star n +_\star D_2{\cdot}_\star n^\star_u \\
= {} & (\Gamma^1_{22})^\star_u {\cdot}_\star f^\star_u +_\star (\Gamma^2_{22})^\star_u {\cdot}_\star f^\star_v +_\star D^\star_{2v}{\cdot}_\star n \\
& +_\star \Gamma^1_{22}{\cdot}_\star \left( \Gamma^1_{11}{\cdot}_\star f^\star_u +_\star \Gamma^2_{11}{\cdot}_\star f^\star_v +_\star D{\cdot}_\star n \right) \\
& +_\star \Gamma^2_{22}{\cdot}_\star \left( \Gamma^1_{12}{\cdot}_\star f^\star_u +_\star \Gamma^2_{12}{\cdot}_\star f^\star_v +_\star D_1{\cdot}_\star n \right) \\
& +_\star D_2{\cdot}_\star \left( a_{11}{\cdot}_\star f^\star_u +_\star a_{12}{\cdot}_\star f^\star_v \right) \\
= {} & \left( (\Gamma^1_{22})^\star_u +_\star \Gamma^1_{22}{\cdot}_\star \Gamma^1_{11} +_\star \Gamma^2_{22}{\cdot}_\star \Gamma^1_{12} +_\star D_2{\cdot}_\star a_{11} \right){\cdot}_\star f^\star_u \\
& +_\star \left( (\Gamma^2_{22})^\star_u +_\star \Gamma^1_{22}{\cdot}_\star \Gamma^2_{11} +_\star \Gamma^2_{22}{\cdot}_\star \Gamma^2_{12} +_\star a_{12}{\cdot}_\star D_2 \right){\cdot}_\star f^\star_v \\
& +_\star \left( D^\star_{2v} +_\star \Gamma^1_{22}{\cdot}_\star D +_\star \Gamma^2_{22}{\cdot}_\star D_1 \right){\cdot}_\star n,
\end{aligned}
$$

i.e.,

$$
\begin{aligned}
f^{\star\star\star}_{v^2 u} = {} & \left( (\Gamma^1_{22})^\star_u +_\star \Gamma^1_{22}{\cdot}_\star \Gamma^1_{11} +_\star \Gamma^2_{22}{\cdot}_\star \Gamma^1_{12} +_\star D_2{\cdot}_\star a_{11} \right){\cdot}_\star f^\star_u \\
& +_\star \left( (\Gamma^2_{22})^\star_u +_\star \Gamma^1_{22}{\cdot}_\star \Gamma^2_{11} +_\star \Gamma^2_{22}{\cdot}_\star \Gamma^2_{12} +_\star a_{12}{\cdot}_\star D_2 \right){\cdot}_\star f^\star_v \\
& +_\star \left( D^\star_{2v} +_\star \Gamma^1_{22}{\cdot}_\star D +_\star \Gamma^2_{22}{\cdot}_\star D_1 \right){\cdot}_\star n.
\end{aligned}
$$

Now, applying (5.6), (5.8) and the last expression, we find the third and fourth multiplicative Gauss equations and the second multiplicative Godazzi-Mainardi equation. This completes the proof.

---

## 5.5 The Multiplicative Darboux Frame

Let $S$ be a multiplicative oriented surface and $(U, f)$ be a multiplicative local representation of $S$ that is multiplicative compatible with the multiplicative orientation of $S$, and $(I, g)$ be a multiplicative parameterized curve whose support lies on $S$. Let also, $\tau$ be the multiplicative unit tangent vector of $g$, $n$ be the multiplicative unit normal of $S$. Take

$$N = n \times_\star \tau.$$

**Definition 5.5:** The multiplicative frame $\{\tau, n, N\}$ will be called the multiplicative Darboux frame of $S$ along the multiplicative curve $g$.

Firstly, we will express the multiplicative vectors $n$ and $N$ in the terms of the multiplicative Frenet frame $\{\tau, \nu, \beta\}$. Let

$$\theta = \angle_\star(\nu, n).$$

We have that

$$\nu = \cos_\star(N, \nu)\cdot_\star N +_\star \sin_\star(N, \nu)\cdot_\star n$$
$$\beta = \cos_\star(N, \beta)\cdot_\star N +_\star \sin_\star(N, \beta)\cdot_\star n.$$

Since

$$\angle_\star(N, \nu) = e^{\frac{\pi}{2}},$$
$$\angle_\star(N, \beta) = \pi -_\star \theta,$$

we find

$$\nu = \sin_\star\theta\cdot_\star N +_\star \cos_\star\theta\cdot_\star n$$
$$\beta = -_\star \cos_\star\theta\cdot_\star N +_\star \sin_\star\theta\cdot_\star n$$

and

$$N = \sin_\star\theta\cdot_\star \nu -_\star \cos_\star\theta\cdot_\star \beta$$
$$n = \cos_\star\theta\cdot_\star \nu +_\star \sin_\star\theta\cdot_\star \beta.$$

Now, we will find representations of $\tau^\star$, $N^\star$ and $n^\star$ in the terms of the multiplicative Darboux frame $\{\tau, n, N\}$. We have, using the multiplicative Frenet formulae,

$$\tau^\star = k\cdot_\star \nu$$
$$= k\cdot_\star \sin_\star\theta\cdot_\star N +_\star k\cdot_\star \cos_\star\theta\cdot_\star n$$

and

$$
\begin{aligned}
N^\star &= \cos{}_\star\theta\cdot{}_\star\ \theta^\star\cdot{}_\star\ \nu\ +_\star\ \sin{}_\star\theta\cdot{}_\star\ \theta^\star\cdot{}_\star\ \beta \\
&\quad +_\star\ \sin{}_\star\theta\cdot{}_\star\ \nu^\star\ -_\star\ \cos{}_\star\theta\cdot{}_\star\ \beta^\star \\
&= \cos{}_\star\theta\cdot{}_\star\ \theta^\star\cdot{}_\star\ \nu\ +_\star\ \sin{}_\star\theta\cdot{}_\star\ \theta^\star\cdot{}_\star\ \beta \\
&\quad +_\star\ \sin{}_\star\theta\cdot{}_\star\ (-_\star k\cdot{}_\star\ \tau\ +_\star\ \kappa\cdot{}_\star\ \beta) \\
&\quad -_\star\ \cos{}_\star\theta\cdot{}_\star\ (-_\star\kappa\cdot{}_\star\ \nu) \\
&= \cos{}_\star\theta\cdot{}_\star\ \theta^\star\cdot{}_\star (\sin{}_\star\theta\cdot{}_\star\ N\ +_\star\ \cos{}_\star\theta\cdot{}_\star\ n) \\
&\quad +_\star\ \sin{}_\star\theta\cdot{}_\star\ \theta^\star\cdot{}_\star (-_\star\cos{}_\star\theta\cdot{}_\star\ N\ +_\star\ \sin{}_\star\theta\cdot{}_\star\ n) \\
&\quad -_\star\ k\cdot{}_\star\ \sin{}_\star\theta\cdot{}_\star\ \tau \\
&\quad +_\star\ \kappa\cdot{}_\star\ \sin{}_\star\theta\cdot{}_\star (-_\star\cos{}_\star\theta\cdot{}_\star\ N\ +_\star\ \sin{}_\star\theta\cdot{}_\star\ n) \\
&\quad +_\star\ \kappa\cdot{}_\star\ \cos{}_\star\theta\cdot{}_\star (\sin{}_\star\theta\cdot{}_\star\ N\ +_\star\ \cos{}_\star\theta\cdot{}_\star\ n) \\
&= \theta^\star\cdot{}_\star\ n\ -_\star\ k\cdot{}_\star\ \sin{}_\star\theta\cdot{}_\star\ \tau\ +_\star\ \kappa\cdot{}_\star\ n \\
&= (\theta^\star\ +_\star\ \kappa)\cdot{}_\star\ n\ -_\star\ k\cdot{}_\star\ \sin{}_\star\theta\cdot{}_\star\ \tau,
\end{aligned}
$$

and

$$
\begin{aligned}
n^\star &= -_\star\ \sin{}_\star\theta\cdot{}_\star\ \theta^\star\cdot{}_\star\ \nu\ +_\star\ \cos{}_\star\theta\cdot{}_\star\ \theta^\star\cdot{}_\star\ \beta \\
&\quad +_\star\ \cos{}_\star\theta\cdot{}_\star\ \nu^\star\ +_\star\ \sin{}_\star\theta\cdot{}_\star\ \beta^\star \\
&= -_\star\ \sin{}_\star\theta\cdot{}_\star\ \theta^\star\cdot{}_\star (\sin{}_\star\theta\cdot{}_\star\ N\ +_\star\ \cos{}_\star\theta\cdot{}_\star\ n) \\
&\quad +_\star\ \cos{}_\star\theta\cdot{}_\star\ \theta^\star\cdot{}_\star (-_\star\cos{}_\star\theta\cdot{}_\star\ N\ +_\star\ \sin{}_\star\theta\cdot{}_\star\ n) \\
&\quad +_\star\ \cos{}_\star\theta\cdot{}_\star (-_\star k\cdot{}_\star\ \tau\ +_\star\ \kappa\cdot{}_\star\ \beta) \\
&\quad +_\star\ \sin{}_\star\theta\cdot{}_\star (-_\star\kappa\cdot{}_\star\ \nu) \\
&= -_\star\theta^\star\cdot{}_\star\ N\ -_\star\ k\cdot{}_\star\ \cos{}_\star\theta'\cdot{}_\star\ \tau \\
&\quad +_\star\ \kappa\cdot{}_\star\ \cos{}_\star\theta\cdot{}_\star (-_\star\cos{}_\star\theta\cdot{}_\star\ N\ +_\star\ \sin{}_\star\theta\cdot{}_\star\ n) \\
&\quad -_\star\ \kappa\cdot{}_\star\ \sin{}_\star\theta\cdot{}_\star (\sin{}_\star\theta\cdot{}_\star\ N\ +_\star\ \cos{}_\star\theta\cdot{}_\star\ n) \\
&= -_\star\theta^\star\cdot{}_\star\ N\ -_\star\ k\cdot{}_\star\ \cos{}_\star\theta\cdot{}_\star\ \tau\ -_\star\ \kappa\cdot{}_\star\ N \\
&= -_\star(\theta^\star\ +_\star\ \kappa)\cdot{}_\star\ N\ -_\star\ k\cdot{}_\star\ \cos{}_\star\theta\cdot{}_\star\ \tau.
\end{aligned}
$$

Therefore, we get the system

$$
\begin{aligned}
\tau^\star &= k\cdot{}_\star\ \sin{}_\star\theta\cdot{}_\star\ N\ +_\star\ k\cdot{}_\star\ \cos{}_\star\theta\cdot{}_\star\ n \\
N^\star &= (\kappa^\star\ +_\star\ \kappa)\cdot{}_\star\ n\ -_\star\ k\cdot{}_\star\ \sin{}_\star\theta\cdot{}_\star\ \tau \\
n^\star &= -_\star(\theta^\star\ +_\star\ \kappa)\cdot{}_\star\ N\ -_\star\ k\cdot{}_\star\ \cos{}_\star\theta\cdot{}_\star\ \tau.
\end{aligned}
$$

Let

$$k_g = k \cdot_\star \sin_\star \theta,$$
$$b = k \cdot_\star \cos_\star \theta,$$
$$c = \theta^\star +_\star \kappa.$$

Thus,

$$\tau^\star = k_g \cdot_\star N +_\star b \cdot_\star n$$
$$N^\star = -_\star k_g \cdot_\star \tau +_\star c \cdot_\star n$$
$$n^\star = k_g \cdot_\star \tau -_\star c \cdot_\star N.$$

**Definition 5.6:** The quantity $k_g$ will be called the multiplicative geodesic curvature or multiplicative tangent curvature of the multiplicative line $g$.

**Definition 5.7:** The quantity $c$ will be called the multiplicative geodesic torsion.

---

## 5.6 The Multiplicative Geodesic Curvature. Multiplicative Geodesic Lines

In this section, we will deduct some expressions for the multiplicative geodesic curvature. We will use the notations by the previous section. Note that

$$k_g = \langle \tau^\star, N \rangle_\star$$
$$= -_\star \langle \tau, N^\star \rangle_\star$$

and

$$k_g = \langle \tau^\star, n \times_\star \tau \rangle_\star.$$

Next,

$$f^\star = \tau \cdot_\star s^\star$$

or

$$\tau = f^\star /_\star s^\star.$$

Hence,

$$\tau^\star = (f^{\star\star} \cdot_\star s^\star -_\star f^\star \cdot_\star s^{\star\star})/_\star (s^\star)^{3\star}$$

and

$$
\begin{aligned}
k_g &= \langle \tau^\star, n \times_\star \tau \rangle_\star \\
&= \langle (f^{\star\star} \cdot_\star s^\star -_\star f^\star \cdot_\star s^{\star\star})/_\star (s^\star)^{3\star}, n \times_\star (f^\star /_\star s^\star) \rangle_\star \\
&= (1_\star /_\star (s^\star)^{4\star}) \cdot_\star \langle f^{\star\star} \cdot_\star s^\star -_\star f^\star \cdot_\star s^{\star\star}, n \times_\star f^\star \rangle_\star \\
&= (1_\star /_\star (s^\star)^{4\star}) \cdot_\star \langle f^{\star\star}, n \times_\star f^\star \rangle_\star.
\end{aligned}
$$

Next, we have

$$f^\star = f_u^\star \cdot_\star u^\star +_\star f_v^\star \cdot_\star v^\star$$

and

$$
\begin{aligned}
f^{\star\star} &= (f_u^{\star\star} \cdot_\star u^\star +_\star f_{uv}^{\star\star} \cdot_\star v^\star) \cdot_\star u^\star \\
&\quad +_\star (f_{uv}^{\star\star} \cdot_\star u^\star +_\star f_v^{\star\star} \cdot_\star v^\star) \cdot_\star v^\star \\
&\quad +_\star f_u^\star \cdot_\star u^{\star\star} +_\star f_v^\star \cdot_\star v^{\star\star} \\
&= f_u^{\star\star} \cdot_\star (u^\star)^{2\star} +_\star e^{2} \cdot_\star f_{uv}^{\star\star} \cdot_\star u^\star \cdot_\star v^\star \\
&\quad +_\star f_v^{\star\star} \cdot_\star (v^\star)^{2\star} +_\star f_u^\star \cdot_\star u^{\star\star} +_\star f_v^\star \cdot_\star v^{\star\star} \\
&= (\Gamma_{11}^1 \cdot_\star f_u^\star +_\star \Gamma_{11}^2 \cdot_\star f_v^\star +_\star D \cdot_\star n) \cdot_\star (u^\star)^{2\star} \\
&\quad +_\star e^{2} \cdot_\star (\Gamma_{12}^1 \cdot_\star f_u^\star +_\star \Gamma_{12}^2 \cdot_\star f_v^\star +_\star D_1 \cdot_\star n) \cdot_\star u^\star \cdot_\star v^\star \\
&\quad +_\star (\Gamma_{22}^1 \cdot_\star f_u^\star +_\star \Gamma_{22}^2 \cdot_\star f_v^\star +_\star D_2 \cdot_\star n) \cdot_\star (v^\star)^{2\star} \\
&\quad +_\star f_u^\star \cdot_\star u^{\star\star} +_\star f_v^\star \cdot_\star v^{\star\star} \\
&= (\Gamma_{11}^1 \cdot_\star f_u^\star +_\star \Gamma_{11}^2 \cdot_\star f_v^\star) \cdot_\star (u^\star)^{2\star} \\
&\quad +_\star e^{2} (\Gamma_{12}^1 \cdot_\star f_u^\star +_\star \Gamma_{12}^2 \cdot_\star f_v^\star) \cdot_\star u^\star \cdot_\star v^\star \\
&\quad +_\star (\Gamma_{22}^1 \cdot_\star f_u^\star +_\star \Gamma_{22}^2 \cdot_\star f_v^\star) \cdot_\star (v^\star)^{2\star} \\
&\quad +_\star \langle \phi_2(f^\star, f^\star), n \rangle_\star +_\star f_u^\star \cdot_\star u^{\star\star} +_\star f_v^\star \cdot_\star v^{\star\star},
\end{aligned}
$$

and

$$k_g = (1_* /_* (s^\star)^{4_*})_* \langle f^{\star\star}, n \times_* f^\star \rangle_*$$
$$= \dot{-}_* (1_* /_* (s^\star)^{4_*})_* \langle f^{\star\star}, f^\star \times_* n \rangle_*$$
$$= \dot{-}_* (1_* /_* (s^\star)^{4_*})_* \langle f^{\star\star} \times_* f^\star, n \rangle_*$$
$$= (1_* /_* (s^\star)^{4_*})_* \langle f^\star \times_* f^{\star\star}, n \rangle_*.$$

Observe that

$$f^\star \times_* f^{\star\star} = (f_u^{\star\cdot}{}_* u^\star +_* f_v^{\star\cdot}{}_* v^\star)$$
$$\times_* ((\Gamma_{11}^1{}_* f_u^\star +_* \Gamma_{11}^2{}_* f_v^\star)_* (u^\star)^{2_*}$$
$$+_* e^{2_*}{}_* (\Gamma_{12}^1{}_* f_u^\star +_* \Gamma_{12}^2{}_* f_v^\star)_* u^{\star\cdot}{}_* v^\star$$
$$+_* (\Gamma_{22}^1{}_* f_u^\star +_* \Gamma_{22}^2{}_* f_v^\star)_* (v^\star)^{2_*}$$
$$+_* f_u^{\star\cdot}{}_* u^{\star\star} +_* f_v^{\star\cdot}{}_* v^{\star\star})$$
$$= (f_u^\star \times_* f_v^\star)_* (u^\star)^{3_*}{}_* \Gamma_{11}^2$$
$$+_* e^{2_*}{}_* \Gamma_{12}^2{}_* (f_u^\star \times_* f_v^\star)_* (u^\star)^{2_*}{}_* v^\star$$
$$+_* \Gamma_{22}^2{}_* (f_u^\star \times_* f_v^\star)_* (v^\star)^{2_*}{}_* u^\star$$
$$\dot{-}_* (f_u^\star \times_* f_v^\star)_* \Gamma_{11}^1{}_* v^{\star\cdot}{}_* (u^\star)^{2_*}$$
$$\dot{-}_* e^{2_*}{}_* (f_u^\star \times_* f_v^\star)_* \Gamma_{12}^1{}_* u^{\star\cdot}{}_* (v^\star)^{2_*}$$
$$\dot{-}_* (f_u^\star \times_* f_v^\star)_* \Gamma_{22}^1{}_* (v^\star)^{3_*}$$
$$+_* (f_u^\star \times_* f_v^\star)_* u^{\star\cdot}{}_* v^{\star\star}$$
$$\dot{-}_* (f_u^\star \times_* f_v^\star)_* u^{\star\star}{}_* v^\star$$
$$= (f_u^\star \times_* f_v^\star)_* ((u^\star)^{3_*}{}_* \Gamma_{11}^2 +_* (e^{2_*}{}_* \Gamma_{12}^2 \dot{-}_* \Gamma_{11}^1)_* v^{\star\cdot}{}_* (u^\star)^{2_*}$$
$$+_* (\Gamma_{22}^2 \dot{-}_* e^{2_*}{}_* \Gamma_{12}^1)_* u^{\star\cdot}{}_* (v^\star)^{2_*}$$
$$+_* u^{\star\cdot}{}_* v^{\star\star} \dot{-}_* u^{\star\star}{}_* v^\star \dot{-}_* \Gamma_{22}^1{}_* (v^\star)^{3_*})$$

and

$$\langle f_u^\star \times_* f_v^\star, n \rangle_* = \langle f_u^\star \times_* f_v^\star, \left( f_u^\star \times_* f_v^\star \right) /_* H \rangle_*$$
$$= H^{2_*} /_* H$$
$$= H.$$

Consequently

$$k_g = H/_\star (s^\star)^{4\star} \cdot_\star ((u^\star)^{3\star} \cdot_\star \Gamma_{11}^2 +_\star (e^{2\cdot}_\star \Gamma_{12}^2 -_\star \Gamma_{11}^1)\cdot_\star v^\star \cdot_\star (u^\star)^{2\star}$$

$$+_\star (\Gamma_{22}^2 -_\star e^{2\cdot}_\star \Gamma_{12}^1)\cdot_\star u^\star \cdot_\star (v^\star)^{2\star}$$

$$+_\star u^\star \cdot_\star v^{\star\star} -_\star u^{\star\star} \cdot_\star v^\star -_\star \Gamma_{22}^1 \cdot_\star (v^\star)^{3\star})$$

$$= H/_\star (s^\star)^{4^\star}$$

$$\cdot_\star \det{}_\star \begin{pmatrix} u^\star & u^{\star\star} +_\star \Gamma_{11}^1 \cdot_\star (u^\star)^{2\star} +_\star e^{2\cdot}_\star \Gamma_{12}^1 \cdot_\star u^\star \cdot_\star v^\star +_\star \Gamma_{22}^1 \cdot_\star (v^\star)^{2\star} \\ v^\star & v^{\star\star} +_\star (u^\star)^{2\star} \cdot_\star \Gamma_{11}^2 +_\star e^{2\cdot}_\star \Gamma_{12}^2 \cdot_\star v^\star \cdot_\star u^\star +_\star \Gamma_{22}^2 \cdot_\star (v^\star)^{2\star} \end{pmatrix}.$$

**Definition 5.8:** The multiplicative line $g$ is said to be multiplicative geodesic line if $k_g = 0_\star$.

If $g$ is a multiplicative geodesic line, then its equations are

$$0_\star = (u^\star)^{3\star} \cdot_\star \Gamma_{11}^2 +_\star (e^{2\cdot}_\star \Gamma_{12}^2 -_\star \Gamma_{11}^1)\cdot_\star v^\star \cdot_\star (u^\star)^{2\star}$$

$$+_\star (\Gamma_{22}^2 -_\star e^{2\cdot}_\star \Gamma_{12}^1)\cdot_\star u^\star \cdot_\star (v^\star)^{2\star}$$

$$+_\star u^\star \cdot_\star v^{\star\star} -_\star u^{\star\star} \cdot_\star v^\star -_\star \Gamma_{22}^1 \cdot_\star (v^\star)^{3\star}$$

and

$$0_\star = u^{\star\star} +_\star \Gamma_{11}^1 \cdot_\star (u^\star)^{2\star} +_\star e^{2\cdot}_\star \Gamma_{12}^1 \cdot_\star u^\star \cdot_\star v^\star +_\star \Gamma_{22}^1 \cdot_\star (v^\star)^{2\star}$$

$$0_\star = v^{\star\star} +_\star (u^\star)^{2\star} \cdot_\star \Gamma_{11}^2 +_\star e^{2\cdot}_\star \Gamma_{12}^2 \cdot_\star v^\star u^\star +_\star \Gamma_{22}^2 \cdot_\star (v^\star)^{2\star}.$$

## 5.7 The Multiplicative Geodesics of the Multiplicative Planes

In the multiplicative Catrtesian coordinates, we have that the multiplicative Christoffel coefficients are equal to $0_\star$. Hence, the equations of the multiplicative geodesics of the multiplicative planes take the form

$$u^{\star\star} = 0_\star$$
$$v^{\star\star} = 0_\star,$$

whereupon

$$u = a_1 \cdot_\star c +_\star b_1$$
$$v = a_2 \cdot_\star s +_\star b_2,$$

where $a_j, b_j \in \mathbb{R}_\star$, $j \in \{1, 2\}$. Thus, the multiplicative geodesics of the multiplicative planes are multiplicative straight lines and only them.

## 5.8 The Multiplicative Geodesics of the Multiplicative Unit Sphere

The multiplicative parameterization of the multiplicative unit sphere is as follows:

$$
\begin{aligned}
x &= \cos_\star \phi \cdot_\star \sin_\star \theta \\
y &= \sin_\star \phi \cdot_\star \sin_\star \theta \\
z &= \cos_\star \theta, \quad \phi \in [1, e^{2\pi}], \quad \theta \in [1, e^\pi].
\end{aligned}
$$

Below, we will suppose that $\phi \in [1, e^{2\pi}], \theta \in [1, e^\pi]$. Then

$$
\begin{aligned}
f(\phi, \theta) &= (\cos_\star \phi \cdot_\star \sin_\star \theta, \sin_\star \phi \cdot_\star \sin_\star \theta, \cos_\star \theta), \\
f_\phi^\star(\phi, \theta) &= (-_\star \sin_\star \phi \cdot_\star \sin_\star \theta, \cos_\star \phi \cdot_\star \sin_\star \theta, 0_\star), \\
f_\theta^\star(\phi, \theta) &= (\cos_\star \phi \cdot_\star \cos_\star \theta, \sin_\star \phi \cdot_\star \cos_\star \theta, -_\star \sin_\star \theta), \\
E(\phi, \theta) &= \langle f_\phi^\star(\phi, \theta), f_\phi^\star(\phi, \theta) \rangle_\star \\
&= (\sin_\star \phi)^{2_\star} \cdot_\star (\sin_\star \theta)^{2_\star} +_\star (\cos_\star \phi)^{2_\star} \cdot_\star (\sin_\star \theta)^{2_\star} \\
&= (\sin_\star \theta)^{2_\star}, \\
F(\phi, \theta) &= \langle f_\phi^\star(\phi, \theta), f_\theta^\star(\phi, \theta) \rangle_\star \\
&= -_\star \sin_\star \phi \cdot_\star \cos_\star \phi \cdot_\star \sin_\star \theta \cdot_\star \cos_\star \theta \\
&\quad +_\star \sin_\star \phi \cdot_\star \cos_\star \phi \cdot_\star \sin_\star \theta \cdot_\star \cos_\star \theta \\
&= 0_\star, \\
G(\phi, \theta) &= \langle f_\theta^\star(\phi, \theta), f_\theta^\star(\phi, \theta) \rangle_\star \\
&= (\cos_\star \phi)^{2_\star} \cdot_\star (\cos_\star \theta)^{2_\star} +_\star (\sin_\star \phi)^{2_\star} \cdot_\star (\cos_\star \theta)^{2_\star} +_\star (\sin_\star \theta)^{2_\star} \\
&= 1_\star, \\
H(\phi, \theta) &= \left( E(\phi, \theta) \cdot_\star G(\phi, \theta) -_\star (F(\phi, \theta))^{2_\star} \right)^{\frac{1}{2}_\star} \\
&= \sin_\star \theta.
\end{aligned}
$$

Hence,

$\Gamma^1_{11}(\phi, \theta) = 0_\star,$

$\Gamma^2_{11}(\phi, \theta) = -_\star (1_\star /_\star e^2) \cdot_\star (\sin_\star \theta)^{2\star} \cdot_\star e^{2\cdot}_\star \sin_\star \theta \cdot_\star \cos_\star \theta /_\star (\sin_\star \theta)^{2\star}$

$\qquad\qquad = -_\star \sin_\star \theta \cdot_\star \cos_\star \theta,$

$\Gamma^1_{12}(\phi, \theta) = 1_\star /_\star e^2 \cdot_\star e^{2\cdot}_\star \sin_\star \theta \cdot_\star \cos_\star \theta /_\star (\sin_\star \theta)^{2\star}$

$\qquad\qquad = \cot_\star \theta,$

$\Gamma^2_{12}(\phi, \theta) = 0_\star,$

$\Gamma^1_{22}(\phi, \theta) = 0_\star,$

$\Gamma^2_{22}(\phi, \theta) = 0_\star.$

Thus, the equations of the multiplicative geodesics of the multiplicative unit sphere are as follows:

$$\phi^{\star\star} +_\star e^{2\cdot}_\star \cot_\star \theta \cdot_\star \phi^\star \cdot_\star \theta^\star = 0_\star$$
$$\theta^{\star\star} -_\star \sin_\star \theta \cdot_\star \cos_\star \theta \cdot_\star (\phi^\star)^{2\star} = 0_\star. \qquad (5.9)$$

Now, suppose that $\theta = \theta(\phi)$. Then

$$\theta^\star = (d^\star \theta /_\star d_\star \phi) \cdot_\star \phi^\star,$$
$$\theta^{\star\star} = (d^2 {}^\star \theta /_\star (d_\star \phi)^{2\star}) /_\star (\phi^\star)^{2\star} +_\star (d^\star \theta /_\star d_\star \phi) \cdot_\star \phi^{\star\star}.$$

Therefore the second equation of the system (5.9) becomes the following form

$$0_\star = (d^{2\star} \theta /_\star d_\star \phi^{2\star}) \cdot_\star (\phi^\star)^{2\star} +_\star (d^\star \theta /_\star d_\star \phi) \cdot_\star \phi^{\star\star}$$
$$-_\star \sin_\star \theta \cdot_\star \cos_\star \theta \cdot_\star (\phi^\star)^{2\star}.$$

Applying the first equation of the system (5.9), we find

$$0_\star = (d^{2\star} \theta /_\star d_\star \phi^{2\star}) \cdot_\star (\phi^\star)^{2\star}$$
$$+_\star (d^\star \theta /_\star d_\star \phi)^{2\star} \cdot_\star (-_\star e^{2\cdot}_\star \cot_\star \theta \cdot_\star (\phi^\star)^{2\star})$$
$$-_\star \sin_\star \theta \cdot_\star \cos_\star \theta \cdot_\star (\phi^\star)^{2\star}$$
$$= (\phi^\star)^{2\star} \cdot_\star (d^{2\star} \theta /_\star d_\star \phi^{2\star} -_\star e^{2\cdot}_\star \cot_\star \theta \cdot_\star (d^\star \theta /_\star d_\star \phi)^{2\star}$$
$$-_\star \sin_\star \theta \cdot_\star \cos_\star \theta).$$

Therefore $\phi = const \in \mathbb{R}_\star$ or

$$0_\star = d^{2\star}\theta /_\star d_\star \phi^{2\star} -_\star e^{2\cdot}_\star \cot_\star \theta \cdot_\star (d^\star \theta /_\star d_\star \phi)^{2\star}$$
$$-_\star \sin_\star \theta \cdot_\star \cos_\star \theta.$$

Set

$$z = \cot_\star \theta.$$

Then

$$d^\star z /_\star d_\star \phi = -_\star (1_\star /_\star (\sin_\star \theta)^{2\star}) \cdot_\star (d^\star \theta /_\star d_\star \phi),$$
$$d^{2\star} z /_\star d_\star \phi^{2\star} = e^{2\cdot}_\star (\cos_\star \theta /_\star (\sin_\star \theta)^{3\star}) \cdot_\star (d^\star \theta /_\star d_\star \phi)^{2\star}$$
$$-_\star (1_\star /_\star (\sin_\star \theta)^{2\star}) \cdot_\star (d^{2\star} \theta /_\star d_\star \phi^{2\star})$$
$$= -_\star (1_\star /_\star (\sin_\star \theta)^{2\star}) \cdot_\star (d^{2\star} \theta /_\star d_\star \phi^{2\star}$$
$$-_\star e^{2\cdot}_\star \cot_\star \theta \cdot_\star (d^\star \theta /_\star d_\star \phi)^{2\star})$$

and the equation takes the form

$$d^{2\star} z /_\star d_\star \phi^{2\star} +_\star z = 0_\star.$$

From here, we obtain

$$z = A_1 \cdot_\star \cos_\star \phi +_\star A_2 \cdot_\star \sin_\star \phi$$
$$= \cot_\star \theta,$$

where $A_1, A_2 \in \mathbb{R}_\star$, or

$$A_1 \cdot_\star \cos_\star \phi \cdot_\star \sin_\star \theta +_\star A_2 \cdot_\star \sin_\star \phi \cdot_\star \sin_\star \theta -_\star \cos_\star \theta = 0_\star.$$

---

## 5.9 Multiplicative Geodesics of Multiplicative Liouville Surfaces

**Definition 5.9:** A multiplicative surface $S$ is said to be a multiplicative Liouville surface if it can be parameterized so that its first multiplicative fundamental form can be written in the following way

$$d_\star s^{2\star} = (U(u) +_\star V(v)) \cdot_\star (d_\star u^{2\star} +_\star d_\star v^{2\star}).$$

Let $S$ be a multiplicative Liouville surface. Then

$$E(u, v) = U(u) +_\star V(v),$$
$$F(u, v) = 0_\star,$$
$$G(u, v) = U(u) +_\star V(v).$$

Hence,

$$H(u, v) = U(u) +_\star V(v),$$
$$E_u^\star(u, v) = G_u^\star(u, v)$$
$$= U^\star(u),$$
$$E_v^\star(u, v) = G_v^\star(u, v)$$
$$= V^\star(v),$$
$$\Gamma_{11}^1(u, v) = (1_\star /_\star e^2) \cdot_\star U^\star(u) \cdot_\star (1_\star /_\star (U(u) +_\star V(v))),$$
$$\Gamma_{11}^2(u, v) = -_\star (1_\star /_\star e^2) \cdot_\star V^\star(v) \cdot_\star (1_\star /_\star (U(u) +_\star V(v))),$$
$$\Gamma_{12}^1(u, v) = (1_\star /_\star e^2) \cdot_\star V^\star(v) \cdot_\star (1_\star /_\star (U(u) +_\star V(v))),$$
$$\Gamma_{12}^2(u, v) = (1_\star /_\star e^2) \cdot_\star U^\star(u) \cdot_\star (1_\star /_\star (U(u) +_\star V(v))),$$
$$\Gamma_{22}^1(u, v) = -\star (1_\star /_\star e^2) \cdot_\star U^\star(u) \cdot_\star (1_\star /_\star (U(u) +_\star V(v))),$$
$$\Gamma_{22}^2(u, v) = (1_\star /_\star e^2) \cdot_\star V^\star(v) \cdot_\star (1_\star /_\star (U(u) +_\star V(v))).$$

Then the equations of the multiplicative geodesics of the multiplicative Liouville surfaces are as follows:

$$0_\star = e^2 \cdot_\star (U(u) +_\star V(v)) \cdot_\star u^{\star\star} +_\star U^\star(u) \cdot_\star (u^\star)^{2\star} +_\star e^2 \cdot_\star V^\star(v) \cdot_\star u^\star \cdot_\star v^\star$$
$$-_\star U^\star(u) \cdot_\star (v^\star)^{2\star},$$
$$0_\star = e^2 \cdot_\star (U(u) +_\star V(v)) \cdot_\star v^{\star\star} -_\star V^\star(v) \cdot_\star (u^\star)^{2\star} +_\star e^2 \cdot_\star U^\star(u) \cdot_\star u^\star \cdot v^\star$$
$$+_\star V^\star(v) \cdot_\star (v^\star)^{2\star}.$$

# 6

## Special Classes of Multiplicative Surfaces

## 6.1 Definition of Multiplicative Ruled Surfaces

Let $I \subseteq \mathbb{R}_\star$, $S$ be a multiplicative oriented surface with a multiplicative local parameterization $(I \times \mathbb{R}_\star, f)$ that is multiplicative compatible with the multiplicative orientation of $S$.

**Definition 6.1:** The multiplicative surface $S$ is said to be multiplicative ruled surface if its multiplicative local parameterization has the form

$$f(u, v) = g(u) +_\star v \cdot_\star b(u), \quad (u, v) \in I \times \mathbb{R}_\star,$$

where $g, b \in \mathscr{C}_\star^2(I)$ and $|b(u)|_\star = 1_\star$, $u \in I$. Below, suppose that $S$ is a multiplicative ruled surface. Then

$$f_u^\star = g^\star +_\star v \cdot_\star b^\star,$$
$$f_v^\star = b,$$
$$f_u^{\star\star} = g^{\star\star} +_\star v \cdot_\star b^{\star\star},$$
$$f_{uv}^{\star\star} = b^\star,$$
$$f_v^{\star\star} = 0_\star.$$

Since $|b|_\star = 1_\star$ on $I$, we have

$$1_\star = \langle b, b \rangle_\star$$

and then

$$0_\star = \langle b^\star, b \rangle_\star +_\star \langle b, b^\star \rangle_\star$$
$$= e^2 \cdot_\star \langle b^\star, b \rangle_\star.$$

DOI: 10.1201/9781003299844-6

Therefore

$$\langle b^\star, b \rangle_\star = 0_\star.$$

## 6.2 The First Multiplicative Fundamental Form of Multiplicative Ruled Surfaces

Here we will use the notations, introduced in Section 6.1. We will find the functions $E$, $F$ and $G$. We have

$$
\begin{aligned}
E &= \langle f_u^\star, f_u^\star \rangle_\star \\
&= \langle g^\star +_\star v \cdot_\star b^\star, g^\star +_\star v \cdot_\star b^\star \rangle_\star \\
&= \langle g^\star, g^\star \rangle_\star +_\star e^{2 \cdot}{}_\star v \cdot_\star \langle g^\star, b^\star \rangle_\star \\
&\quad +_\star v^{2 \star \cdot}{}_\star \langle b^\star, b^\star \rangle_\star, \\
F &= \langle f_u^\star, f_v^\star \rangle_\star \\
&= \langle g^\star +_\star v \cdot_\star b^\star, b \rangle_\star \\
&= \langle g^\star, b \rangle_\star +_\star v \cdot_\star \langle b^\star, b \rangle_\star \\
&= \langle g^\star, b \rangle_\star, \\
G &= \langle f_v^\star, f_v^\star \rangle_\star \\
&= \langle b, b \rangle_\star \\
&= 1_\star.
\end{aligned}
$$

Therefore the first multiplicative fundamental form of $S$ can be written as follows:

$$
\begin{aligned}
d_\star s^{2\star} = \; &(\langle g^\star, g^\star \rangle_\star +_\star e^{2 \cdot}{}_\star v \cdot_\star \langle g^\star, b^\star \rangle_\star \\
&+_\star v^{2 \star \cdot}{}_\star \langle b^\star, b^\star \rangle_\star) \cdot_\star d_\star u^{2\star} \\
&+_\star e^{2 \cdot}{}_\star \langle g^\star, b \rangle_{\star \cdot \star} d_\star u \cdot_\star d_\star v +_\star d_\star v^{2\star}.
\end{aligned}
$$

## 6.3 The Multiplicative Tangent Plane of Multiplicative Ruled Surfaces

Here we will use the notations in Section 6.1. In this section, we will deduct the equations of the multiplicative tangent plane of the multiplicative ruled surface $S$. We have

$$N = f_u^\star \times_\star f_v^\star$$
$$= (g^\star +_\star v \cdot_\star b^\star) \times_\star b$$
$$= g^\star \times_\star b +_\star v \cdot_\star (b^\star \times_\star b).$$

Hence, the equations of the multiplicative tangent plane of $S$ are as follows:

$$0_\star = \langle R -_\star f, N \rangle_\star$$
$$= \langle R -_\star g_\star +_\star v \cdot_\star b, g^\star \times_\star b +_\star v \cdot_\star (b^\star \times_\star b)) \rangle_\star$$
$$= \langle R, g^\star \times_\star b \rangle_\star +_\star v \cdot_\star \langle R, b^\star \times_\star b \rangle_\star$$
$$-_\star \langle g, g^\star \times_\star b \rangle_\star -_\star v \cdot_\star \langle g, b^\star \times b \rangle_\star$$

or

$$0_\star = \langle R \times_\star g^\star, b \rangle_\star +_\star v \cdot_\star \langle R \times_\star b^\star, b \rangle_\star$$
$$-_\star \langle g \times_\star g^\star, b \rangle_\star -_\star v \cdot_\star \langle g \times_\star b^\star, b \rangle_\star$$
$$= \langle R \times_\star g^\star +_\star v \cdot_\star (R \times_\star b^\star) -_\star (g \times_\star g^\star) -_\star v \cdot_\star (g \times_\star b^\star), b \rangle_\star.$$

Note that

$$\langle N, b \rangle_\star = 0_\star.$$

## 6.4 The Multiplicative Gaussian Curvature of Multiplicative Ruled Surfaces

Here we will use the notations in Section 6.1. In this section, we will find the multiplicative Gaussian curvature of the multiplicative ruled surface $S$. For this aim, we will find $D$, $D_1$ and $D_2$. We have

$$D = (1_\star /_\star H) \cdot_\star \langle f_u^\star \times_\star f_v^\star, f_u^{\star\star} \rangle_\star$$
$$= (1_\star /_\star H) \cdot_\star \langle (g^\star +_\star v \cdot_\star b^\star) \times_\star b, g^{\star\star} +_\star v \cdot_\star b^{\star\star} \rangle_\star,$$
$$D_1 = (1_\star /_\star H) \cdot_\star \langle f_u^\star \times_\star f_v^\star, f_{uv}^{\star\star} \rangle_\star$$
$$= (1_\star /_\star H) \cdot_\star \langle (g^\star +_\star v \cdot_\star b^\star) \times_\star b, b^\star \rangle_\star$$
$$= (1_\star /_\star H) \cdot_\star \langle g^\star \times_\star b, b^\star \rangle_\star,$$
$$D_2 = (1_\star /_\star H) \cdot_\star \langle f_u^\star \times_\star f_v^\star, f_v^{\star\star} \rangle_\star$$
$$= 0_\star.$$

Consequently the multiplicative Gaussian curvature $K_t$ of $S$ is

$$K_t = -_\star D_1^{2\star} /_\star H^{2\star}$$
$$= -_\star (1_\star /_\star H^{4\star}) \cdot_\star (\langle g^\star \times_\star b, b^\star \rangle_\star)^{2\star}.$$

## 6.5 Multiplicative Minimal Surfaces

Let $S$ be a multiplicative oriented surface with a multiplicative local parameterization $(U, f)$ that is multiplicative compatible with the multiplicative orientation of $S$.

**Definition 6.2:** The multiplicative surface $S$ is said to be multiplicative minimal if its multiplicative mean curvature $K_m = 0_\star$.
  In other words, the multiplicative surface $S$ is multiplicative minimal if and only if

$$D\cdot_\star G -_\star e^{2\cdot}_\star D_1\cdot_\star F +_\star D_2\cdot_\star E = 0_\star.$$

**Definition 6.3:** A multiplicative local parameterization $(U, f)$ of $S$ is said to be multiplicative isothermic if the multiplicative coordinate lines are multiplicative orthogonal, i.e.,

$$F = 0_\star, \quad E = G.$$

**Theorem 6.1:** *A multiplicative surface $S$ is multiplicative minimal if and only if its multiplicative asymptotic directions are multiplicative orthogonal.*

*Proof.* We choose a multiplicative isothermic representation of $S$. Then

$$E = G = \lambda^{2\star}, \quad F = 0_\star.$$

Hence, the multiplicative minimality condition is equivalent to the condition

$$E\cdot_\star (D +_\star D_2) = 0_\star,$$

whereupon

$$D +_\star D_2 = 0_\star.$$

The equation of the multiplicative asymptotic directions is

$$D \cdot_\star t^2_\star +_\star e^2 \cdot_\star D_1 \cdot_\star t +_\star D_2 = 0_\star$$

and the condition they to be multiplicative orthogonal is

$$t_1 \cdot_\star t_2 = -_\star 1_\star,$$

whereupon

$$D = -_\star D_2$$

or

$$D +_\star D_2 = 0_\star.$$

This completes the proof.

# 7

## Multiplicative Differential Forms

### 7.1 Algebra of Multiplicative Differential Forms

We will work in $\mathbb{R}_\star^3$ and we will use the coordinates $x$, $y$ and $z$. This is really for convenience. In the end of this chapter, we will briefly indicate what happens in $\mathbb{R}_\star^n$.

**Definition 7.1:**

1. A 0-multiplicative differential form is a function $f\colon \mathbb{R}_\star^3 \to \mathbb{R}_\star$.
2. A 1-multiplicative differential form is

$$\phi = f\cdot_\star d_\star x +_\star g\cdot_\star d_\star y +_\star h\cdot_\star d_\star h,$$

where $f, g, h\colon \mathbb{R}_\star^3 \to \mathbb{R}_\star$ are given functions.

3. A 2-multiplicative differential form is

$$\phi = f\cdot_\star d_\star x\cdot_\star d_\star y +_\star g\cdot_\star d_\star y\cdot_\star d_\star z +_\star h\cdot_\star d_\star z\cdot_\star d_\star x,$$

where $f, g, h\colon \mathbb{R}_\star^3 \to \mathbb{R}_\star$ are given functions.

4. A 3-multiplicative differential form is

$$\phi = f\cdot_\star d_\star x\cdot_\star d_\star y\cdot_\star d_\star z,$$

where $f\colon \mathbb{R}_\star^3 \to \mathbb{R}_\star$ is a given function.

In fact,

1. For 1-multiplicative differential form, we have the representation

$$\phi = e^{\frac{\log f}{x}dx + \frac{\log g}{y}dy + \frac{\log h}{z}dz}.$$

DOI: 10.1201/9781003299844-7

2. For 2-multiplicative differential form, we have the representation

$$\phi = e^{\frac{\log f}{xy}dxdy + \frac{\log g}{yz}dydz + \frac{\log h}{xz}dxdz}.$$

3. For 3-multiplicative differential form, we have the representation

$$\phi = e^{\frac{\log f}{xyz}dxdydz}.$$

**Example 7.1:** The form

$$\phi = x +_\star y +_\star z,$$

is a 0-multiplicative differential form.

**Example 7.2:** The form

$$\phi = x \cdot_\star d_\star x +_\star y \cdot_\star d_\star y +_\star z \cdot_\star d_\star z$$

is a 1-multiplicative differential form.

**Example 7.3:** The form

$$\phi = \sin_\star(x +_\star y +_\star z) \cdot_\star d_\star x \cdot_\star d_\star y \cdot_\star d_\star z$$

is a 3-multiplicativce differential form.

**Example 7.4:** The form

$$\phi = d_\star y \cdot_\star d_\star x +_\star d_\star x \cdot_\star d_\star z$$

is a 2-multiplicative differential form.

**Definition 7.2:** Let

$$\phi_1 = f_1 \cdot_\star d_\star x +_\star g_1 \cdot_\star d_\star y +_\star h_1 \cdot_\star d_\star z,$$
$$\phi_2 = f_2 \cdot_\star d_\star x +_\star g_2 \cdot_\star d_\star y +_\star h_2 \cdot_\star d_\star z$$

be two 1-multiplicative differential forms. Then we define

$$\phi_1 +_\star \phi_2 = (f_1 +_\star f_2) \cdot_\star d_\star x +_\star (g_1 +_\star g_2) \cdot_\star d_\star y +_\star (h_1 +_\star h_2) \cdot_\star d_\star z$$

and

$$\phi_1 -_\star \phi_2 = (f_1 -_\star f_2) \cdot_\star d_\star x +_\star (g_1 -_\star g_2) \cdot_\star d_\star y +_\star (h_1 -_\star h_2) \cdot_\star d_\star z.$$

**Definition 7.3:** Let

$$\phi_1 = f_1 \cdot_\star d_\star x \cdot_\star d_\star y +_\star g_1 \cdot_\star d_\star y \cdot_\star d_\star z +_\star h_1 \cdot_\star d_\star z \cdot_\star d_\star x,$$
$$\phi_2 = f_2 \cdot_\star d_\star x \cdot_\star d_\star y +_\star g_2 \cdot_\star d_\star y \cdot_\star d_\star z +_\star h_2 \cdot_\star d_\star z \cdot_\star d_\star x$$

be two 2-multiplicative differential forms. Then we define

$$\phi_1 +_\star \phi_2 = (f_1 +_\star f_2) \cdot_\star d_\star x \cdot_\star d_\star y +_\star (g_1 +_\star g_2) \cdot_\star d_\star y \cdot_\star d_\star z +_\star (h_1 +_\star h_2)$$
$$\cdot_\star d_\star z \cdot_\star d_\star x,$$

and

$$\phi_1 -_\star \phi_2 = (f_1 -_\star f_2) \cdot_\star d_\star x \cdot_\star d_\star y +_\star (g_1 -_\star g_2) \cdot_\star d_\star y \cdot_\star d_\star z +_\star (h_1 -_\star h_2)$$
$$\cdot_\star d_\star z \cdot_\star d_\star x.$$

**Definition 7.4:** Let

$$\phi_1 = f_1 \cdot_\star d_\star x \cdot_\star d_\star y \cdot_\star d_\star z,$$
$$\phi_2 = f_2 \cdot_\star d_\star x \cdot_\star d_\star y \cdot_\star d_\star z$$

be two 3-multiplicative differential forms. Then we define

$$\phi_1 +_\star \phi_2 = (f_1 +_\star f_2) \cdot_\star d_\star x \cdot_\star d_\star y \cdot_\star d_\star z$$

and

$$\phi_1 -_\star \phi_2 = (f_1 -_\star f_2) \cdot_\star d_\star x \cdot_\star d_\star y \cdot_\star d_\star z.$$

**Remark 7.1:** Note that the multiplicative addition or multiplicative subtraction makes a sense only for $k$-multiplicative differential forms, $k \in \{0, 1, 2, 3\}$, not for a $k$-multiplicative differential form and for an $l$-multiplicative differential form, $k \neq l$, $k, l \in \{0, 1, 2, 3\}$.

**Example 7.5:** Let

$$\phi = \left(x^2{}_\star +_\star e^2{}_\star x{}_\star{}_\star y\right){}_\star\, d_\star x +_\star (x{}_\star y -_\star e){}_\star\, d_\star y +_\star x^3{}_\star{}_\star\, d_\star z,$$

$$\psi = \left(x^2{}_\star -_\star e^2{}_\star x{}_\star{}_\star y\right){}_\star\, d_\star x +_\star d_\star y +_\star x^3{}_\star{}_\star\, d_\star z.$$

Then

$$\phi +_\star \psi = e^2{}_\star x^2{}_\star{}_\star\, d_\star x +_\star x{}_\star y{}_\star\, d_\star y +_\star e^2{}_\star x^3{}_\star{}_\star\, d_\star z$$

and

$$\phi -_\star \psi = e^4{}_\star x{}_\star y{}_\star\, d_\star x +_\star (x{}_\star y -_\star e^2){}_\star\, d_\star y.$$

**Exercise 7.1:** Let

$$\phi = (x -_\star y +_\star e^3{}_\star y^2{}_\star){}_\star\, d_\star x{}_\star\, d_\star y -_\star d_\star z{}_\star\, d_\star x,$$

$$\psi = (x^2{}_\star +_\star y -_\star e^2{}_\star y){}_\star\, d_\star x{}_\star\, d_\star y +_\star (x -_\star e^3{}_\star x^2{}_\star{}_\star y){}_\star\, d_\star x{}_\star\, d_\star y$$

$$+_\star e^3{}_\star\, d_\star z{}_\star\, d_\star x.$$

Find

$$\phi +_\star \psi, \quad \phi_\star -_\star \psi.$$

**Definition 7.5:** Let $m, f, g, h\colon \mathbb{R}^3_\star \to \mathbb{R}_\star$ and

$$\phi = f{}_\star\, d_\star x +_\star g{}_\star\, d_\star y +_\star h{}_\star\, d_\star z.$$

Then, we define $m{}_\star\, \phi$ as follows:

$$m{}_\star\, \phi = m{}_\star f{}_\star\, d_\star x +_\star m{}_\star g{}_\star\, d_\star y +_\star m{}_\star h{}_\star\, d_\star z.$$

**Definition 7.6:** Let $m, f, g, h\colon \mathbb{R}^3_\star \to \mathbb{R}_\star$ and

$$\phi = f{}_\star\, d_\star x{}_\star\, d_\star y +_\star g{}_\star\, d_\star y{}_\star\, d_\star z +_\star h{}_\star\, d_\star z{}_\star\, d_\star x.$$

Then, we define $m{}_\star\, \phi$ as follows:

$$m{}_\star\, \phi = m{}_\star f{}_\star\, d_\star x{}_\star\, d_\star y +_\star m{}_\star g{}_\star\, d_\star y{}_\star\, d_\star z +_\star m{}_\star h{}_\star\, d_\star z{}_\star\, d_\star x.$$

**Definition 7.7:** Let $m, f\colon \mathbb{R}^3_\star \to \mathbb{R}_\star$ and

$$\phi = f \cdot_\star d_\star x \cdot_\star d_\star y \cdot_\star d_\star z.$$

Then, we define $m \cdot_\star \phi$ as follows:

$$m \cdot_\star \phi = m \cdot_\star f \cdot_\star d_\star x \cdot_\star d_\star y \cdot_\star d_\star z.$$

**Example 7.6:** Let

$$\phi = e^2 \cdot_\star x^{3_\star} \cdot_\star d_\star x \cdot_\star d_\star y +_\star x \cdot_\star d_\star y \cdot_\star d_\star z$$
$$+_\star x^{2_\star} \cdot_\star y^{4_\star} \cdot_\star d_\star z \cdot_\star d_\star x.$$

Then

$$e^3 \cdot_\star \phi = e^6 \cdot_\star x^{3_\star} \cdot_\star d_\star x \cdot_\star d_\star y +_\star e^3 \cdot_\star x \cdot_\star d_\star y \cdot_\star d_\star z$$
$$+_\star e^3 \cdot_\star x^{2_\star} \cdot_\star y^{4_\star} \cdot_\star d_\star z \cdot_\star d_\star x.$$

Next,

$$(1_\star /_\star x) \cdot_\star \phi = e^2 \cdot_\star x^{2_\star} \cdot_\star d_\star x \cdot_\star d_\star y +_\star d_\star y \cdot_\star d_\star z$$
$$+_\star x \cdot_\star y^{4_\star} \cdot_\star d_\star z \cdot_\star d_\star x.$$

**Exercise 7.2:** Find

1. $e^6 \cdot_\star \phi$.
2. $\sin_\star x \cdot_\star \phi$.
3. $(\cos_\star x /_\star \log_\star(e +_\star x^{2_\star})) \cdot_\star \phi$,

where

1.

$$\phi = (x +_\star y^{2_\star} -_\star x^{3_\star}) \cdot_\star d_\star x \cdot_\star d_\star y +_\star x^{6_\star} \cdot_\star d_\star y \cdot_\star d_\star z$$
$$+_\star (x^{4_\star} +_\star x^{2_\star} \cdot_\star y) \cdot_\star d_\star z \cdot_\star d_\star x.$$

2.

$$\phi = (x -_\star e^2 \cdot_\star y^{2_\star}) \cdot_\star d_\star x +_\star d_\star y +_\star d_\star z.$$

For the multiplicative multiplication of multiplicative differential forms, we introduce the following rules.

$$d_\star x \cdot_\star d_\star x = 0_\star,$$
$$d_\star y \cdot_\star d_\star y = 0_\star,$$
$$d_\star z \cdot_\star d_\star z = 0$$

and

$$d_\star x \cdot_\star d_\star y = -_\star d_\star y \cdot_\star d_\star x,$$
$$d_\star x \cdot_\star d_\star z = -_\star d_\star z \cdot_\star d_\star x,$$
$$d_\star z \cdot_\star d_\star y = -_\star d_\star y \cdot_\star d_\star z.$$

**Example 7.7:** We have

$$d_\star x \cdot_\star d_\star y \cdot_\star d_\star x +_\star d_\star x \cdot_\star d_\star y +_\star e^2 \cdot_\star d_\star y \cdot_\star d_\star x$$
$$= -_\star d_\star x \cdot_\star d_\star x \cdot_\star d_\star y +_\star d_\star x \cdot_\star d_\star y -_\star e^2 \cdot_\star d_\star x \cdot_\star d_\star y$$
$$= d_\star x \cdot_\star d_\star y -_\star e^2 \cdot_\star d_\star x \cdot_\star d_\star y$$
$$= -_\star d_\star x \cdot_\star d_\star y.$$

**Example 7.8:** We have

$$d_\star x \cdot_\star (d_\star x +_\star d_\star y +_\star d_\star z) = d_\star x \cdot_\star d_\star x +_\star d_\star x \cdot_\star d_\star y +_\star d_\star x \cdot_\star d_\star z$$
$$= d_\star x \cdot_\star d_\star y +_\star d_\star x \cdot_\star d_\star z$$
$$= d_\star x \cdot_\star d_\star y -_\star d_\star z \cdot_\star d_\star x.$$

**Example 7.9:** Let

$$\phi = x^3 \cdot_\star y \cdot_\star d_\star x +_\star y \cdot_\star d_\star y,$$
$$\psi = x^4 \cdot_\star d_\star x +_\star x \cdot_\star d_\star y +_\star z^2 \cdot_\star d_\star z$$

and

$$\rho = x \cdot_\star y \cdot_\star z \cdot_\star d_\star z \cdot_\star d_\star x.$$

Then

$$\phi \cdot_\star \psi = \left( x^{3\star} \cdot_\star y \cdot_\star d_\star x +_\star y \cdot_\star d_\star y \right)$$
$$\cdot_\star \left( x^{4\star} \cdot_\star d_\star x +_\star x \cdot_\star d_\star y +_\star z^{P2\star} \cdot_\star d_\star z \right)$$
$$= x^{7\star} \cdot_\star y \cdot_\star d_\star x \cdot_\star d_\star x +_\star x^{4\star} \cdot_\star y \cdot_\star d_\star y \cdot_\star d_\star x$$
$$+_\star x^{4\star} \cdot_\star y \cdot_\star d_\star x \cdot_\star d_\star y +_\star y \cdot_\star x \cdot_\star d_\star y \cdot_\star d_\star y$$
$$+_\star x^{3\star} \cdot_\star y \cdot_\star z^{2\star} \cdot_\star d_\star x \cdot_\star d_\star z +_\star z^{2\star} \cdot_\star y \cdot_\star d_\star y \cdot_\star d_\star z$$
$$= -_\star x^{4\star} \cdot_\star y \cdot_\star d_\star x \cdot_\star d_\star y +_\star x^{4\star} \cdot_\star y \cdot_\star d_\star x \cdot_\star d_\star y$$
$$-_\star x^{3\star} \cdot_\star y \cdot_\star z^{2\star} \cdot_\star d_\star z \cdot_\star d_\star x +_\star z^{2\star} \cdot_\star y \cdot_\star d_\star y \cdot_\star d_\star z$$
$$= -_\star x^{3\star} \cdot_\star y \cdot_\star z^{2\star} \cdot_\star d_\star z \cdot_\star d_\star x +_\star z^{2\star} \cdot_\star y \cdot_\star d_\star y \cdot_\star d_\star z.$$

Next,

$$\phi \cdot_\star \rho = (x^{3\star} \cdot_\star y \cdot_\star d_\star x +_\star y \cdot_\star d_\star y) \cdot_\star x \cdot_\star y \cdot_\star z \cdot_\star d_\star z \cdot_\star d_\star x$$
$$= x^{4\star} \cdot_\star y^{2\star} \cdot_\star z \cdot_\star d_\star x \cdot_\star d_\star z \cdot_\star d_\star x$$
$$+_\star x \cdot_\star y^{2\star} \cdot_\star z \cdot_\star d_\star y \cdot_\star d_\star z \cdot_\star d_\star x$$
$$= x \cdot_\star y^{2\star} \cdot_\star z \cdot_\star d_\star x \cdot_\star d_\star y \cdot_\star d_\star z.$$

**Exercise 7.3:** Let

$$\phi_1 = x \cdot_\star d_\star x +_\star y \cdot_\star d_\star y +_\star z \cdot_\star d_\star z,$$
$$\phi_2 = d_\star x \cdot_\star d_\star y,$$
$$\phi_3 = d_\star x \cdot_\star d_\star y -_\star d_\star y \cdot_\star d_\star z +_\star d_\star z \cdot_\star d_\star x.$$

Find

1. $e^{2} \cdot_\star \phi_3 \cdot_\star \phi_2$.
2. $\phi_1 +_\star y \cdot_\star \phi_4$.
3. $\phi_1 \cdot_\star \phi_2$.
4. $\phi_1 \cdot_\star \phi_3$.
5. $\phi_3 \cdot_\star \phi_4$.
6. $\phi_4 \cdot_\star \phi_1$

**Theorem 7.1:** *Let $\phi$ be a k-multiplicative differential form and k is odd. Then*

$$\phi^{2\star} = 0_\star.$$

*Proof.*

1. Let

$$\phi = f \cdot_\star d_\star x +_\star g \cdot_\star d_\star y +_\star h \cdot_\star d_\star z,$$

where $f, g, h\colon \mathbb{R}^3_\star \to \mathbb{R}_\star$. Then

$$
\begin{aligned}
\phi^{2\star} &= (f \cdot_\star d_\star x +_\star g \cdot_\star d_\star y +_\star h \cdot_\star d_\star z) \\
&\quad \cdot_\star (f \cdot_\star d_\star x +_\star g \cdot_\star d_\star y +_\star h \cdot_\star d_\star z) \\
&= f^{2\star} \cdot_\star d_\star x \cdot_\star d_\star x +_\star f \cdot_\star g \cdot_\star d_\star x \cdot_\star d_\star y +_\star f \cdot_\star h \cdot_\star d_\star x \cdot_\star d_\star z \\
&\quad +_\star g \cdot_\star f \cdot_\star d_\star y \cdot_\star d_\star x +_\star g^{2\star} \cdot_\star d_\star y \cdot_\star d_\star y +_\star g \cdot_\star h \cdot_\star d_\star y \cdot_\star d_\star z \\
&\quad +_\star h \cdot_\star f \cdot_\star d_\star z \cdot_\star d_\star x +_\star h \cdot_\star g \cdot_\star d_\star z \cdot_\star d_\star y +_\star h^{2\star} \cdot_\star d_\star z \cdot_\star d_\star z \\
&= f \cdot_\star g \cdot_\star d_\star x \cdot_\star d_\star y -_\star f \cdot_\star g \cdot_\star d_\star x \cdot_\star d_\star y \\
&\quad +_\star f \cdot_\star h \cdot_\star d_\star x \cdot_\star d_\star z -_\star f \cdot_\star h \cdot_\star d_\star x \cdot_\star d_\star z \\
&\quad +_\star g \cdot_\star h \cdot_\star d_\star y \cdot_\star d_\star z -_\star g \cdot_\star h \cdot_\star d_\star y \cdot_\star d_\star z \\
&= 0_\star.
\end{aligned}
$$

2. Let

$$\phi = f \cdot_\star d_\star x \cdot_\star d_\star y \cdot_\star d_\star z,$$

where $f\colon \mathbb{R}^3_\star \to \mathbb{R}_\star$. Then

$$
\begin{aligned}
\phi^{2\star} &= \phi \cdot_\star \phi \\
&= (f \cdot_\star d_\star x \cdot_\star d_\star y \cdot_\star d_\star z) \cdot_\star (f \cdot_\star d_\star x \cdot_\star d_\star y \cdot_\star d_\star z) \\
&= f^{2\star} \cdot_\star d_\star x \cdot_\star d_\star y \cdot_\star d_\star z \cdot_\star d_\star x \cdot_\star d_\star y \cdot_\star d_\star z \\
&= 0_\star.
\end{aligned}
$$

This completes the proof.

**Theorem 7.2:** *Let*

$$
\begin{aligned}
\phi_1 &= f_1 \cdot_\star d_\star x +_\star g_1 \cdot_\star d_\star y +_\star h_1 \cdot_\star d_\star z, \\
\phi_2 &= f_2 \cdot_\star d_\star x +_\star g_2 \cdot_\star d_\star y +_\star h_2 \cdot_\star d_\star z.
\end{aligned}
$$

*Then*

$$\phi_1 \star \phi_2 = (f_1 \cdot \star g_2 \to_\star g_1 \cdot \star f_2) \cdot_\star d_\star x \cdot_\star d_\star y$$
$$+_\star (h_1 \cdot \star f_2 \to_\star f_1 \cdot \star h_2) \cdot_\star d_\star z \cdot_\star d_\star x$$
$$+_\star (g_1 \cdot \star h_2 \to_\star h_1 \cdot \star g_2) \cdot_\star d_\star z \cdot_\star d_\star y.$$

*Proof.* We have

$$\phi_1 \star \phi_2 = (f_1 \cdot \star d_\star x +_\star g_1 \cdot \star d_\star y +_\star h_1 \cdot \star d_\star z)$$
$$\cdot_\star (f_2 \cdot \star d_\star x +_\star g_2 \cdot \star d_\star y +_\star h_2 \cdot \star d_\star z)$$
$$= f_1 \cdot \star f_2 \cdot_\star d_\star x \cdot_\star d_\star x +_\star f_1 \cdot \star g_2 \cdot_\star d_\star x \cdot_\star d_\star y$$
$$+_\star f_1 \cdot \star h_2 \cdot_\star d_\star x \cdot_\star d_\star z +_\star g_1 \cdot \star f_2 \cdot_\star d_\star y \cdot_\star d_\star x$$
$$+_\star g_1 \cdot \star g_2 \cdot_\star d_\star y \cdot_\star d_\star y +_\star g_1 \cdot \star h_2 \cdot_\star d_\star y \cdot_\star d_\star z$$
$$+_\star h_1 \cdot \star f_2 \cdot_\star d_\star z \cdot_\star d_\star x +_\star h_1 \cdot \star g_2 \cdot_\star d_\star z \cdot_\star d_\star y$$
$$+_\star h_1 \cdot \star h_2 \cdot_\star d_\star z \cdot_\star d_\star z$$
$$= (f_1 \cdot \star g_2 \to_\star g_1 \cdot \star f_2) \cdot_\star d_\star x \cdot_\star d_\star y$$
$$+_\star (h_1 \cdot \star f_2 \to_\star f_1 \cdot \star h_2) \cdot_\star d_\star z \cdot_\star d_\star x$$
$$+_\star (g_1 \cdot \star h_2 \to_\star h_1 \cdot \star g_2) \cdot_\star d_\star z \cdot_\star d_\star y.$$

This completes the proof.

**Theorem 7.3:** *Let*

$$\phi_1 = f_1 \cdot_\star d_\star x +_\star g_1 \cdot_\star d_\star y +_\star h_1 \cdot_\star d_\star z,$$
$$\phi_2 = f_2 \cdot_\star d_\star x \cdot_\star d_\star y +_\star g_2 \cdot_\star d_\star y \cdot_\star d_\star z$$
$$+_\star h_2 \cdot_\star d_\star z \cdot_\star d_\star x.$$

*Then*

$$\phi_1 \star \phi_2 = (f_1 \cdot \star g_2 +_\star g_1 \cdot \star h_2 +_\star h_1 \cdot \star f_2) \cdot_\star d_\star x \cdot_\star d_\star y \cdot_\star d_\star z.$$

*Proof.* We have

$$\phi_1 \cdot_\star \phi_2 = (f_1 \cdot_\star d_\star x +_\star g_1 \cdot_\star d_\star y +_\star h_1 \cdot_\star d_\star z)$$
$$\cdot_\star (f_2 \cdot_\star d_\star x \cdot_\star d_\star y +_\star g_2 \cdot_\star d_\star y \cdot_\star d_\star z$$
$$+_\star h_2 \cdot_\star d_\star z \cdot_\star d_\star x)$$
$$= f_1 \cdot_\star f_2 \cdot_\star d_\star x \cdot_\star d_\star x \cdot_\star d_\star y$$
$$+_\star f_1 \cdot_\star g_2 \cdot_\star d_\star x \cdot_\star d_\star y \cdot_\star d_\star z$$
$$+_\star f_1 \cdot_\star h_2 \cdot_\star d_\star x \cdot_\star d_\star z \cdot_\star d_\star x$$
$$+_\star g_1 \cdot_\star f_2 \cdot_\star d_\star y \cdot_\star d_\star x \cdot_\star d_\star y$$
$$+_\star g_1 \cdot_\star g_2 \cdot_\star d_\star y \cdot_\star d_\star y \cdot_\star d_\star z$$
$$+_\star g_1 \cdot_\star h_2 \cdot_\star d_\star y \cdot_\star d_\star z \cdot_\star d_\star x$$
$$+_\star h_1 \cdot_\star f_2 \cdot_\star d_\star z \cdot_\star d_\star x \cdot_\star d_\star y$$
$$+_\star h_1 \cdot_\star g_2 \cdot_\star d_\star z \cdot_\star d_\star y \cdot_\star d_\star z$$
$$+_\star h_1 \cdot_\star h_2 \cdot_\star d_\star z \cdot_\star d_\star z \cdot_\star d_\star x$$
$$= (f_1 \cdot_\star g_2 +_\star g_1 \cdot_\star h_2 +_\star h_1 \cdot_\star f_2) \cdot_\star d_\star x \cdot_\star d_\star y \cdot_\star d_\star z.$$

This completes the proof.

## 7.2 Multiplicative Exterior Differentiation

In this section, we introduce multiplicative exterior differentiation of multiplicative differential forms.

**Definition 7.8:** Let $f$ be a 0-multiplicative differential form. Then its multiplicative exterior derivative is defined by

$$d_\star \phi = f_x^\star \cdot_\star d_\star x +_\star f_y^\star \cdot_\star d_\star y +_\star f_z^\star \cdot_\star d_\star z.$$

**Example 7.10:** Let

$$f(x, y, z) = \frac{x^2 + y^2}{z}, \quad (x, y, z) \in \mathbb{R}^3_\star.$$

Then

$$f_x(x, y, z) = \frac{2x}{z},$$
$$f_y(x, y, z) = \frac{2y}{z},$$
$$f_z(x, y, z) = -\frac{x^2 + y^2}{z^2}, \quad (x, y, z) \in \mathbb{R}^3_\star.$$

Hence,

$$f_x^\star(x, y, z) = e^{x\frac{f_x(x,y,z)}{f(x,y,z)}}$$

$$= e^{x\frac{\frac{2x}{z}}{\frac{x^2+y^2}{z}}}$$

$$= e^{\frac{2x^2}{x^2+y^2}},$$

$$f_y^\star(x, y, z) = e^{y\frac{f_y(x,y,z)}{f(x,y,z)}}$$

$$= e^{y\frac{\frac{2y}{z}}{\frac{x^2+y^2}{z}}}$$

$$= e^{\frac{2y^2}{x^2+y^2}},$$

$$f_z^\star(x, y, z) = e^{z\frac{f_z(x,y,z)}{f(x,y,z)}}$$

$$= e^{-z\frac{\frac{x^2+y^2}{z^2}}{\frac{x^2+y^2}{z}}}$$

$$= e^{-1}, \quad (x, y, z) \in \mathbb{R}_\star^3.$$

Therefore

$$d_\star f(x, y, z) = f_x^\star(x, y, z) \cdot_\star d_\star x +_\star f_y^\star(x, y, z) \cdot_\star d_\star y +_\star f_z^\star(x, y, z) \cdot_\star d_\star z$$

$$= e^{\frac{2x^2}{x^2+y^2}} \cdot_\star d_\star x +_\star e^{\frac{2y^2}{x^2+y^2}} \cdot_\star d_\star y +_\star e^{-1} \cdot_\star d_\star z,$$

$(x, y, z) \in \mathbb{R}_\star^3.$

**Example 7.11:** Let

$$f(x, y, z) = e^x + e^y + e^z, \quad (x, y, z) \in \mathbb{R}_\star^3.$$

Then

$$f_x^\star(x, y, z) = e^x,$$

$$f_y^\star(x, y, z) = e^y,$$

$$f_z^\star(x, y, z) = e^z, \quad (x, y, z) \in \mathbb{R}_\star^3.$$

Therefore

$$d_\star f(x, y, z) = f_x^\star(x, y, z) \cdot_\star d_\star x +_\star f_y^\star(x, y, z) \cdot_\star d_\star y +_\star f_z(x, y, z) \cdot_\star d_\star z$$
$$= e^x \cdot_\star d_\star x +_\star e^y \cdot_\star d_\star y +_\star e^z \cdot_\star d_\star z,$$

$$(x, y, z) \in \mathbb{R}_\star^3.$$

**Exercise 7.4:** Let

$$f(x, y, z) = x + \frac{x^2 + y^2}{y^2 + z^2 + 1} + z^3, \quad (x, y, z) \in \mathbb{R}_\star^3.$$

Find $d_\star f$.

**Definition 7.9:** Let $\phi$ be a $k$-multiplicative differential form. Then its multiplicative exterior derivative $d_\star \phi$ is a $(k + 1)$-multiplicative differential form obtained from $\phi$ by applying $d_\star$ to each of the functions included in $\phi$.

1. Let

$$\phi = f \cdot_\star d_\star x +_\star g \cdot_\star d_\star y +_\star h \cdot_\star d_\star z,$$

where $f, g, h: \mathbb{R}_\star^3 \to \mathbb{R}_\star$ are given functions. Then

$$d_\star \phi = (d_\star f) \cdot_\star d_\star x +_\star (d_\star g) \cdot_\star d_\star y +_\star (d_\star h) \cdot_\star d_\star z$$
$$= (f_x^\star \cdot_\star d_\star x +_\star f_y^\star \cdot_\star d_\star y +_\star f_z^\star \cdot_\star d_\star z) \cdot_\star d_\star x$$
$$+_\star (g_x^\star \cdot_\star d_\star x +_\star g_y^\star \cdot_\star d_\star y +_\star g_z^\star \cdot_\star d_\star z) \cdot_\star d_\star y$$
$$+_\star (h_x^\star \cdot_\star d_\star x +_\star h_y^\star \cdot_\star d_\star y +_\star h_z^\star \cdot_\star d_\star z) \cdot_\star d_\star z$$
$$= f_y^\star \cdot_\star d_\star y \cdot_\star d_\star x +_\star f_z^\star \cdot_\star d_\star z \cdot_\star d_\star x$$
$$+_\star g_x^\star \cdot_\star d_\star x \cdot_\star d_\star y +_\star g_z^\star \cdot_\star d_\star z \cdot_\star d_\star y$$
$$+_\star h_x^\star \cdot_\star d_\star x \cdot_\star d_\star z +_\star h_y^\star \cdot_\star d_\star y \cdot_\star d_\star z$$
$$= (g_x^\star -_\star f_y^\star) \cdot_\star d_\star x \cdot_\star d_\star y +_\star (f_z^\star -_\star h_x^\star) \cdot_\star d_\star z \cdot_\star d_\star x$$
$$+_\star (h_y^\star -_\star g_z^\star) \cdot_\star d_\star y \cdot_\star d_\star z.$$

**Example 7.12:** Let

$$\phi = e^{y^2+z^3} \cdot_\star d_\star x +_\star e^{x+z^3} \cdot_\star d_\star y +_\star e^{x+y^2} \cdot_\star d_\star z,$$

$(x, y, z) \in \mathbb{R}^3_\star$. We have

$$f(x, y, z) = e^{y^2+z^3},$$
$$g(x, y, z) = e^{x+z^3},$$
$$h(x, y, z) = e^{x+y^2}, \quad (x, y, z) \in \mathbb{R}^3_\star.$$

Hence,

$$f^\star_y(x, y, z) = e^{2y^2},$$
$$f^\star_z(x, y, z) = e^{3z^3},$$
$$g^\star_x(x, y, z) = e^x,$$
$$g^\star_z(x, y, z) = e^{3z^3},$$
$$h^\star_x(x, y, z) = e^x,$$
$$h_y(x, y, z) = e^{2y^2}, \quad (x, y, z) \in \mathbb{R}^3_\star,$$

and

$$d_\star\phi = (e^x -_\star e^{2y^2}) \cdot_\star d_\star x \cdot_\star d_\star y +_\star (e^{3z^3} -_\star e^x) \cdot_\star d_\star z \cdot_\star d_\star x$$
$$+_\star (e^{2y^2} -_\star e^{3z^3}) \cdot_\star d_\star y \cdot_\star d_\star z$$
$$= e^{x-2y^2} \cdot_\star d_\star x \cdot_\star d_\star y +_\star e^{3z^3-x} \cdot_\star d_\star z \cdot_\star d_\star x$$
$$+_\star e^{2y^2-3z^3} \cdot_\star d_\star y \cdot_\star d_\star z.$$

**Exercise 7.5:** Let

$$\phi = e^{x+y+z} \cdot_\star d_\star x +_\star e^{x^2+y^2+z^2} \cdot_\star d_\star y +_\star e^{x+2y+3z} \cdot_\star d_\star z.$$

Find $d_\star\phi$.

2. Let

$$\phi = f \cdot_\star d_\star x \cdot_\star d_\star y +_\star g \cdot_\star d_\star y \cdot_\star d_\star z +_\star h \cdot_\star d_\star z \cdot_\star d_\star x,$$

where $f, g, h: \mathbb{R}^3_\star \to \mathbb{R}_\star$. Then

$$
\begin{aligned}
d_\star \phi &= (f^\star_x \cdot_\star d_\star x +_\star f^\star_y \cdot_\star d_\star y +_\star f^\star_z \cdot_\star d_\star z) \cdot_\star d_\star x \cdot_\star d_\star y \\
&+_\star (g^\star_x \cdot_\star d_\star x +_\star g^\star_y \cdot_\star d_\star y +_\star g^\star_z \cdot_\star d_\star z) \cdot_\star d_\star y \cdot_\star d_\star z \\
&+_\star (h^\star_x \cdot_\star d_\star x +_\star h^\star_y \cdot_\star d_\star y +_\star h^\star_z \cdot_\star d_\star z) \cdot_\star d_\star z \cdot_\star d_\star x \\
&= f^\star_z \cdot_\star d_\star x \cdot_\star d_\star y \cdot_\star d_\star z +_\star g^\star_x \cdot_\star d_\star x \cdot_\star d_\star y \cdot_\star d_\star z \\
&+_\star h^\star_y \cdot_\star d_\star x \cdot_\star d_\star y \cdot_\star d_\star z \\
&= (f^\star_z +_\star g^\star_x +_\star h^\star_y) \cdot_\star d_\star x \cdot_\star d_\star y \cdot_\star d_\star z.
\end{aligned}
$$

**Example 7.13:** Let

$$
\begin{aligned}
\phi &= e^{x^2-y^2-z^2} \cdot_\star d_\star x \cdot_\star d_\star y +_\star e^{x+z} \cdot_\star d_\star y \cdot_\star d_\star z \\
&+_\star e^{y^3} \cdot_\star d_\star z \cdot_\star d_\star x.
\end{aligned}
$$

Here

$$
\begin{aligned}
f(x, y, z) &= e^{x^2-y^2-z^2}, \\
g(x, y, z) &= e^{x+z}, \\
h(x, y, z) &= e^{y^3}, \quad (x, y, z) \in \mathbb{R}^3_\star,
\end{aligned}
$$

and

$$
\begin{aligned}
f^\star_z(x, y, z) &= e^{-2z^2}, \\
g^\star_x(x, y, z) &= e^{x}, \\
h^\star_y(x, y, z) &= e^{3y^3}, \quad (x, y, z) \in \mathbb{R}^3_\star.
\end{aligned}
$$

Therefore

$$
\begin{aligned}
d_\star \phi &= (e^{-2z^2} +_\star e^{x} +_\star e^{3y^3}) \cdot_\star d_\star x \cdot_\star d_\star y \cdot_\star d_\star z \\
&= e^{-2z^2+x+3y^3} \cdot_\star d_\star x \cdot_\star d_\star y \cdot_\star d_\star z.
\end{aligned}
$$

**Exercise 7.6:** Let

$$\phi = e^{\frac{x+y+z}{1+x^2+y^2+z^2}} \cdot_{\star} (d_{\star}x \cdot_{\star} d_{\star}y +_{\star} d_{\star}y \cdot_{\star} d_{\star}z +_{\star} d_{\star}z \cdot_{\star} d_{\star}x),$$

$(x, ,y, z) \in \mathbb{R}_{\star}^3$. Find $d_{\star}\phi$.

3. Let

$$\phi = f \cdot_{\star} d_{\star}x \cdot_{\star} d_{\star}y \cdot_{\star} d_{\star}z,$$

where $f: \mathbb{R}_{\star}^3 \to \mathbb{R}_{\star}$. Then

$$d_{\star}\phi = (f_x^{\star} \cdot_{\star} d_{\star}x +_{\star} f_y^{\star} \cdot_{\star} d_{\star}y +_{\star} f_z^{\star} \cdot_{\star} d_{\star}z) \cdot_{\star} d_{\star}x \cdot_{\star} d_{\star}y \cdot_{\star} d_{\star}z$$
$$= 0_{\star}.$$

## 7.3 Properties of the Multiplicative Exterior Differentiation

**Theorem 7.4:** *The multiplicative exterior differentiation is a linear operation.*

*Proof.*

1. Let $\phi_1, \phi_2$ be 0-multiplicative differential forms that are multiplicative differentiable functions and $a_1, a_2 \in \mathbb{R}_{\star}$. Then

$$d_{\star}\phi_1 = \phi_{1x}^{\star} \cdot_{\star} d_{\star}x +_{\star} \phi_{1y}^{\star} \cdot_{\star} d_{\star}y +_{\star} \phi_{1z}^{\star} \cdot_{\star} d_{\star}z,$$

$$d_{\star}(a_1 \cdot_{\star} \phi_1) = (a_1 \cdot_{\star} \phi_{1x}^{\star}) \cdot_{\star} d_{\star}x +_{\star} (a_1 \cdot_{\star} \phi_{1y}^{\star}) \cdot_{\star} d_{\star}y +_{\star} (a_1 \cdot_{\star} \phi_{1z}^{\star}) \cdot_{\star} d_{\star}z$$

$$= a_1 \cdot_{\star} (\phi_{1x}^{\star} \cdot_{\star} d_{\star}x +_{\star} \phi_{1y}^{\star} \cdot_{\star} d_{\star}y +_{\star} \phi_{1z}^{\star} \cdot_{\star} d_{\star}z)$$

$$= a_1 \cdot_{\star} d_{\star}\phi_1,$$

$$d_{\star}\phi_2 = \phi_{2x}^{\star} \cdot_{\star} d_{\star}x +_{\star} \phi_{2y}^{\star} \cdot_{\star} d_{\star}y +_{\star} \phi_{2z}^{\star} \cdot_{\star} d_{\star}z,$$

$$d_{\star}(a_2 \cdot_{\star} \phi_2) = (a_2 \cdot_{\star} \phi_{2x}^{\star}) \cdot_{\star} d_{\star}x +_{\star} (a_2 \cdot_{\star} \phi_{2y}^{\star}) \cdot_{\star} d_{\star}y +_{\star} (a_2 \cdot_{\star} \phi_{2z}^{\star}) \cdot_{\star} d_{\star}z$$

$$= a_2 \cdot_{\star} (\phi_{2x}^{\star} \cdot_{\star} d_{\star}x +_{\star} \phi_{2y}^{\star} \cdot_{\star} d_{\star}y +_{\star} \phi_{2z}^{\star} \cdot_{\star} d_{\star}z)$$

$$= a_2 \cdot_{\star} d_{\star}\phi_2$$

and

$$d_\star(a_1 \cdot_\star \phi_1 +_\star a_2 \cdot_\star \phi_2) = (a_1 \cdot_\star \phi_{1x}^\star +_\star a_2 \cdot_\star \phi_{2x}^\star) \cdot_\star d_\star x$$

$$+_\star (a_1 \cdot_\star \phi_{1y} +_\star a_2 \cdot_\star \phi_{2y}) \cdot_\star d_\star y$$

$$+_\star (a_1 \cdot_\star \phi_{1z}^\star +_\star a_2 \cdot_\star \phi_{2z}^\star) \cdot_\star d_\star z$$

$$= (a_1 \cdot_\star \phi_{1x}^\star) \cdot_\star d_\star x +_\star (a_1 \cdot_\star \phi_{1y}^\star) \cdot_\star d_\star y +_\star (a_1 \cdot_\star \phi_{1z}^\star) \cdot_\star d_\star z$$

$$+_\star a_2 \cdot_\star (\phi_{2x}^\star \cdot_\star d_\star x +_\star \phi_{2y}^\star \cdot_\star d_\star y +_\star \phi_{2z}^\star \cdot_\star d_\star z)$$

$$= a_1 \cdot_\star d_\star \phi_1 +_\star a_2 \cdot_\star d_\star \phi_2.$$

2. Let $\phi_1$ and $\phi_2$ be two 1-multiplicative differential forms, i.e.,

$$\phi_1 = f_1 \cdot_\star d_\star x +_\star g_1 \cdot_\star d_\star y +_\star h_1 \cdot_\star d_\star z,$$
$$\phi_2 = f_2 \cdot_\star d_\star x +_\star g_2 \cdot_\star d_\star y +_\star h_2 \cdot_\star d_\star z,$$

where $f_1, f_2, g_1, g_2, h_1, h_2 \colon \mathbb{R}^3_\star \to \mathbb{R}_\star$ are given multiplicative differentiable functions. Let also, $a_1, a_2 \in \mathbb{R}_\star$. Then

$$d_\star \phi_1 = (g_{1x}^\star -_\star f_{1y}^\star) \cdot_\star d_\star x \cdot_\star d_\star y$$

$$+_\star (f_{1z}^\star -_\star h_{1x}^\star) \cdot_\star d_\star z \cdot_\star d_\star x$$

$$+_\star (h_{1y}^\star -_\star g_{1z}^\star) \cdot_\star d_\star y \cdot_\star d_\star z,$$

$$d_\star(a_1 \cdot_\star \phi_1) = (a_1 \cdot_\star g_{1x}^\star -_\star a_1 \cdot_\star f_{1y}^\star) \cdot_\star d_\star x \cdot_\star d_\star y$$

$$+_\star (a_1 \cdot_\star f_{1z}^\star -_\star a_1 \cdot_\star h_{1x}^\star) \cdot_\star d_\star z \cdot_\star d_\star x$$

$$+_\star (a_1 \cdot_\star h_{1y}^\star -_\star a_1 \cdot_\star g_{1z}^\star) \cdot_\star d_\star y \cdot_\star d_\star z$$

$$= a_1 \cdot_\star ((g_{1x}^\star -_\star f_{1y}^\star) \cdot_\star d_\star x \cdot_\star d_\star y$$

$$+_\star (f_{1z}^\star -_\star h_{1x}^\star) \cdot_\star d_\star z \cdot_\star d_\star x$$

$$+_\star (h_{1y}^\star -_\star g_{1z}^\star) \cdot_\star d_\star y \cdot_\star d_\star z)$$

$$= a_1 \cdot_\star d_\star \phi_1,$$

and

$$d_\star \phi_2 = (g_{2x}^\star \multimap_\star f_{2y}^\star) \cdot_\star d_\star x \cdot_\star d_\star y$$
$$+_\star (f_{2z}^\star \multimap_\star h_{2x}^\star) \cdot_\star d_\star z \cdot_\star d_\star x$$
$$+_\star (h_{2y}^\star \multimap_\star g_{2z}^\star) \cdot_\star d_\star y \cdot_\star d_\star z,$$

$$d_\star (a_2 \cdot_\star \phi_2) = (a_2 \cdot_\star g_{2x}^\star \multimap_\star a_2 \cdot_\star f_{2y}^\star) \cdot_\star d_\star x \cdot_\star d_\star y$$
$$+_\star (a_2 \cdot_\star f_{2z}^\star \multimap_\star a_2 \cdot_\star h_{2x}^\star) \cdot_\star d_\star z \cdot_\star d_\star x$$
$$+_\star (a_2 \cdot_\star h_{2y}^\star \multimap_\star a_2 \cdot_\star g_{2z}^\star) \cdot_\star d_\star y \cdot_\star d_\star z$$
$$= a_2 \cdot_\star ((g_{2x}^\star \multimap_\star f_{2y}^\star) \cdot_\star d_\star x \cdot_\star d_\star y$$
$$+_\star (f_{2z}^\star \multimap_\star h_{2x}^\star) \cdot_\star d_\star z \cdot_\star d_\star x$$
$$+_\star (h_{2y}^\star \multimap_\star g_{2z}^\star) \cdot_\star d_\star y \cdot_\star d_\star z)$$
$$= a_2 \cdot_\star d_\star \phi_2.$$

Consequently

$$a_1 \cdot_\star \phi_1 +_\star a_2 \cdot_\star \phi_2 = (a_1 \cdot_\star f_1 +_\star a_2 \cdot_\star f_2) \cdot_\star d_\star x +_\star (a_1 \cdot_\star g_1 +_\star a_2 \cdot_\star g_2) \cdot_\star d_\star y$$
$$+_\star (a_1 \cdot_\star h_1 +_\star a_2 \cdot_\star h_2) \cdot_\star d_\star z$$

and

$$d_\star (a_1 \cdot_\star \phi_1 +_\star a_2 \cdot_\star \phi_2) = (a_1 \cdot_\star g_{1x}^\star +_\star a_2 \cdot_\star g_{2x}^\star \multimap_\star a_1 \cdot_\star f_{1y}^\star \multimap_\star a_2 \cdot_\star f_{2y}^\star) \cdot_\star d_\star x \cdot_\star d_\star y$$
$$+_\star (a_1 \cdot_\star f_{1z}^\star +_\star a_2 \cdot_\star f_{2z}^\star \multimap_\star a_1 \cdot_\star h_{1x}^\star \multimap_\star a_2 \cdot_\star h_{2x}^\star) \cdot_\star d_\star z \cdot_\star d_\star x$$
$$+_\star (a_1 \cdot_\star h_{1y}^\star +_\star a_2 \cdot_\star h_{2y}^\star \multimap_\star a_1 \cdot_\star g_{1z}^\star \multimap_\star a_2 \cdot_\star g_{2z}^\star) \cdot_\star d_\star y \cdot_\star d_\star z$$
$$= a_1 \cdot_\star ((g_{1x}^\star \multimap_\star f_{1y}^\star) \cdot_\star d_\star x \cdot_\star d_\star y$$
$$+_\star (f_{1z}^\star \multimap_\star h_{1x}^\star) \cdot_\star d_\star z \cdot_\star d_\star x$$
$$+_\star (h_{1y}^\star \multimap_\star g_{1z}^\star) \cdot_\star d_\star y \cdot_\star d_\star z)$$
$$+_\star a_2 \cdot_\star ((g_{2x}^\star \multimap_\star f_{2y}^\star) \cdot_\star d_\star x \cdot_\star d_\star y$$
$$+_\star (f_{2z}^\star \multimap_\star h_{2x}^\star) \cdot_\star d_\star z \cdot_\star d_\star x$$
$$+_\star (h_{2y}^\star \multimap_\star g_{2z}^\star) \cdot_\star d_\star y \cdot_\star d_\star z)$$
$$= a_1 \cdot_\star d_\star \phi_1 +_\star a_2 \cdot_\star d_\star \phi_2.$$

3. Let $\phi_1$ and $\phi_2$ be two 2-multiplicative differential forms, i.e.,

$$\phi_1 = f_1 \cdot_\star d_\star x \cdot_\star d_\star y +_\star g_1 \cdot_\star d_\star y \cdot_\star d_\star z +_\star h_1 \cdot_\star d_\star z \cdot_\star d_\star y,$$
$$\phi_2 = f_2 \cdot_\star d_\star x \cdot_\star d_\star y +_\star g_2 \cdot_\star d_\star y \cdot_\star d_\star z +_\star h_2 \cdot_\star d_\star z \cdot_\star d_\star y,$$

where $f_1, f_2, g_1, g_2, h_1, h_2 \colon \mathbb{R}^3_\star \to \mathbb{R}_\star$ are given multiplicative differentiable functions. Let also, $a_1, a_2 \in \mathbb{R}_\star$. Then

$$d_\star \phi_1 = (f^\star_{1z} +_\star g^\star_{1x} +_\star h^\star_{1y}) \cdot_\star d_\star x \cdot_\star d_\star y \cdot_\star d_\star z,$$
$$d_\star (a_1 \cdot_\star \phi_1) = (a_1 \cdot_\star f^\star_{1z} +_\star a_1 \cdot_\star g^\star_{1x} +_\star a_1 \cdot_\star h^\star_{1y}) \cdot_\star d_\star x \cdot_\star d_\star y \cdot_\star d_\star z$$
$$= a_1 \cdot_\star (f^\star_{1z} +_\star g^\star_{1x} +_\star h^\star_{1y}) \cdot_\star d_\star x \cdot_\star d_\star y \cdot_\star d_\star z$$
$$= a_1 \cdot_\star d_\star \phi_1$$

and

$$d_\star \phi_2 = (f^\star_{2z} +_\star g^\star_{2x} +_\star h^\star_{2y}) \cdot_\star d_\star x \cdot_\star d_\star y \cdot_\star d_\star z,$$
$$d_\star (a_2 \cdot_\star \phi_2) = (a_2 \cdot_\star f^\star_{2z} +_\star a_2 \cdot_\star g^\star_{2x} +_\star a_2 \cdot_\star h^\star_{2y}) \cdot_\star d_\star x \cdot_\star d_\star y \cdot_\star d_\star z$$
$$= a_2 \cdot_\star (f^\star_{2z} +_\star g^\star_{2x} +_\star h^\star_{2y}) \cdot_\star d_\star x \cdot_\star d_\star y \cdot_\star d_\star z$$
$$= a_2 \cdot_\star d_\star \phi_2.$$

Consequently

$$a_1 \cdot_\star \phi_1 +_\star a_2 \cdot_\star \phi_2 = a_1 \cdot_\star f_1 \cdot_\star d_\star x \cdot_\star d_\star y +_\star a_1 \cdot_\star g_1 \cdot_\star d_\star y \cdot_\star d_\star z +_\star a_1 \cdot_\star h_1 \cdot_\star d_\star z$$
$$\cdot_\star d_\star x$$
$$+_\star a_2 \cdot_\star f_2 \cdot_\star d_\star x \cdot_\star d_\star y +_\star a_2 \cdot_\star g_2 \cdot_\star d_\star y \cdot_\star d_\star z$$
$$+_\star a_2 \cdot_\star h_2 \cdot_\star d_\star z \cdot_\star d_\star x$$
$$= (a_1 \cdot_\star f_1 +_\star a_2 \cdot_\star f_2) \cdot_\star d_\star x \cdot_\star d_\star y +_\star (a_1 \cdot_\star g_1 +_\star a_2 \cdot_\star g_2) \cdot_\star d_\star y$$
$$\cdot_\star d_\star z$$
$$+_\star (a_1 \cdot_\star h_1 +_\star a_2 \cdot_\star h_2) \cdot_\star d_\star z \cdot_\star d_\star x$$

and

$$d_\star(a_1 \cdot_\star \phi_1 +_\star a_2 \cdot_\star \phi_2) = (a_1 \cdot_\star f_{1z}^\star +_\star a_2 \cdot_\star f_{2z}^\star +_\star a_1 \cdot_\star g_{1x}^\star +_\star a_2 \cdot_\star g_{2x}^\star +_\star a_1$$
$$\cdot_\star h_{1y}^\star +_\star a_2 \cdot_\star h_{2y}^\star) \cdot_\star d_\star x \cdot_\star d_\star y \cdot_\star d_\star z$$
$$= a_1 \cdot_\star (f_{1z}^\star +_\star g_{1x}^\star +_\star h_{1y}^\star) \cdot_\star d_\star x \cdot_\star d_\star y \cdot_\star d_\star z$$
$$+_\star a_2 \cdot_\star (f_{2z}^\star +_\star g_{2x}^\star +_\star h_{2y}^\star) \cdot_\star d_\star x \cdot_\star d_\star y \cdot_\star d_\star z$$
$$= a_1 \cdot_\star d_\star \phi_1 +_\star a_2 \cdot_\star d_\star \phi_2.$$

This completes the proof.

**Theorem 7.5:** *Let $\phi_1$ and $\phi_2$ be 0-multiplicative differential forms. Then*

$$d_\star(\phi_1 \cdot_\star \phi_2) = \phi_1 \cdot_\star d_\star \phi_2 +_\star \phi_2 \cdot_\star d_\star \phi_1.$$

*Proof.* We have

$$d_\star \phi_1 = \phi_{1x}^\star \cdot_\star d_\star x +_\star \phi_{1y}^\star \cdot_\star d_\star y +_\star \phi_{1z}^\star \cdot_\star d_\star z,$$
$$d_\star \phi_2 = \phi_{2x}^\star \cdot_\star d_\star x +_\star \phi_{2y}^\star \cdot_\star d_\star y +_\star \phi_{2z}^\star \cdot_\star d_\star z$$

and

$$d_\star(\phi_1 \cdot_\star \phi_2) = (\phi_1 \cdot_\star \phi_2)_x^\star d_\star x +_\star (\phi_1 \cdot_\star \phi_2)_y^\star \cdot_\star d_\star y +_\star (\phi_1 \cdot_\star \phi_2)_z^\star \cdot_\star d_\star z$$
$$= (\phi_1 \cdot_\star \phi_{2x}^\star +_\star \phi_{1x}^\star \cdot_\star \phi_2) \cdot_\star d_\star x$$
$$+_\star (\phi_1 \cdot_\star \phi_{2y}^\star +_\star \phi_{1y}^\star \cdot_\star \phi_2) \cdot_\star d_\star y$$
$$+_\star (\phi_1 \cdot_\star \phi_{2z}^\star +_\star \phi_{1z}^\star \cdot_\star \phi_2) \cdot_\star d_\star z$$
$$= \phi_1 \cdot_\star (\phi_{2x}^\star \cdot_\star d_\star x +_\star \phi_{2y}^\star \cdot_\star d_\star y +_\star \phi_{2z}^\star \cdot_\star d_\star z)$$
$$+_\star \phi_2 \cdot_\star (\phi_{1x}^\star \cdot_\star d_\star x +_\star \phi_{1y}^\star \cdot_\star d_\star y +_\star \phi_{1z}^\star \cdot_\star d_\star z)$$
$$= \phi_1 \cdot_\star d_\star \phi_2 +_\star \phi_2 \cdot_\star d_\star \phi_1.$$

This completes the proof.

**Theorem 7.6:** *Let $\phi_1$ be a 0-multiplicative differential form and $\phi_2$ be a 1-multiplicative form*

$$\phi_2 = f \cdot_\star d_\star x +_\star g \cdot_\star d_\star y +_\star h \cdot_\star d_\star z.$$

*Then*

$$d_\star(\phi_1 \cdot_\star \phi_2) = \phi_1 \cdot_\star d_\star \phi_2 +_\star \phi_2 \cdot_\star d_\star \phi_1.$$

*Proof.* We have

$$d_\star \phi_2 = (g_x^\star \rightarrow_\star f_y^\star) \cdot_\star d_\star x \cdot_\star d_\star y$$
$$+_\star (f_z^\star \rightarrow_\star h_x^\star) \cdot_\star d_\star z \cdot_\star d_\star x$$
$$+_\star (h_y^\star \rightarrow_\star g_z^\star) \cdot_\star d_\star y \cdot_\star d_\star z$$

and

$$\phi_1 \cdot_\star \phi_2 = (\phi_1 \cdot_\star f) \cdot_\star d_\star x +_\star (\phi_1 \cdot_\star g) \cdot_\star d_\star y +_\star (\phi_1 \cdot_\star h) \cdot_\star d_\star z.$$

Hence,

$$d_\star(\phi_1 \cdot_\star \phi_2) = (\phi_{1x}^\star \cdot_\star g +_\star \phi_1 \cdot_\star g_x^\star \rightarrow_\star \phi_1 \cdot_\star f_y^\star \rightarrow_\star \phi_{1y}^\star \cdot_\star f) \cdot_\star d_\star x \cdot_\star d_\star y$$
$$+_\star (\phi_{1z}^\star \cdot_\star f +_\star \phi_1 \cdot_\star f_z^\star \rightarrow_\star \phi_1 \cdot_\star h_x^\star \rightarrow_\star \phi_{1x}^\star \cdot_\star h) \cdot_\star d_\star z \cdot_\star d_\star x$$
$$+_\star (\phi_{1y}^\star \cdot_\star h +_\star \phi_1 \cdot_\star h_y^\star \rightarrow_\star \phi_1 \cdot_\star g_z^\star \rightarrow_\star \phi_{1z}^\star \cdot_\star g) \cdot_\star d_\star y \cdot_\star d_\star z$$
$$= \phi_1 \cdot_\star (g_x^\star \rightarrow_\star f_y^\star) \cdot_\star d_\star x \cdot_\star d_\star y +_\star \phi_1 \cdot_\star (f_z^\star \rightarrow_\star h_x^\star) \cdot_\star d_\star z \cdot_\star d_\star x$$
$$+_\star \phi_1 \cdot_\star (h_y^\star \rightarrow_\star g_z^\star) \cdot_\star d_\star y \cdot_\star d_\star z +_\star (\Phi_{1x}^\star \cdot_\star g \rightarrow_\star \Phi_{1y}^\star \cdot_\star f) \cdot_\star d_\star x \cdot_\star d_\star y$$
$$+_\star (\phi_{1z}^\star \cdot_\star f \rightarrow_\star \phi_{1x}^\star \cdot_\star h) \cdot_\star d_\star z \cdot_\star d_\star x +_\star (\phi_{1y}^\star \cdot_\star h \rightarrow_\star \phi_{1z}^\star \cdot_\star g) \cdot_\star d_\star y \cdot_\star d_\star z$$
$$= \phi_1 \cdot_\star d_\star \phi_2 +_\star (\Phi_{1x}^\star \cdot_\star g \rightarrow_\star \Phi_{1y}^\star \cdot_\star f) \cdot_\star d_\star x \cdot_\star d_\star y$$
$$+_\star (\phi_{1z}^\star \cdot_\star f \rightarrow_\star \phi_{1x}^\star \cdot_\star h) \cdot_\star d_\star z \cdot_\star d_\star x +_\star (\phi_{1y}^\star \cdot_\star h \rightarrow_\star \phi_{1z}^\star \cdot_\star g) \cdot_\star d_\star y \cdot_\star d_\star z.$$

Next,

$$d_\star \phi_1 = \phi_{1x} \cdot_\star d_\star x +_\star \phi_{1y} \cdot_\star d_\star y +_\star \phi_{1z} \cdot_\star d_\star z$$

and

$$
\begin{aligned}
d_\star \phi_1 \cdot_\star \phi_2 &= (\phi_{1x}^\star \cdot_\star d_\star x +_\star \phi_{1y}^\star \cdot_\star d_\star y +_\star \phi_{1z}^\star \cdot_\star d_\star z) \\
&\quad \cdot_\star (f \cdot_\star d_\star x +_\star g \cdot_\star d_\star y +_\star h \cdot_\star d_\star z) \\
&= \phi_{1x}^\star \cdot_\star f \cdot_\star d_\star x \cdot_\star d_\star x +_\star \phi_{1x}^\star \cdot_\star g \cdot_\star d_\star x \cdot_\star d_\star y \\
&\quad +_\star \phi_{1x}^\star \cdot_\star h \cdot_\star d_\star x \cdot_\star d_\star z +_\star \phi_{1y}^\star \cdot_\star f \cdot_\star d_\star y \cdot_\star d_\star x \\
&\quad +_\star \phi_{1y}^\star \cdot_\star g \cdot_\star d_\star y \cdot_\star d_\star y +_\star \phi_{1y}^\star \cdot_\star h \cdot_\star d_\star y \cdot_\star d_\star z \\
&\quad +_\star \phi_{1z}^\star \cdot_\star f \cdot_\star d_\star z \cdot_\star d_\star x +_\star \phi_{1z}^\star \cdot_\star g \cdot_\star d_\star z \cdot_\star d_\star y \\
&\quad +_\star \phi_{1z}^\star \cdot_\star h \cdot_\star d_\star z \cdot_\star d_\star z \\
&= (\phi_{1z}^\star \cdot_\star f -_\star \phi_{1x}^\star \cdot_\star h) \cdot_\star d_\star z \cdot_\star d_\star x \\
&\quad +_\star (\phi_{1y}^\star \cdot_\star h -_\star \phi_{1z}^\star \cdot_\star g) \cdot_\star d_\star y \cdot_\star d_\star z \\
&\quad +_\star (\phi_{1x}^\star \cdot_\star g -_\star \phi_{1y}^\star \cdot_\star f) \cdot_\star d_\star x \cdot_\star d_\star y.
\end{aligned}
$$

Consequently

$$
d_\star(\phi_1 \cdot_\star \phi_2) = d_\star \phi_1 \cdot_\star \phi_2 +_\star \phi_1 \cdot_\star d_\star \phi_2.
$$

This completes the proof.

**Theorem 7.7:** *Let $\phi_1$ be a 0-multiplicative differential form and*

$$
\phi_2 = f \cdot_\star d_\star x \cdot_\star d_\star y +_\star g \cdot_\star d_\star y \cdot_\star d_\star z +_\star h \cdot_\star d_\star z \cdot_\star d_\star x.
$$

*Then*

$$
d_\star(\phi_1 \cdot_\star \phi_2) = d_\star \phi_1 \cdot_\star \phi_2 +_\star \phi_1 \cdot_\star d_\star \phi_2.
$$

*Proof.* We have

$$
d_\star \phi_2 = (f_z^\star +_\star g_x^\star +_\star h_y^\star) \cdot_\star d_\star x \cdot_\star d_\star y \cdot_\star d_\star z
$$

and

$$
\begin{aligned}
\phi_1 \cdot_\star \phi_2 &= (\phi_1 \cdot_\star f) \cdot_\star d_\star x \cdot_\star d_\star y +_\star (\phi_1 \cdot_\star g) \cdot_\star d_\star y \cdot_\star d_\star z \\
&\quad +_\star (\phi_1 \cdot_\star h) \cdot_\star d_\star z \cdot_\star d_\star x.
\end{aligned}
$$

Hence,

$$d_\star(\phi_1 \cdot_\star \phi_2) = ((\phi_1 \cdot_\star f)_z^\star +_\star (\phi_1 \cdot_\star g)_x^\star +_\star (\phi_1 \cdot_\star h)_y^\star) \cdot_\star d_\star x \cdot_\star d_\star y \cdot_\star d_\star z$$

$$= (\phi_{1z}^\star \cdot_\star f +_\star \phi_1 \cdot_\star f_z^\star +_\star \phi_{1x}^\star \cdot_\star g +_\star \phi_1 \cdot_\star g_x^\star +_\star \phi_{1y}^\star \cdot_\star h$$

$$+_\star \phi_1 \cdot_\star h_y^\star) \cdot_\star d_\star x \cdot_\star d_\star y \cdot_\star d_\star z$$

$$= \phi_1 \cdot_\star (f_z^\star +_\star g_x^\star +_\star h_y^\star) \cdot_\star d_\star x \cdot_\star d_\star y \cdot_\star d_\star z$$

$$+_\star (\phi_{1z}^\star \cdot_\star f +_\star \phi_{1x}^\star \cdot_\star g +_\star \phi_{1y}^\star \cdot_\star h) \cdot_\star d_\star x \cdot_\star d_\star y \cdot_\star d_\star z$$

$$= \phi_1 \cdot_\star d_\star \phi_2 +_\star (\phi_{1z}^\star \cdot_\star f +_\star \phi_{1x}^\star \cdot_\star g +_\star \phi_{1y}^\star \cdot_\star h) \cdot_\star d_\star x \cdot_\star d_\star y \cdot_\star d_\star z.$$

Also,

$$d_\star \phi_1 \cdot_\star \phi_2 = (\phi_{1x}^\star \cdot_\star d_\star x +_\star \phi_{1y}^\star \cdot_\star d_\star y +_\star \phi_{1z}^\star \cdot_\star d_\star z)$$

$$\cdot_\star (f \cdot_\star d_\star x \cdot_\star d_\star y +_\star g \cdot_\star d_\star y \cdot_\star d_\star z +_\star h \cdot_\star d_\star z \cdot_\star d_\star x)$$

$$= \phi_{1x}^\star \cdot_\star f \cdot_\star d_\star x \cdot_\star d_\star x \cdot_\star d_\star y$$

$$+_\star \phi_{1x}^\star \cdot_\star g \cdot_\star d_\star x \cdot_\star d_\star y \cdot_\star d_\star z$$

$$+_\star \phi_{1x}^\star \cdot_\star h \cdot_\star d_\star x \cdot_\star d_\star z \cdot_\star d_\star x$$

$$+_\star \phi_{1y}^\star \cdot_\star f \cdot_\star d_\star y \cdot_\star d_\star x \cdot_\star d_\star y$$

$$+_\star \phi_{1y}^\star \cdot_\star g \cdot_\star d_\star y \cdot_\star d_\star y \cdot_\star d_\star z$$

$$+_\star \phi_{1y}^\star \cdot_\star h \cdot_\star d_\star y \cdot_\star d_\star z \cdot_\star d_\star x$$

$$+_\star \phi_{1z}^\star \cdot_\star f \cdot_\star d_\star z \cdot_\star d_\star x \cdot_\star d_\star y$$

$$+_\star \phi_{1z}^\star \cdot_\star g \cdot_\star d_\star z \cdot_\star d_\star y \cdot_\star d_\star z$$

$$+_\star \phi_{1z}^\star \cdot_\star h \cdot_\star d_\star z \cdot_\star d_\star z \cdot_\star d_\star x$$

$$= (\phi_{1x}^\star \cdot_\star g +_\star \phi_{1y}^\star \cdot_\star h +_\star \phi_{1z}^\star \cdot_\star f) \cdot_\star d_\star x \cdot_\star d_\star y \cdot_\star d_\star z.$$

Consequently

$$d_\star(\phi_1 \cdot_\star \phi_2) = d_\star \phi_1 \cdot_\star \phi_2 +_\star \phi_1 \cdot_\star d_\star \phi_2.$$

This completes the proof.

**Theorem 7.8:** *Let*

$$\phi_1 = f_1 \cdot {}_\star d_\star x +_\star g_1 \cdot {}_\star d_\star y +_\star h_1 \cdot {}_\star d_\star z,$$
$$\phi_2 = f_2 \cdot {}_\star d_\star x +_\star g_2 \cdot {}_\star d_\star y +_\star h_2 \cdot {}_\star d_\star z.$$

*Then*

$$d_\star(\phi_1 \cdot_\star \phi_2) = d_\star \phi_1 \cdot_\star \phi_2 -_\star \phi_1 \cdot_\star d_\star \phi_2.$$

*Proof.* We have

$$\begin{aligned}
\phi_1 \cdot_\star \phi_2 = {} & (f_1 \cdot_\star g_2 -_\star g_1 \cdot_\star f_2) \cdot_\star d_\star x \cdot_\star d_\star y \\
& +_\star (h_1 \cdot_\star f_2 -_\star f_1 \cdot_\star h_2) \cdot_\star d_\star z \cdot_\star d_\star x \\
& +_\star (g_1 \cdot_\star h_2 -_\star h_1 \cdot_\star g_2) \cdot_\star d_\star y \cdot_\star d_\star z.
\end{aligned}$$

Hence,

$$\begin{aligned}
d_\star(\phi_1 \cdot_\star \phi_2) = {} & ((f_1 \cdot_\star g_2 -_\star g_1 \cdot_\star f_2)_z^\star +_\star (h_1 \cdot_\star f_2 -_\star f_1 \cdot_\star h_2)_y^\star \\
& -_\star (g_1 \cdot_\star h_2 -_\star h_1 \cdot_\star g_2)_x^\star) \cdot_\star d_\star x \cdot_\star d_\star y \cdot_\star d_\star z \\
= {} & (f_{1z}^\star \cdot_\star g_2 +_\star f_1 \cdot_\star g_{2z}^\star -_\star g_{1z}^\star \cdot_\star f_2 -_\star g_1 \cdot_\star f_{2z}^\star \\
& +_\star h_{1y}^\star \cdot_\star f_2 +_\star h_1 \cdot_\star f_{2y}^\star -_\star f_{1y}^\star \cdot_\star h_2 -_\star f_1 \cdot_\star h_{2y}^\star \\
& +_\star g_{1x}^\star \cdot_\star h_2 +_\star g_1 \cdot_\star h_{2x}^\star -_\star h_{1x}^\star \cdot_\star g_2 -_\star h_1 \cdot_\star g_{2x}^\star) \cdot_\star d_\star x \cdot_\star d_\star y \\
& \cdot_\star d_\star z
\end{aligned}$$

and

$$\begin{aligned}
d_\star \phi_1 = {} & (g_{1x}^\star -_\star f_{1y}^\star) \cdot_\star d_\star x \cdot_\star d_\star y \\
& +_\star (f_{1z}^\star -_\star h_{1x}^\star) \cdot_\star d_\star z \cdot_\star d_\star x \\
& +_\star (h_{1y}^\star -_\star g_{1z}^\star) \cdot_\star d_\star y \cdot_\star d_\star z,
\end{aligned}$$

and

$$d_\star \phi_1 \star \phi_2 = ((g_{1x}^\star \rightarrow_\star f_{1y}^\star) \cdot_\star d_\star x \cdot_\star d_\star y$$

$$+_\star (f_{1z}^\star \rightarrow_\star h_{1x}^\star) \cdot_\star d_\star z \cdot_\star d_\star x$$

$$+_\star (h_{1y}^\star \rightarrow_\star g_{1z}^\star) \cdot_\star d_\star y \cdot_\star d_\star z)$$

$$\cdot_\star (f_2 \cdot_\star d_\star x +_\star g_2 \cdot_\star d_\star y +_\star h_2 \cdot_\star d_\star z)$$

$$= f_2 \cdot_\star (h_{1y}^\star \rightarrow_\star g_{1z}^\star) \cdot_\star d_\star y \cdot_\star d_\star z \cdot_\star d_\star x$$

$$+_\star g_2 \cdot_\star (f_{1z}^\star \rightarrow_\star h_{1x}^\star) \cdot_\star d_\star z \cdot_\star d_\star x \cdot_\star d_\star y$$

$$+_\star h_2 \cdot_\star (g_{1x}^\star \rightarrow_\star f_{1y}^\star) \cdot_\star d_\star x \cdot_\star d_\star y \cdot_\star d_\star z$$

$$= (f_2 \cdot_\star (h_{1y}^\star \rightarrow_\star g_{1z}^\star) +_\star g_2 \cdot_\star (f_{1z}^\star \rightarrow_\star h_{1x}^\star)$$

$$+_\star h_2 \cdot_\star (g_{1x}^\star \rightarrow_\star f_{1y}^\star)) \cdot_\star d_\star x \cdot_\star d_\star y \cdot_\star d_\star z.$$

Moreover,

$$d_\star \phi_2 = (g_{2x}^\star \rightarrow_\star f_{2y}^\star) \cdot_\star d_\star x \cdot_\star d_\star y$$

$$+_\star (f_{2z}^\star \rightarrow_\star h_{2x}^\star) \cdot_\star d_\star z \cdot_\star d_\star x$$

$$+_\star (h_{2y}^\star \rightarrow_\star g_{2z}^\star) \cdot_\star d_\star y \cdot_\star d_\star z,$$

and

$$\phi_1 \star d_\star \phi_2 = (f_1 \cdot_\star d_\star x +_\star g_1 \cdot_\star d_\star y +_\star h_1 \cdot_\star d_\star z)$$

$$\cdot_\star ((g_{2x}^\star \rightarrow_\star f_{2y}^\star) \cdot_\star d_\star x \cdot_\star d_\star y$$

$$+_\star (f_{2z}^\star \rightarrow_\star h_{2x}^\star) \cdot_\star d_\star z \cdot_\star d_\star x$$

$$+_\star (h_{2y}^\star \rightarrow_\star g_{2z}^\star) \cdot_\star d_\star y \cdot_\star d_\star z)$$

$$= (f_1 \cdot_\star (h_{2y}^\star \rightarrow_\star g_{2z}^\star) +_\star g_1 \cdot_\star (f_{2z}^\star \rightarrow_\star h_{2x}^\star)$$

$$+_\star h_1 \cdot_\star (g_{2x}^\star \rightarrow_\star f_{2y}^\star)) \cdot_\star d_\star x \cdot_\star d_\star y \cdot_\star d_\star z.$$

Therefore

$$d_\star \phi_1 {\cdot}_\star \phi_2 \to_\star \phi_1 {\cdot}_\star d_\star \phi_2$$

$$
\begin{aligned}
&= (f_{1z}^\star {\cdot}_\star g_2 +_\star f_1 {\cdot}_\star g_{2z}^\star \to_\star g_{1z}^\star {\cdot}_\star f_2 \to_\star g_1 {\cdot}_\star f_{2z}^\star \\
&\quad +_\star h_{1y}^\star {\cdot}_\star f_2 +_\star h_1 {\cdot}_\star f_{2y}^v \to_\star f_{1y}^\star {\cdot}_\star h_2 \to_\star f_1 {\cdot}_\star h_{2y}^\star \\
&\quad +_\star g_{1x}^\star {\cdot}_\star h_2 +_\star g_1 {\cdot}_\star h_{2x}^\star \to_\star h_{1x}^\star {\cdot}_\star g_2 \to_\star h_1 {\cdot}_\star g_{2x}^\star ) {\cdot}_\star d_\star x \\
&\quad {\cdot}_\star d_\star y {\cdot}_\star d_\star z \\
&= d_\star (\phi_1 {\cdot}_\star \phi_2).
\end{aligned}
$$

$$
\begin{aligned}
d_\star \phi_1 &= (g_{1x}^\star \to_\star f_{1y}^\star) {\cdot}_\star d_\star x {\cdot}_\star d_\star y \\
&\quad +_\star (f_{1z}^\star \to_\star h_{1x}^\star) {\cdot}_\star d_\star z {\cdot}_\star d_\star x \\
&\quad +_\star (h_{1y}^\star \to_\star g_{1z}^\star) {\cdot}_\star d_\star y {\cdot}_\star d_\star z \\
&= (f_2 {\cdot}_\star (h_{1y}^\star \to_\star g_{1z}^\star) \to_\star g_2 {\cdot}_\star (f_{1z}^\star \to_\star h_{1x}^\star) \\
&\quad +_\star h_2 {\cdot}_\star (g_{1x}^\star \to_\star f_{1y}^\star) \to_\star f_1^\star {\cdot}_\star (h_{2y}^\star \to_\star g_{2z}^\star) \\
&\quad \to_\star g_1 {\cdot}_\star (f_{2z}^\star \to_\star h_{2x}^\star) \to_\star h_1 {\cdot}_\star (g_{2x}^\star \to_\star f_{2y}^\star)) {\cdot}_\star d_\star x {\cdot}_\star d_\star y {\cdot}_\star d_\star z \\
&= d_\star (\phi_1 {\cdot}_\star \phi_2).
\end{aligned}
$$

This completes the proof.

**Theorem 7.9:** *Let $\phi$ be a 0-multiplicative differential form. Then*

$$d_\star (d_\star \phi) = 0_\star.$$

*Proof.* We have

$$d_\star \phi = \phi_x^\star {\cdot}_\star d_\star x +_\star \phi_y^\star {\cdot}_\star d_\star y +_\star \phi_z^\star {\cdot}_\star d_\star z$$

and

$$
\begin{aligned}
d_\star (d_\star \phi) &= d_\star (\phi_x^\star {\cdot}_\star d_\star x +_\star \phi_y^\star {\cdot}_\star d_\star y +_\star \phi_z^\star {\cdot}_\star d_\star z) \\
&= (\phi_{yx}^\star \to_\star \phi_{xy}^\star) {\cdot}_\star d_\star x {\cdot}_\star d_\star y \\
&\quad +_\star (\phi_{xz}^\star \to_\star \phi_{zx}^\star) {\cdot}_\star d_\star z {\cdot}_\star d_\star x \\
&\quad +_\star (\phi_{zy}^\star \to_\star \phi_{yx}^\star) {\cdot}_\star d_\star y {\cdot}_\star d_\star z \\
&= 0_\star.
\end{aligned}
$$

This completes the proof.

**Theorem 7.10:** *Let $\phi$ be a 1-multiplicative differential form,*

$$\phi = f_\star \, d_\star x +_\star \, g_\star \, d_\star y +_\star \, h_\star \, d_\star z.$$

*Then*

$$d_\star (d_\star \phi) = 0_\star.$$

*Proof.* We have

$$
\begin{aligned}
d_\star \phi = &(g_x^\star -_\star f_y^\star) \cdot_\star d_\star x \cdot_\star d_\star y \\
&+_\star (f_z^\star -_\star h_x^\star) \cdot_\star d_\star z \cdot_\star d_\star x \\
&+_\star (h_y^\star -_\star g_z^\star) \cdot_\star d_\star y \cdot_\star d_\star z.
\end{aligned}
$$

Then

$$
\begin{aligned}
d_\star (d_\star \phi) \; = \; & d_\star ((g_x^\star -_\star f_y^\star) \cdot_\star d_\star x \cdot_\star d_\star y \\
& +_\star (f_z^\star -_\star h_x^\star) \cdot_\star d_\star z \cdot_\star d_\star x \\
& +_\star (h_y^\star -_\star g_z^\star) \cdot_\star d_\star y \cdot_\star d_\star z) \\
= \; & ((g_x^\star -_\star f_y^\star)_z^\star \\
& +_\star (f_z^\star -_\star h_x^\star)_y^\star \\
& +_\star (h_y^\star -_\star g_z^\star)_x^\star) \cdot_\star d_\star x \cdot_\star d_\star y \cdot_\star d_\star z \\
= \; & (g_{xz}^\star -_\star f_{yz}^\star +_\star f_{zx}^\star -_\star h_{xy}^\star \\
& +_\star h_{yx}^\star -_\star g_{zx}^\star) \cdot_\star d_\star x \cdot_\star d_\star y \cdot_\star d_\star z \\
= \; & 0_\star.
\end{aligned}
$$

This completes the proof.

---

## 7.4 Multiplicative Closed Differential Forms. Multiplicative Exact Differential Forms

**Definition 7.10:** A multiplicative differential form $\phi$ is said to be multiplicative closed differential form if $d_\star \phi = 0_\star$.

**Example 7.14:** Let

$$\phi(x, y, z) = e^{5(\log x)^4 (\log y)^2 (\log z)^3} \cdot_\star d_\star x$$
$$+_\star \; e^{2(\log x)^5 \log y (\log z)^3} \cdot_\star d_\star y$$
$$+_\star \; e^{3(\log x)^5 (\log y)^2 (\log z)^2} \cdot_\star d_\star z, \quad (x, y, z) \in \mathbb{R}^3_\star.$$

Here

$$f(x, y, z) = e^{5(\log x)^4 (\log y)^2 (\log z)^3},$$
$$g(x, y, z) = e^{2(\log x)^5 (\log y)(\log z)^3},$$
$$h(x, y, z) = e^{3(\log x)^5 (\log y)^2 (\log z)^2}, \quad (x, y, z) \in \mathbb{R}^3_\star.$$

Then

$$f^\star_y(x, y, z) = e^{10(\log x)^4 \log y (\log z)^3},$$
$$f^\star_z(x, y, z) = e^{15(\log x)^4 (\log y)^2 (\log z)^2},$$
$$g^\star_x(x, y, z) = e^{10(\log x)^4 (\log y)(\log z)^3},$$
$$g^\star_z(x, y, z) = e^{6(\log x)^5 \log y (\log z)^2},$$
$$h^\star_x(x, y, z) = e^{15(\log x)^4 (\log y)^2 (\log z)^2},$$
$$h^\star_y(x, y, z) = e^{6(\log x)^5 \log y (\log z)^2}, \quad (x, y, z) \in \mathbb{R}^3_\star.$$

Hence,

$$d_\star \phi(x, y, z) = (g^\star_x(x, y, z) -_\star f^\star_y(x, y, z)) \cdot_\star d_\star x \cdot_\star d_\star y$$
$$+_\star (f^\star_z(x, y, z) -_\star h^\star_x(x, y, z)) \cdot_\star d_\star z \cdot_\star d_\star x$$
$$+_\star (h^\star_y(x, y, z) -_\star g^\star_z(x, y, z)) \cdot_\star d_\star y \cdot_\star d_\star z$$
$$= \left( e^{10(\log x)^4 \log y (\log z)^3} -_\star e^{10(\log x)^4 \log y (\log z)^3} \right) \cdot_\star d_\star x \cdot_\star d_\star y$$
$$+_\star \left( e^{15(\log x)^4 (\log y)^2 (\log z)^2} -_\star e^{15(\log x)^4 (\log y)^2 (\log z)^2} \right) \cdot_\star d_\star z \cdot_\star d_\star x$$
$$+_\star \left( e^{6(\log x)^5 \log y (\log z)^2} -_\star e^{6(\log x)^5 \log y (\log z)^2} \right) \cdot_\star d_\star y \cdot_\star d_\star x$$
$$= 0_\star.$$

Thus, $\phi$ is a multiplicative closed differential form on $\mathbb{R}^3_\star$.

**Example 7.15:** Let

$$\phi(x, y, z) = (x/_\star z -_\star e^{2\cdot}_\star z)\cdot_\star d_\star y\cdot_\star d_\star z$$
$$+_\star \left(x^{2\cdot}\cdot_\star e^z -_\star y/_\star z\right)\cdot_\star d_\star z\cdot_\star d_\star x, \quad (x, y, z) \in \mathbb{R}^3_\star.$$

We have

$$g(x, y, z) = x/_\star z -_\star e^{2\cdot}_\star z$$
$$= e^{\frac{\log x}{\log z} - 2\log z},$$
$$h(x, y, z) = x^{2\cdot}\cdot_\star e^z -_\star y/_\star z$$
$$= e^{z(\log x)^2 - \frac{\log y}{\log z}}, \quad (x, y, z) \in \mathbb{R}^3_\star,$$

and

$$g_x(x, y, z) = \frac{1}{x\log z} e^{\frac{\log x}{\log z} - 2\log z},$$
$$h_y(x, y, z) = -\frac{1}{y\log z} e^{z(\log x)^2 - \frac{\log y}{\log z}}, \quad (x, y, z) \in \mathbb{R}^3_\star,$$

and

$$g^\star_x(x, y, z) = e^{x\frac{g_x(x,y,z)}{g(x,y,z)}}$$
$$= e^{\frac{1}{\log z}},$$
$$h^\star_y(x, y, z) = e^{y\frac{h_y(x,y,z)}{h(x,y,z)}}$$
$$= e^{-\frac{1}{\log z}}, \quad (x, y, z) \in \mathbb{R}^3_\star.$$

Therefore

$$d_\star\phi(x, y, z) = (g^\star_x(x, y, z) +_\star h^\star_y(x, y, z))\cdot_\star d_\star x\cdot_\star d_\star y\cdot_\star d_\star z$$
$$= (e^{\frac{1}{\log z}} +_\star e^{-\frac{1}{\log z}})\cdot_\star d_\star x\cdot_\star d_\star y\cdot_\star d_\star z$$
$$= 0_\star.$$

Thus, $\phi$ is a multiplicative closed differential form.

**Exercise 7.7:** Prove that

$$\phi(x, y, z) = (e^2 \cdot_\star x \cdot_\star y \cdot_\star z^{3\star} +_\star y^{2\star} +_\star e^4 \cdot_\star z +_\star e^2) \cdot_\star d_\star x$$
$$+_\star (x^{2\star} \cdot_\star z^{3\star} +_\star e^2 \cdot_\star x \cdot_\star y +_\star e^2 \cdot_\star z^{3\star} -_\star e) \cdot_\star d_\star y$$
$$+_\star (e^3 \cdot_\star x^{2\star} \cdot_\star y \cdot_\star z^{2\star} +_\star e^6 \cdot_\star y \cdot_\star z^{2\star} +_\star e^4 \cdot_\star x -_\star e^4 \cdot_\star z) \cdot_\star d_\star z$$

is a multiplicative closed differential form on $\mathbb{R}^3_\star$.

**Definition 7.11:** A multiplicative differential form $\phi$ is said to be a multiplicative exact differential form if there is a multiplicative closed differential form $\psi$ so that

$$\phi = -d_\star \psi.$$

**Example 7.16:** Let $\phi$ be as in Example 7.14. Let also,

$$\psi(x, y, z) = e^{(\log x)^5 (\log y)^2 (\log z)^3}, \quad (x, y, z) \in \mathbb{R}^3_\star.$$

Then

$$\psi_x^\star(x, y, z) = e^{5(\log x)^4 (\log y)^2 (\log z)^3},$$
$$\psi_y^\star(x, y, z) = e^{2(\log x)^5 (\log y)(\log z)^3},$$
$$\psi_z^\star(x, y, z) = e^{3(\log x)^5 (\log y)^2 (\log z)^2}, \quad (x, y, z) \in \mathbb{R}^3_\star.$$

Thus,

$$\phi(x, y, z) = d_\star \psi(x, y, z), \quad (x, y, z) \in \mathbb{R}^3_\star.$$

Consequently $\phi$ is a multiplicative exact differential form.

**Example 7.17:** Let $\phi$ be as in Example 7.5. Let also,

$$\psi(x, y, z) = x^{2\star} \cdot_\star e^z \cdot_\star d_\star x +_\star z^{2\star} \cdot_\star d_\star y +_\star ((x \cdot_\star y)/_\star z) \cdot_\star d_\star z,$$

$(x, y, z) \in \mathbb{R}^3_\star$. Set

$$f_1(x, y, z) = x^2 \cdot_\star e^z$$
$$= e^{(\log x)^2 z},$$
$$g_1(x, y, z) = z^{2\star}$$
$$= e^{(\log z)^2},$$
$$h_1(x, y, z) = (x \cdot_\star y)/_\star z$$
$$= e^{\frac{\log x \log y}{\log z}}, \quad (x, y, z) \in \mathbb{R}^3_\star.$$

We have

$$g^\star_{1x}(x, y, z) = 0_\star,$$
$$g^\star_{1z}(x, y, z) = e^{2\log z},$$
$$f^\star_{1y}(x, y, z) = 0_\star,$$
$$f^\star_{1z}(x, y, z) = e^{(\log x)^2 z},$$
$$h^\star_{1x}(x, y, z) = e^{\frac{\log y}{\log z}},$$
$$h^\star_{1y}(x, y, z) = e^{\frac{\log x}{\log z}}, \quad (x, y, z) \in \mathbb{R}^3_\star.$$

Therefore

$$\phi = d_\star \psi.$$

Thus, $\phi$ is a multiplicative exact differential form.

**Theorem 7.11:** *Any multiplicative exact differential form is a multiplicative closed differential form.*

*Proof.* Let $\phi$ be a multiplicative exact differential form. Then there is a multiplicative differential form $\psi$ so that

$$\phi = d_\star \psi.$$

Hence,

$$d_\star \phi = d_\star(d_\star \psi)$$
$$= 0_\star.$$

Thus, $\phi$ is a multiplicative closed differential form. This completes the proof.

**Theorem 7.12:** *Let $\phi$ be any multiplicative differential form and $\psi$ be a multiplicative closed differential form. Then*

$$d_\star(\phi +_\star \psi) = d_\star\phi.$$

*Proof.* Since $\psi$ is a multiplicative closed differential form, we have

$$d_\star\psi = 0_\star.$$

Hence,

$$d_\star(\phi +_\star \psi) = d_\star\phi +_\star d_\star\psi$$
$$= d_\star\phi.$$

This completes the proof.

**Theorem 7.13:** *Let $\phi_1$ and $\phi_2$ be two multiplicative differential forms so that*

$$d_\star\phi_1 = d_\star\phi_2.$$

*Then*

$$\phi_2 = \phi_1 +_\star \psi,$$
*where $\psi$ is a multiplicative closed differential form.*

*Proof.* Let

$$\psi = \phi_2 -_\star \phi_1.$$

Then

$$d_\star\psi = d_\star(\phi_2 -_\star \phi_1)$$
$$= d_\star\phi_2 -_\star d_\star\phi_1$$
$$= 0_\star.$$

Thus, $\psi$ is a multiplicative closed differential form. This completes the proof.

## 7.5 Multiplicative Gradient, Multiplicative Curl and Multiplicative Divergence

Suppose that

$$e_1 = (1_\star, 0_\star, 0_\star),$$
$$e_2 = (0_\star, 1_\star, 0_\star),$$
$$e_3 = (0_\star, 0_\star, 1_\star).$$

Let also, $R$ be a region in $\mathbb{R}^3_\star$.

**Definition 7.12:** A multiplicative vector field on the region $R$ is a multiplicative vector-valued function

$$F(x, y, z) = f(x, y, z) \cdot_\star e_1 +_\star g(x, y, z) \cdot_\star e_2 +_\star h(x, y, z) \cdot_\star e_3,$$

where $f, g, h \colon \mathbb{R}^3_\star \to \mathbb{R}_\star$ are given functions.

**Example 7.18:** Let

$$f(x, y, z) = x^{2_\star},$$
$$g(x, y, z) = y,$$
$$h(x, y, z) = e^z, \quad (x, y, z) \in \mathbb{R}^3_\star.$$

Then

$$F(x, y, z) = x^{2_\star} \cdot_\star e_1 +_\star y \cdot_\star e_2 +_\star e^z \cdot_\star e_3, \quad (x, y, z) \in \mathbb{R}^3_\star,$$

is a multiplicative vector field.

**Definition 7.13:** Let $f \in \mathscr{C}^1_\star(R)$. Then the multiplicative gradient of $f$ is defined by

$$\mathrm{grad}_\star f(x, y, z) = f^\star_x(x, y, z) \cdot_\star e_1 +_\star f^\star_y(x, y, z) \cdot_\star e_2 +_\star f^\star_z(x, y, z)$$
$$\cdot_\star e_3, \quad (x, y, z) \in \mathbb{R}^3_\star.$$

**Example 7.19:** Let

$$f(x, y, z) = e^{x^2 + y^2 + z^2}, \quad (x, y, z) \in \mathbb{R}^3_\star.$$

Then

$$f_x^\star (x, y, z) = e^{2x^2},$$
$$f_y^\star (x, y, z) = e^{2y^2},$$
$$f_z^\star (x, y, z) = e^{2z^2}, \quad (x, y, z) \in \mathbb{R}_\star^3.$$

Then

$$\mathrm{grad}_\star f (x, y, z) = e^{2x^2} \cdot_\star e_1 +_\star e^{2y^2} \cdot_\star e_2 +_\star e^{2z^2} \cdot_\star e_3, \quad (x, y, z) \in \mathbb{R}_\star^3.$$

**Example 7.20:** Let

$$f (x, y, z) = \left(x^{2\star} +_\star y^{2\star}\right)/_\star z, \quad (x, y, z) \in \mathbb{R}_\star^3.$$

We have

$$f (x, y, z) = e^{\frac{(\log x)^2 + (\log y)^2}{\log z}}, \quad (x, y, z) \in \mathbb{R}_\star^3.$$

Then

$$f_x^\star (x, y, z) = e^{2\frac{\log x}{\log z}},$$
$$f_y^\star (x, y, z) = e^{2\frac{\log y}{\log z}},$$
$$f_z^\star (x, y, z) = e^{-\frac{(\log x)^2 + (\log y)^2}{(\log z)^2}}, \quad (x, y, z) \in \mathbb{R}_\star^3.$$

Therefore

$$\mathrm{grad}_\star f (x, y, z) = e^{2\frac{\log x}{\log z}} \cdot_\star e_1 +_\star e^{2\frac{\log y}{\log z}} \cdot_\star e_2$$
$$+_\star e^{-\frac{(\log x)^2 + (\log y)^2}{(\log z)^2}} \cdot_\star e_3, \quad (x, y, z) \in \mathbb{R}_\star^3.$$

**Exercise 7.8:** Let

$$f (x, y, z) = \left(x^{3\star} -_\star e^2 \cdot_\star y^{2\star} +_\star x \cdot_\star y\right)\Big|_\star z^{2\star}, \quad (x, y, z) \in \mathbb{R}_\star^3.$$

Find

$$\text{grad}_\star f(x, y, z), \quad (x, y, z) \in \mathbb{R}^3_\star.$$

**Definition 7.14:** Let

$$F(x, y, z) = f(x, y, z) \cdot_\star e_1 +_\star g(x, y, z) \cdot_\star e_2 +_\star h(x, y, z) \cdot_\star e_3, \qquad (7.1)$$

$(x, y, z) \in R$, where $f, g, h \in \mathscr{C}^1_\star(R)$. We define the multiplicative curl of $F$ as follows:

$$\text{curl}_\star F = (h^\star_y -_\star g^\star_z) \cdot_\star e_1 +_\star (f^\star_z -_\star h^\star_x) \cdot_\star e_2$$
$$+_\star (g^\star_x -_\star f^\star_y) \cdot_\star e_3.$$

**Example 7.21:** Let

$$F(x, y, z) = (x^{2\star} +_\star y^{2\star} \cdot_\star z) \cdot_\star e_1 +_\star (y^{2\star} -_\star e^{3\star} \cdot_\star x \cdot_\star z) \cdot_\star e_2$$
$$+_\star (x^{4\star} +_\star y^{3\star} -_\star z^{2\star}) \cdot_\star e_3,$$

$(x, y, z) \in \mathbb{R}^3_\star$. Here

$$\begin{aligned}
f(x, y, z) &= x^{2\star} +_\star e^{3\star} \cdot_\star z \\
&= e^{(\log x)^2 + (\log y)^3 \log z}, \\
g(x, y, z) &= y^{2\star} -_\star e^{3\star} \cdot_\star x \cdot_\star z \\
&= e^{(\log y)^2 - 3 \log x \log z}, \\
h(x, y, z) &= x^{4\star} +_\star y^{3\star} -_\star z^{2\star} \\
&= e^{(\log x)^4 + (\log y)^3 - (\log z)^2}, \quad (x, y, z) \in \mathbb{R}^3_\star.
\end{aligned}$$

Then

$$\begin{aligned}
f^\star_x(x, y, z) &= e^{2 \log x}, \\
f^\star_y(x, y, z) &= e^{3(\log y)^2 \log z}, \\
f^\star_z(x, y, z) &= e^{(\log y)^3}, \\
g^\star_x(x, y, z) &= e^{-3 \log z}, \\
g^\star_y(x, y, z) &= e^{2 \log y}, \\
g^\star_z(x, y, z) &= e^{-3 \log x}, \\
h^\star_x(x, y, z) &= e^{4(\log x)^3}, \\
h^\star_y(x, y, z) &= e^{3(\log y)^2}, \\
h^\star_z(x, y, z) &= e^{-2 \log z}, \quad (x, y, z) \in \mathbb{R}^3_\star.
\end{aligned}$$

Consequently

$$\text{curl}_\star F(x, y, z) = \left(e^{3(\log y)^2} -_\star e^{-3\log x}\right)\cdot_\star e_1$$
$$+_\star \left(e^{(\log y)^3} -_\star e^{4(\log x)^3}\right)\cdot_\star e_2$$
$$+_\star \left(e^{-3\log z} -_\star e^{3(\log y)^2 \log z}\right)\cdot_\star e_3$$
$$= e^{-\frac{(\log y)^2}{\log x}}\cdot_\star e_1 +_\star e^{\frac{(\log y)^3}{4(\log x)^3}}\cdot_\star e_2$$
$$+_\star e^{-\frac{1}{(\log y)^2}}\cdot_\star e_3, \quad (x, y, z) \in \mathbb{R}^3_\star.$$

**Definition 7.15:** Let $F$ be given by (7.1). Then its multiplicative divergence is defined by

$$\div_\star F(x, y, z) = f^\star_x(x, y, z) +_\star g^\star_y(x, y, z) +_\star h^\star_z(x, y, z),$$

$$(x, y, z) \in \mathbb{R}^3_\star.$$

**Example 7.22:** Let $F$ be defined as in Example7.21. Then

$$\div_\star F(x, y, z) = e^{2\log x} +_\star e^{2\log y} +_\star e^{-2\log z}$$
$$= e^{-8\log x \log y \log z}, \quad (x, y, z) \in \mathbb{R}^3_\star.$$

**Exercise 7.9:** Let

$$F(x, y, z) = \left(x^{2_\star} +_\star y^{2_\star}\cdot_\star z\right)\cdot_\star e_1 +_\star x\cdot_\star y\cdot_\star z\cdot_\star e_2 +_\star \left(x^{2_\star} -_\star y^{3_\star}\right)\cdot_\star e_3,$$

$(x, y, z) \in \mathbb{R}^3_\star.$ Find

1. $\text{curl}_\star F(x, y, z)$, $(x, y, z) \in \mathbb{R}^3_\star.$
2. $\div_\star F(x, y, z)$, $(x, y, z) \in \mathbb{R}^3_\star.$

## 7.6 Multiplicative Differential Forms in $\mathbb{R}^n_\star$

**Definition 7.16:** A 0-multiplicative differential form $\phi$ is a function $f(x_1, \ldots, x_n)$, $(x_1, \ldots, x_n) \in \mathbb{R}^n_\star.$

**Definition 7.17:** A $k$-multiplicative differential form is a multiplicative sum of terms of the form

$$f(x_1, \ldots, x_n) \cdot_\star d_\star x_{j_1} \cdot_\star \cdots \cdot_\star d_\star x_{j_k}.$$

Multiplicative addition of multiplicative differential forms are defined in the usual way. The multiplicative multiplication of multiplicative differential forms ia subject to the following rules

$$d_\star x_j \cdot_\star d_\star x_k = -_\star d_\star x_k \cdot_\star d_\star x_j, \quad j, k \in \{1, \ldots, n\},$$
$$d_\star x_j \cdot_\star d_\star x_j = 0_\star, \quad j \in \{1, \ldots, n\}.$$

**Definition 7.18:** For a 0-multiplicative differential form $\phi$, we define

$$d_\star \phi = \phi^\star_{x_1} \cdot_\star d_\star x_1 +_\star \cdots +_\star \phi^\star_{x_n} \cdot_\star d_\star x_n.$$

**Definition 7.19:** Let $\phi$ be a $k$-multiplicative differential form. Its multiplicative exterior derivative $d_\star \phi$ is the $(k + 1)$-multiplicative differential form obtained from $\phi$ by applying $d_\star$ to each function involved in $\phi$.

Let

$$e_1 = (1_\star, 0_\star, \ldots, 0_\star),$$
$$e_2 = (0_\star, 1_\star, \ldots, 0_\star),$$
$$\vdots$$
$$e_n = (0_\star, 0_\star, \ldots, 1_\star).$$

**Definition 7.20:** A multiplicative vector field on $\mathbb{R}^n_\star$ is a multiplicative vector-valued function of the form

$$f_1 \cdot_\star e_1 +_\star f_2 \cdot_\star e_2 +_\star \cdots +_\star f_n \cdot_\star e)n,$$

where $f_j \colon \mathbb{R}^n_\star \to \mathbb{R}_\star, j \in \{1, \ldots, n\}$ are given functions.

## 7.7 Advanced Practical Problems

Find $\phi +_\star \psi$ and $\phi -_\star \psi$, where

1.

$$\phi = (x +_\star y)\cdot_\star d_\star x -_\star (e^2\cdot_\star x -_\star y^2\star)\cdot_\star d_\star y +_\star (x^4\star -_\star y^3\star)\cdot_\star d_\star z,$$
$$\psi = \sin_\star x\cdot_\star d_\star x -_\star \cos_\star (x +_\star y +_\star z)\cdot_\star d_\star y -_\star (x -_\star e^4\cdot_\star x)\cdot_\star d_\star z.$$

2.

$$\phi = (x +_\star y)\cdot_\star d_\star x\cdot_\star d_\star y -_\star d_\star y\cdot_\star d_\star z +_\star (x -_\star y)d_\star z\cdot_\star d_\star x,$$
$$\psi = d_\star x\cdot_\star d_\star y +_\star d_\star y\cdot_\star d_\star z +_\star d_\star z\cdot_\star d_\star x.$$

3.

$$\phi = (\sin_\star x -_\star \cos_\star y +_\star z)\cdot_\star d_\star x\cdot_\star d_\star y\cdot_\star d_\star z,$$
$$\psi = \log_\star (e +_\star x^2\star +_\star y^2\star)\cdot_\star d_\star x\cdot_\star d_\star y\cdot_\star d_\star z.$$

**Problem 7.2:** Find

1. $e^3\cdot_\star \phi.$
2. $(x +_\star y)\cdot_\star \phi.$
3. $(x^2\star -_\star y^3\star)\cdot_\star \phi.$
4. $x\cdot_\star \phi +_\star e^2\cdot_\star \phi,$

where

1.

$$\phi = \left(e^x +_\star \sin_\star x\cdot_\star \cos (e^2\cdot_\star x) -_\star y^2\star -_\star z\right)\cdot_\star d_\star x$$
$$+_\star x\cdot_\star y^3\star z\cdot_\star d_\star y -_\star \cos_\star x\cdot_\star d_\star z.$$

2.

$$\phi = ((e +_\star x)/_\star (e^2 +_\star x^2\star))\cdot_\star d_\star x\cdot_\star d_\star y +_\star (x^2\star +_\star x\cdot_\star y)\cdot_\star d_\star y\cdot_\star d_\star z$$
$$+_\star x\cdot_\star y\cdot_\star z\cdot_\star d_\star z\cdot_\star d_\star y.$$

3.

$$\phi = (e +_\star x +_\star x^2\star +_\star x^3\star)/_\star (e +_\star x^2\star)\cdot_\star d_\star x\cdot_\star d_\star y\cdot_\star d_\star z.$$

**Problem 7.3:** Let

$$\phi_1 = e^2 \cdot_\star x^{2\star} \cdot_\star d_\star x +_\star (x +_\star y) \cdot_\star d_\star y,$$

$$\phi_2 = -_\star x \cdot_\star d_\star x +_\star (x -_\star e^2 \cdot_\star y) \cdot_\star d_\star y,$$

$$\phi_3 = x^{3\star} \cdot_\star d_\star x +_\star y \cdot_\star z \cdot_\star d_\star y -_\star (x^{2\star} +_\star y^{2\star} +_\star z^{2\star}) \cdot_\star d_\star z,$$

$$\phi_4 = y^{2\star} \cdot_\star z \cdot_\star d_\star x -_\star x \cdot_\star z \cdot_\star d_\star y +_\star (e^2 \cdot_\star x +_\star e) \cdot_\star d_\star z,$$

$$\phi_5 = x \cdot_\star d_\star x +_\star y^{2\star} \cdot_\star d_\star y +_\star z^{2\star} \cdot_\star d_\star z,$$

$$\phi_6 = d_\star x +_\star e^2 \cdot_\star d_\star y +_\star e^3 \cdot_\star d_\star z,$$

$$\psi_1 = (x^{2\star} -_\star y^{2\star}) \cdot_\star d_\star x \cdot_\star d_\star y,$$

$$\psi_2 = (x -_\star y) \cdot_\star d_\star x \cdot_\star d_\star y.$$

Find

1. $\phi_1 \cdot_\star \phi_2$.
2. $\phi_2 \cdot_\star \phi_1$.
3. $\phi_2 \cdot_\star \phi_3$.
4. $\phi_2 \cdot_\star \phi_4$.
5. $\phi_4 \cdot_\star \phi_6$.
6. $\phi_6 \cdot_\star \phi_5$.
7. $y \cdot_\star \psi_1 +_\star x^{3\star} \cdot_\star \psi_2$.
8. $e^2 \cdot_\star y \cdot_\star \psi_1 +_\star \psi_2$.
9. $\phi_4 \cdot_\star \psi_2$.
10. $\psi_2 \cdot_\star \phi_3 +_\star \psi_1 \cdot_\star \phi_4$.

**Problem 7.4:** Let

$$f(x, y, z) = \sin_\star x +_\star \cos_\star y +_\star \tan_\star z, \quad (x, y, z) \in \mathbb{R}^3_\star.$$

Find $d_\star f$.

**Problem 7.5:** Let

$$\phi = e^{y^4 + z^3} \cdot_\star (d_\star x +_\star d_\star y +_\star d_\star z), \quad (x, y, z) \in \mathbb{R}^3_\star.$$

Find $d_\star \phi$.

**Problem 7.6:** Let

$$\phi = e^{x-y+z} \cdot_\star d_\star x \cdot_\star d_\star y +_\star e^{x+y+z} \cdot_\star d_\star y \cdot_\star d_\star z$$
$$+_\star e^{x+y-z} \cdot_\star d_\star z \cdot_\star d_\star x.$$

Find $d_\star \phi$.

**Problem 7.7:** Prove that

$$\phi(x, y) = (e^2 \cdot_\star x \cdot_\star y^{3\star} +_\star e^4 \cdot_\star x^{3\star}) \cdot_\star d_\star x$$
$$+_\star (e^3 \cdot_\star x^{2\star} \cdot_\star y^{2\star} +_\star e^2 \cdot_\star y) \cdot_\star d_\star y, \quad (x, y) \in \mathbb{R}_\star^2,$$

is a multiplicative closed differential form on $\mathbb{R}_\star^2$.

**Problem 7.8:** Let

$$f(x, y, z) = e^x \cdot_\star (x^{2\star} +_\star y^{2\star} +_\star z^{2\star}), \quad (x, y, z) \in \mathbb{R}_\star^3.$$

Find

$$\text{grad}_\star f(x, y, z), \quad (x, y, z) \in \mathbb{R}_\star^3.$$

**Problem 7.9:** Let

$$F(x, y, z) = e^{x^2 - 4xyz + z^3 + y^4}, \quad (x, y, z) \in \mathbb{R}_\star^3.$$

Find

1. $\text{curl}_\star F(x, y, z), (x, y, z) \in \mathbb{R}_\star^3.$
2. $\div_\star F(x, y, z), (x, y, z) \in \mathbb{R}_\star^3.$

# 8

## The Multiplicative Nature Connection

### 8.1 The Multiplicative Directional Derivative

Suppose that $p, v \in \mathbb{R}_\star^n$ and $f : \mathbb{R}_\star^n \to \mathbb{R}_\star$, $f \in \mathscr{C}_\star^1(\mathbb{R}_\star^n)$. Define

$$F(t) = f(p +_\star t \cdot_\star v), \quad t \in \mathbb{R}_\star.$$

**Definition 8.1:** The multiplicative directional derivative of the function $f$ at the point $p$ in the multiplicative direction $v$ is defined to be the number

$$F^\star(0_\star)$$

and we will use the notation

$$(\partial_\star f /_\star \partial_\star v)(p) = F^\star(0_\star).$$

We have

$$F(t) = f(p_1 +_\star t \cdot_\star v_1, p_2 +_\star t \cdot_\star v_2, \ldots, p_n +_\star t \cdot_\star v_n)$$

and

$$
\begin{aligned}
F^\star(t) = \; & f_{x_1}^\star(p_1 +_\star t \cdot_\star v_1, p_2 +_\star t \cdot_\star v_2, \ldots, p_n +_\star t \cdot_\star v_n) \cdot_\star v_1 \\
& +_\star f_{x_2}^\star(p_1 +_\star t \cdot_\star v_1, p_2 +_\star t \cdot_\star v_2, \ldots, p_n +_\star t \cdot_\star v_n) \cdot_\star v_2 \\
& +_\star \cdots \\
& +_\star f_{x_n}^\star(p_1 +_\star t \cdot_\star v_1, p_2 +_\star t \cdot_\star v_2, \ldots, p_n +_\star t \cdot_\star v_n) \cdot_\star v_n.
\end{aligned}
$$

Thus,

$$
\begin{aligned}
\partial_\star f /_\star \partial_\star v(p) = \; & f_{x_1}^\star(p_1, p_2, \ldots, p_n) \cdot_\star v_1 \\
& +_\star f_{x_2}^\star(p_1, p_2, \ldots, p_n) \cdot_\star v_2 \\
& +_\star \cdots \\
& +_\star f_{x_n}^\star(p_1, p_2, \ldots, p_n) \cdot_\star v_n.
\end{aligned}
$$

DOI: 10.1201/9781003299844-8

**Example 8.1:** Let

$$f(x_1, x_2, x_3) = e^{x_1^3} +_\star e^{x_2^2} +_\star e^{x_3^4}, \quad (x_1, x_2, x_3) \in \mathbb{R}_\star^3.$$

Let also,

$$v = (2, 3, 4),$$
$$p = (1, 1, 1).$$

Then

$$f_{x_1}^\star (x_1, x_2, x_3) = e^{3x_1^3},$$
$$f_{x_2}^\star (x_1, x_2, x_3) = e^{2x_2^2},$$
$$f_{x_3}^\star (x_1, x_2, x_3) = e^{4x_3^4}, \quad (x_1, x_2, x_3) \in \mathbb{R}_\star^3,$$

and

$$f_{x_1}^\star (1, 1, 1) = e^3,$$
$$f_{x_2}^\star (1, 1, 1) = e^2,$$
$$f_{x_3}^\star (1, 1, 1) = e^4$$

and

$$\begin{aligned}
\partial_\star f /_\star \, \partial_\star v (1, 1, 1) &= e^{3} \cdot_\star 2 +_\star e^{2} \cdot_\star 2 +_\star e^{4} \cdot_\star 4 \\
&= e^{3 \log 2} +_\star e^{2 \log 2} +_\star e^{4 \log 4} \\
&= e^{3 \log 2} +_\star e^{2 \log 2} +_\star e^{8 \log 2} \\
&= e^{13 \log 2}.
\end{aligned}$$

**Exercise 8.1:** Let

$$f(x_1, x_2, x_3) = x_1^{2\star} +_\star x_3^{4\star} +_\star e^{x_1 + x_2 + x_3}, \quad (x_1, x_2, x_3) \in \mathbb{R}_\star^3,$$
$$p = (2, 5, 6),$$
$$v = (1, 2, 1).$$

Find

$$(\partial_\star f /_\star \partial_\star v)(p).$$

**Theorem 8.1:** *Let $f, g \in \mathscr{C}^1_\star(\mathbb{R}^n_\star)$ and $\alpha, \beta \in \mathbb{R}_\star$. Then*

$$(\partial_\star(\alpha\cdot_\star f +_\star \beta\cdot_\star g)/_\star \partial_\star v)(p)$$
$$= \alpha\cdot_\star(\partial_\star f/_\star \partial_\star v)(p) +_\star \beta\cdot_\star(\partial_\star g/_\star \partial_\star v)(p).$$

*Proof.* We have

$$(\alpha\cdot_\star f +_\star \beta\cdot_\star g)/_\star \partial_\star v(p)$$
$$= \sum_{\star j=1}^{n} (\alpha\cdot_\star f +_\star \beta\cdot_\star g)^\star_{x_j}(p)\cdot_\star v_j$$
$$= \sum_{\star j=1}^{n} (\alpha\cdot_\star f^\star_{x_j} +_\star \beta\cdot_\star g^\star_{x_j}(p))\cdot_\star v_j = \sum_{\star j=1}^{n} (\alpha\cdot_\star f^\star_{x_j}(p) +_\star \beta\cdot_\star g^\star_{x_j}(p))\cdot_\star v_j$$
$$= \sum_{\star j=1}^{n} \alpha\cdot_\star f^\star_{x_j}(p)\cdot_\star v_j +_\star \beta\cdot_\star g^\star_{x_j}(p)\cdot_\star v_j$$
$$= \alpha\cdot_\star \sum_{\star j=1}^{n} f^\star_{x_j}(p)\cdot_\star v_j +_\star \beta\cdot_\star \sum_{\star j=1}^{n} g^\star_{x_j}(p)\cdot_\star v_j$$
$$= \alpha\cdot_\star(\partial_\star f/_\star \partial_\star v)(p) +_\star \beta\cdot_\star(\partial_\star g/_\star \partial_\star v)(p).$$

This completes the proof.

**Theorem 8.2:** *Let $f, g \in \mathscr{C}^1_\star(\mathbb{R}^n_\star)$. Then*

$$(\partial_\star(fg)/_\star \partial_\star v)(p) = f(p)\cdot_\star(\partial_\star g/_\star \partial_\star v)(p)$$
$$+_\star g(p)\cdot_\star(\partial_\star f/_\star \partial_\star v)(p).$$

*Proof.* We have

$$(\partial_\star(f\cdot_\star g)/_\star \partial_\star v)(p) = \sum_{\star j=1}^{n} (f\cdot_\star g)^\star_{x_j}(p)$$
$$= \sum_{\star j=1}^{n} (f^\star_{x_j}(p)\cdot_\star g(p)\cdot_\star v_j +_\star f(p)\cdot_\star g^\star_{x_j}(p)\cdot_\star v_j)$$
$$= g(p)\cdot_\star \sum_{\star j=1}^{n} f^\star_{x_j}(p)\cdot_\star v_j +_\star f(p)\cdot_\star \sum_{\star j=1}^{n} g^\star_{x_j}(p)\cdot_\star v_j$$
$$= g(p)\cdot_\star(\partial_\star f/_\star \partial_\star v)(p) +_\star f(p)\cdot_\star(\partial_\star g/_\star \partial_\star v)(p).$$

This completes the proof.

## 8.2 Multiplicative Tangent Spaces

Let $p \in \mathbb{R}_\star^n$ and $T_{\star p}(\mathbb{R}_\star^n)$ be the multiplicative tangent space that have $p$ as a point of application.

**Definition 8.2:** The set

$$\{\partial_\star /_\star \partial_\star t_1, \ldots, \partial_\star /_\star \partial_\star t_n\}$$

is said to be multiplicative basis of $T_{\star p}(\mathbb{R}_\star^n)$.

Any vector field $V$ can be represented by

$$V = \overset{n}{\underset{\star j=1}{\Sigma}} V_j \cdot_\star \partial_\star /_\star \partial_\star t_j.$$

**Definition 8.3:** Let

$$V_1 = \overset{n}{\underset{\star j=1}{\Sigma}} V_{1j} \cdot_\star \partial_\star /_\star \partial_\star t_j,$$

$$V_2 = \overset{n}{\underset{\star j=1}{\Sigma}} V_{2j} \cdot_\star \partial_\star /_\star \partial_\star t_j,$$

where $V_{kj} : \mathbb{R}_\star^n \to \mathbb{R}_\star$, $k \in \{1, 2\}$, $j \in \{1, \ldots, n\}$. Let also, $f : \mathbb{R}_\star^n \to \mathbb{R}_\star$. We define

$$a \cdot_\star V_1 +_\star b \cdot_\star V_2 = \overset{n}{\underset{\star j=1}{\Sigma}} (a \cdot_\star V_{1j} +_\star b \cdot_\star V_{2j}) \cdot_\star \partial_\star /_\star \partial_\star t_j$$

and

$$f \cdot_\star V_1 = \overset{n}{\underset{\star j=1}{\Sigma}} (f \cdot_\star V_{1j}) \cdot_\star \partial_\star /_\star \partial_\star t_j.$$

**Example 8.2:** Let

$$V_1 = (3 +_\star t_1 +_\star t_2) \cdot_\star \partial_\star /_\star \partial_\star t_1 +_\star t_1^{2\star} \cdot_\star \partial_\star /_\star \partial_\star t_2,$$
$$V_2 = (t_1 -_\star t_2) \cdot_\star \partial_\star /_\star \partial_\star t_1 +_\star \partial_\star /_\star \partial_\star t_2.$$

Then

$$V_1 +_\star V_2 = (3 +_\star e^2 \cdot_\star t_1) \cdot_\star \partial_\star/_\star \partial_\star t_1 +_\star (t_1^{2\star} +_\star 1_\star) \cdot_\star \partial_\star/_\star \partial_\star t_2$$

and

$$t_1 \cdot_\star V_1 \;=\; (3 \cdot_\star t_1 +_\star t_1^{2\star} +_\star t_1 \cdot_\star t_2) \cdot_\star \partial_\star/_\star \partial_\star t_1$$
$$+_\star t_1^{3\star} \cdot_\star \partial_\star/_\star \partial_\star t_2.$$

**Exercise 8.2:** Let

$$V_1 = t_1 \cdot_\star \partial_\star/_\star \partial_\star t_1 +_\star t_1 \cdot_\star \partial_\star/_\star \partial_\star t_2,$$

$$V_2 = (t_1 -_\star t_2) \cdot_\star \partial_\star/_\star \partial_\star t_1 +_\star t_1 \cdot_\star \partial_\star/_\star \partial_\star t_2.$$

Find

1. $V_1 +_\star V_2$.
2. $V_1 -_\star V_2$.
3. $(t_1 +_\star t_2) \cdot_\star V_1$.
4. $t_2 \cdot_\star V_2$.

## 8.3 The Multiplicative Covariant Derivative

Let $p \in \mathbb{R}_\star^n$ and $v \in T_{\star p}(\mathbb{R}_\star^n)$, and

$$W = \sum_{\star j=1}^{n} w_j \cdot_\star \partial_\star/_\star \partial_\star t_j.$$

**Definition 8.4:** The multiplicative covariant derivative of $W$ with respect to $v$ we define as follows:

$$\Delta_\star W = \sum_{\star j=1}^{n} (\partial_\star w_j/_\star \partial_\star v)(p) \cdot_\star (\partial_\star/_\star \partial_\star t_j)(p).$$

**Theorem 8.3:** *Let*

$$W_1 = \sum_{\star j=1}^{n} w_{1j} \cdot_\star \partial_\star/_\star \partial_\star t_j,$$

$$W_2 = \sum_{\star} w_{2j} \cdot_\star \partial_\star/_\star \partial_\star t_j,$$

*and* $a, b \in \mathbb{R}_\star$. *Then*

$$\Delta_{\star v}(a \cdot_\star W_1 +_\star b \cdot_\star W_2) = a \cdot_\star \Delta_{\star v} W_1 +_\star b \cdot_\star \Delta_{\star v} W_2.$$

*Proof.* We have

$$\Delta_{\star v} W_1 = \sum_{\star j=1}^{n} (\partial_\star w_{1j}/_\star \partial_\star v) \cdot_\star (\partial_\star /_\star \partial_\star t_j),$$

$$\Delta_{\star v} W_2 = \sum_{\star j=1}^{n} (\partial_\star w_{2j}/_\star \partial_\star v) \cdot_\star (\partial_\star /_\star \partial_\star t_j)$$

and

$$a \cdot_\star W_1 +_\star b \cdot_\star W_2 = \sum_{\star j=1}^{n} (a \cdot_\star w_{1j} +_\star b \cdot_\star w_{2j}) \cdot_\star (\partial_\star /_\star \partial_\star t_j).$$

Then

$$
\begin{aligned}
\Delta_{\star v}(a \cdot_\star W_1 +_\star b \cdot_\star W_2) &= \sum_{\star j=1}^{n} \partial_\star(a \cdot_\star w_{1j} +_\star b \cdot_\star w_{2j})/_\star \partial_\star v \cdot_\star (\partial_\star /_\star \partial_\star t_j) \\
&= \sum_{\star j=1}^{n} (a \cdot_\star \partial_\star w_{1j}/_\star \partial_\star v +_\star b \cdot_\star \partial_\star w_{2j}/_\star \partial_\star v) \cdot_\star (\partial_\star /_\star \partial_\star t_j) \\
&= \sum_{\star j=1}^{n} (a \cdot_\star (\partial_\star w_{1j}/_\star \partial_\star v) \cdot_\star (\partial_\star /_\star \partial_\star t_j) \\
&\quad +_\star b \cdot_\star (\partial_\star w_{2j}/_\star \partial_\star v) \cdot_\star (\partial_\star /_\star \partial_\star t_j)) \\
&= \sum_{\star j=1}^{n} a \cdot_\star (\partial_\star w_{1j}/_\star \partial_\star v) \cdot_\star (\partial_\star /_\star \partial_\star t_j) \\
&\quad +_\star \sum_{\star j=1}^{n} b \cdot_\star (\partial_\star w_{2j}/_\star \partial_\star v) \cdot_\star (\partial_\star /_\star \partial_\star t_j) \\
&= a \cdot_\star \sum_{\star j=1}^{n} (\partial_\star w_{1j}/_\star \partial_\star v) \cdot_\star (\partial_\star /_\star \partial_\star t_j) \\
&\quad +_\star b \cdot_\star \sum_{\star j=1}^{n} (\partial_\star w_{2j}/_\star \partial_\star v) \cdot_\star (\partial_\star /_\star \partial_\star t_j) \\
&= a \cdot_\star \Delta_{\star v} W_1 +_\star b \cdot_\star \Delta_{\star v} W_2.
\end{aligned}
$$

This completes the proof.

**Theorem 8.4:** *Let*

$$W_1 = \sum_{\star j=1}^{n} w_{1j} \cdot_\star \partial_\star /_\star \partial_\star t_j$$

*and* $f : \mathbb{R}_\star^n \to \mathbb{R}_\star$, $f \in \mathscr{C}_\star^1(\mathbb{R}_\star^n)$. *Then*

$$\Delta_{\star v}(f \cdot_\star W_1) = f \cdot_\star \Delta_{\star v} W_1 +_\star W_1 \cdot_\star \Delta_{\star v} f.$$

*Proof.* We have

$$\Delta_{\star v}W_1 = \sum_{\star j=1}^{n} (\partial_{\star} w_{1j}/_{\star}\partial_{\star}t_j)\cdot_{\star}(\partial_{\star}/_{\star}\partial_{\star}t_j)$$

and

$$f\cdot_{\star} W_1 = \sum_{\star j=1}^{n} (f\cdot_{\star}w_{1j})(\partial_{\star}/_{\star}\partial_{\star}t_j).$$

Then

$$\Delta_{\star v}(f\cdot_{\star} W_1) = \sum_{\star j=1}^{n} \partial_{\star}(f\cdot_{\star}w_{1j})/_{\star}\partial_{\star}v\cdot_{\star}(\partial_{\star}/_{\star}\partial_{\star}t_j)$$

$$= \sum_{\star j=1}^{n} (f\cdot_{\star}\partial_{\star}w_{1j}/_{\star}\partial_{\star}v +_{\star} w_{1j}\cdot_{\star}\partial_{\star}f/_{\star}\partial_{\star}v)\cdot_{\star}(\partial_{\star}/_{\star}\partial_{\star}t_j)$$

$$= \sum_{\star j=1}^{n} (f\cdot_{\star}(\partial_{\star}w_{1j}/_{\star}\partial_{\star}v)\cdot_{\star}(\partial_{\star}/_{\star}\partial_{\star}t_j)$$

$$+_{\star} W_1\cdot_{\star}(\partial_{\star}f/_{\star}\partial_{\star}v)\cdot_{\star}(\partial_{\star}/_{\star}\partial_{\star}t_j))$$

$$= \sum_{\star j=1}^{n} f\cdot_{\star}(\partial_{\star}w_{1j}/_{\star}\partial_{\star}v)\cdot_{\star}(\partial_{\star}/_{\star}\partial_{\star}t_j)$$

$$+_{\star} \sum_{\star j=1}^{n} W_1\cdot_{\star}(\partial_{\star}f/_{\star}\partial_{\star}v)\cdot_{\star}(\partial_{\star}/_{\star}\partial_{\star}t_j)$$

$$= f\cdot_{\star} \sum_{\star j=1}^{n} (\partial_{\star}w_{1j}/_{\star}\partial_{\star}v)\cdot_{\star}(\partial_{\star}/_{\star}\partial_{\star}t_j)$$

$$+_{\star} W_1\cdot_{\star} \sum_{\star j=1}^{n} (\partial_{\star}f/_{\star}\partial_{\star}v)\cdot_{\star}(\partial_{\star}/_{\star}\partial_{\star}t_j)$$

$$= f\cdot_{\star} \Delta_{\star v}W_1 +_{\star} W_1\cdot_{\star} \Delta_{\star v}f.$$

This completes the proof.

## 8.4 The Multiplicative Lie Brackets

Suppose that

$$V_1 = \sum_{\star j=1}^{n} V_{1j}\cdot_{\star}\partial_{\star}/_{\star}\partial_{\star}t_j,$$

$$V_2 = \sum_{\star j=1}^{n} V_{2j}\cdot_{\star}\partial_{\star}/_{\star}\partial_{\star}t_j,$$

where $V_{kj} : \mathbb{R}^3_\star \to \mathbb{R}_\star$, $V_{kj} \in \mathscr{C}^1_\star(\mathbb{R}^n_\star)$, $k \in \{1, 2\}$, $j \in \{1, ..., n\}$.

**Definition 8.5:** Define

$$V_1[V_2]_\star(f) = \sum_{\star j=1}^{n} V_{1j}\cdot_\star \partial_\star \left( \sum_{\star k=1}^{n} V_{2k}\cdot_\star \partial_\star f /_\star \partial_\star t_k \right) /_\star \partial_\star t_j.$$

**Example 8.3:** Let

$$V_1(t_1, t_2) = e^{t_1^2 + t_2^2}\cdot_\star \partial_\star /_\star \partial_\star t_1 +_\star e^{t_1 t_2}\cdot_\star \partial_\star /_\star \partial_\star t_2,$$
$$V_2(t_1, t_2) = e^{t_1^2 - t_2^2}\cdot_\star \partial_\star /_\star \partial_\star t_1 +_\star e^{t_1 + t_2}\cdot_\star \partial_\star /_\star \partial_\star t_2,$$
$$f(t_1, t_2) = e^{t_1^3 + t_2^3}, \quad (t_1, t_2) \in \mathbb{R}^2_\star.$$

Here

$$V_{11}(t_1, t_2) = e^{t_1^2 + t_2^2},$$
$$V_{12}(t_1, t_2) = e^{t_1 t_2},$$
$$V_{21}(t_1, t_2) = e^{t_1^2 - t_2^2},$$
$$V_{22}(t_1, t_2) = e^{t_1 + t_2}, \quad (t_1, t_2) \in \mathbb{R}^2_\star.$$

We have

$$V^\star_{11_{t_1}}(t_1, t_2) = e^{2t_1^2},$$
$$V^\star_{11_{t_2}}(t_1, t_2) = e^{2t_2^2},$$
$$V^\star_{12_{t_1}}(t_1, t_2) = e^{t_1 t_2},$$
$$V^\star_{12_{t_2}}(t_1, t_2) = e^{t_1 t_2},$$
$$V^\star_{21_{t_1}}(t_1, t_2) = e^{2t_1^2},$$
$$V^\star_{21_{t_2}}(t_1, t_2) = e^{-2t_2^2},$$
$$V^\star_{22_{t_1}}(t_1, t_2) = e^{t_1},$$
$$V^\star_{22_{t_2}}(t_1, t_2) = e^{t_2},$$
$$f^\star_{t_1}(t_1, t_2) = e^{3t_1^3},$$
$$f^v_{t_2}(t_1, t_2) = e^{3t_2^3}, \quad (t_1, t_2) \in \mathbb{R}^2_\star.$$

Therefore

$$\sum_{\star k=1}^{2} V_{2k}\,(t_1,\,t_2) \cdot_\star \partial_\star f /_\star \partial_\star t_k\,(t_1,\,t_2) = V_{21}\,(t_1,\,t_2) \cdot_\star f_{t_1}^\star\,(t_1,\,t_2)$$

$$+_\star V_{22}\,(t_1,\,t_2) \cdot_\star f_{t_2}^\star\,(t_1,\,t_2)$$

$$= e^{t_1^2 - t_2^2} \cdot_\star e^{3t_1^3} +_\star e^{t_1 + t_2} \cdot_\star e^{3t_2^3}$$

$$= e^{t_1^2 - t_2^2 + 3t_1^3} +_\star e^{t_1 + t_2 + 3t_2^3}$$

$$= e^{t_1 + t_2 + t_1^2 - t_2^2 + 3t_1^3 + 3t_2^3}, \quad (t_1,\,t_2) \in \mathbb{R}_\star^2.$$

Let

$$h\,(t_1,\,t_2) = e^{t_1 + t_2 + t_1^2 - t_2^2 + 3t_1^3 + 3t_2^3}, \quad (t_1,\,t_2) \in \mathbb{R}_\star^2.$$

Then

$$h_{t_1}^\star\,(t_1,\,t_2) = e^{t_1 + 2t_1^2 + 9t_1^3},$$

$$h_{t_2}^\star\,(t_1,\,t_2) = e^{t_2 - 2t_2^2 + 9t_2^3}, \quad (t_1,\,t_2) \in \mathbb{R}_\star^2.$$

Hence,

$$V_1[V_2]_\star\,(f)(t_1,\,t_2) = V_{11}\,(t_1,\,t_2) \cdot_\star h_{t_1}^\star\,(t_1,\,t_2)$$

$$+_\star V_{12}\,(t_1,\,t_2) \cdot_\star h_{t_2}^\star\,(t_1,\,t_2)$$

$$= e^{t_1^2 + t_2^2} \cdot_\star e^{t_1 + 2t_1^2 + 9t_1^3}$$

$$+_\star e^{t_1 t_2} \cdot_\star e^{t_2 - 2t_2^2 + 9t_2^3}$$

$$= e^{t_1 + 3t_1^2 + t_2^2 + 9t_1^3} +_\star e^{t_1 t_2 + t_2 - 2t_2^2 + 9t_2^3}$$

$$= e^{t_1 + t_2 + t_1 t_2 + 3t_1^2 - t_2^2 + 9t_1^3 + 9t_2^3}, \quad (t_1,\,t_2) \in \mathbb{R}_\star^2.$$

Next,

$$V_{11}\,(t_1,\,t_2) \cdot_\star f_{t_1}^\star\,(t_1,\,t_2) +_\star V_{12}\,(t_1,\,t_2) \cdot_\star f_{t_2}^\star\,(t_1,\,t_2) = e^{t_1^2 + t_2^2} \cdot_\star e^{3t_1^3} +_\star e^{t_1 t_2} \cdot_\star e^{3t_2^3}$$

$$= e^{t_1^2 + 3t_1^3 + t_2^2} +_\star e^{t_1 t_2 + 3t_2^3}$$

$$= e^{t_1 t_2 + t_1^2 + 3t_1^3 + t_2^2 + 3t_2^3}, \quad (t_1,\,t_2) \in \mathbb{R}_\star^2.$$

Let

$$g(t_1, t_2) = e^{t_1 t_2 + t_1^2 + 3t_1^3 + t_2^2 + 3t_2^3}, \quad (t_1, t_2) \in \mathbb{R}_\star^2.$$

Then

$$g_{t_1}^\star(t_1, t_2) = e^{t_1 t_2 + 2t_1^2 + 9t_1^3},$$
$$g_{t_2}^\star(t_1, t_2) = e^{t_1 t_2 + 2t_2^2 + 9t_2^3}, \quad (t_1, t_2) \in \mathbb{R}_\star^2.$$

Therefore

$$V_2[V_1](f)(t_1, t_2) = V_{21}(t_1, t_2) \cdot_\star g_{t_1}^\star(t_1, t_2) +_\star V_{22}(t_1, t_2) \cdot_\star g_{t_2}^\star(t_1, t_2)$$
$$= e^{t_1^2 - t_2^2} \cdot_\star e^{t_1 t_2 + 2t_1^2 + 9t_1^3}$$
$$+_\star e^{t_1 + t_2} \cdot_\star e^{t_1 t_2 + 2t_2^2 + 2t_2^3}$$
$$= e^{3t_1^2 + t_1 t_2 - t_2^2 + 9t_1^3} +_\star e^{t_1 + t_2 + t_1 t_2 + 2t_2^2 + 9t_2^3}$$
$$= e^{t_1 + t_2 + 2t_1 t_2 + 3t_1^2 + t_2^2 + 9t_1^3 + 9t_2^3}, \quad (t_1, t_2) \in \mathbb{R}_\star^2.$$

**Exercise 8.3:** Let

$$V_1(t_1, t_2) = e^{t_1^3 - t_2^2} \cdot_\star \partial_\star /_\star \partial_\star t_1 +_\star e^{t_1 t_2^3} \cdot_\star \partial_\star /_\star \partial_\star t_2,$$
$$V_2(t_1, t_2) = e^{t_1^2 t_2^2} \cdot_\star \partial_\star /_\star \partial_\star t_1 +_\star e^{t_1 - 3t_2 + 4t_1^2} \cdot_\star \partial_\star /_\star \partial_\star t_2,$$
$$V_3(t_1, t_2) = e^{t_1^4 + t_2^5} \cdot_\star \partial_\star /_\star \partial_\star t_1 +_\star e^{t_1^6 - t_2^3} \cdot_\star \partial_\star /_\star \partial_\star t_2,$$
$$V_4(t_1, t_2) = e^{t_1^2} \cdot_\star \partial_\star /_\star \partial_\star t_1 +_\star e^{t_2^2} \cdot_\star \partial_\star /_\star \partial_\star t_2,$$
$$f(t_1, t_2) = t_1 t_2^4,$$

$(t_1, t_2) \in \mathbb{R}_\star^2$. Find

1. $V_1[V_2]_\star(f)$.
2. $V_1[V_3]_\star(f)$.
3. $V_1[V_4]_\star(f)$.
4. $V_2[V_3]_\star(f)$.
5. $V_2[V_4]_\star(f)$.
6. $V_3[V_4]_\star(f)$.

**Theorem 8.5:** *Let* $V_j$, $j \in \{1, ..., 4\}$, *be multiplicative vector fields such that*

$$V_j = \sum_{j=1}^{k} V_{jk} \cdot_\star \partial_\star /_\star \partial_\star t_k, \quad j \in \{1, ..., 4\}.$$

*Let also,* $a, b, c, d \in \mathbb{R}_\star$. *Then*

$$(a \cdot_\star V_1 +_\star b \cdot_\star V_2)[c \cdot_\star V_3 +_\star d \cdot_\star V_4]_\star = a \cdot_\star c \cdot_\star (V_1[V_3]_\star) +_\star a \cdot_\star d \cdot_\star (V_1[V_4]_\star)$$
$$+_\star b \cdot_\star c \cdot_\star (V_2[V_3]_\star) +_\star b \cdot_\star d \cdot_\star (V_2[V_4]_\star).$$

*Proof.* We have

$$a \cdot_\star V_1 = a \cdot_\star \left( \sum_{\star k=1}^{n} V_{1k} \cdot_\star \partial_\star /_\star \partial_\star t_k \right)$$
$$= \sum_{\star k=1}^{n} (a \cdot_\star V_{1k}) \cdot_\star \partial_\star /_\star \partial_\star t_k,$$

$$b \cdot_\star V_2 = b \cdot_\star \left( \sum_{\star k=1}^{n} V_{2k} \cdot_\star \partial_\star /_\star \partial_\star t_k \right)$$
$$= \sum_{\star k=1}^{n} (b \cdot_\star V_{2k}) \cdot_\star \partial_\star /_\star \partial_\star t_k,$$

$$c \cdot_\star V_3 = c \cdot_\star \left( \sum_{\star k=1}^{n} V_{3k} \cdot_\star \partial_\star /_\star \partial_\star t_k \right)$$
$$= \sum_{\star k=1}^{n} (c \cdot_\star V_{3k}) \cdot_\star \partial_\star /_\star \partial_\star t_k,$$

$$d \cdot_\star V_4 = d \cdot_\star \left( \sum_{\star k=1}^{n} V_{4k} \cdot_\star \partial_\star /_\star \partial_\star t_k \right)$$
$$= \sum_{\star k=1}^{n} (d \cdot_\star V_{4k}) \cdot_\star \partial_\star /_\star \partial_\star t_k$$

and

$$a \cdot_\star V_1 +_\star b \cdot_\star V_2 = \sum_{\star k=1}^{n} (a \cdot_\star V_{1k}) \cdot_\star \partial_\star /_\star \partial_\star t_k$$
$$+_\star \sum_{\star k=1}^{n} (b \cdot_\star V_{2k}) \cdot_\star \partial_\star /_\star \partial_\star t_k$$
$$= \sum_{\star k=1}^{n} (a \cdot_\star V_{1k} +_\star b \cdot_\star V_{2k}) \cdot_\star \partial_\star /_\star \partial_\star t_k,$$

$$c \cdot_\star V_1 +_\star d \cdot_\star V_2 = \sum_{\star k=1}^{n} (c \cdot_\star V_{3k}) \cdot_\star \partial_\star /_\star \partial_\star t_k$$

$$+_\star \sum_{\star k=1}^{n} (d \cdot_\star V_{4k}) \cdot_\star \partial_\star /_\star \partial_\star t_k$$

$$= \sum_{\star k=1}^{n} c \cdot_\star V_{3k} +_\star d \cdot_\star V_{4k}) \cdot_\star \partial_\star /_\star \partial_\star t_k.$$

Hence,

$$(a \cdot_\star V_1 +_\star b \cdot_\star V_2)[c \cdot_\star V_3 +_\star d \cdot_\star V_4]_\star$$

$$= \sum_{\star k=1}^{n} (a \cdot_\star V_{1j} +_\star b \cdot_\star V_{2j}) \cdot_\star \partial_\star \sum_{\star k=1}^{n} (c \cdot_\star V_{3k} +_\star d \cdot_\star V_{4k}) \cdot_\star \partial_\star /_\star \partial_\star t_k /_\star \partial_\star t_j$$

$$= a \cdot_\star c \cdot_\star \sum_{\star j=1}^{n} V_{1j} \cdot_\star \partial_\star \sum_{\star k=1}^{n} V_{3k} \cdot_\star \partial_\star /_\star \partial_\star t_k /_\star \partial_\star t_j$$

$$+_\star a \cdot_\star d \cdot_\star \sum_{\star j=1}^{n} V_{1j} \cdot_\star \partial_\star \sum_{\star k=1}^{n} V_{4k} \cdot_\star \partial_\star /_\star \partial_\star t_k /_\star \partial_\star t_j$$

$$+_\star b \cdot_\star c \cdot_\star \sum_{\star j=1}^{n} V_{2j} \cdot_\star \partial_\star \sum_{\star k=1}^{n} V_{3k} \cdot_\star \partial_\star /_\star \partial_\star t_k /_\star \partial_\star t_j$$

$$+_\star b \cdot_\star d \cdot_\star \sum_{\star j=1}^{n} V_{2j} \cdot_\star \partial_\star \sum_{\star k=1}^{n} V_{4k} \cdot_\star \partial_\star /_\star \partial_\star t_k /_\star \partial_\star t_j$$

$$= a \cdot_\star c \cdot_\star (V_1[V_3]_\star) +_\star a \cdot_\star d \cdot_\star (V_1[V_4]_\star)$$

$$+_\star b \cdot_\star c \cdot_\star (V_2[V_3]_\star) +_\star b \cdot_\star d \cdot_\star (V_2[V_4]_\star).$$

This completes the proof.

**Definition 8.6:** Let $V_j$, $j \in \{1, 2\}$, be multiplicative vector fields. The equation

$$[V_1, V_2]_\star = V_1[V_2]_\star -_\star V_2[V_1]_\star$$

will be called the multiplicative Lie brackets.

**Example 8.4:** Let $V_1$, $V_2$ and $f$ be as in Example 8.3. Then

$$[V_1, V_2]_\star = e^{t_1 + t_2 + t_1 t_2 + 3t_1^2 - t_2^2 + 9t_1^3 + 9t_2^3}$$

$$-_\star e^{t_1 + t_2 + 2t_1 t_2 + 3t_1^2 + t_2^2 + 9t_1^3 + 9t_2^3}$$

$$= e^{-t_1 t_2 - 2t_2^2}, \quad (t_1, t_2) \in \mathbb{R}_\star^2.$$

**Exercise 8.4:** Let $V_1$ and $V_2$ be multiplicative vector fields. Prove that

$$[V_1, V_2]_\star = -_\star [V_2, V_1]_\star.$$

## 8.5 Advanced Practical Problems

**Problem 8.1:** Let

$$f(x_1, x_2, x_3) = (x_1^3 +_\star x_2^4 +_\star x_3^5)/_\star (1 +_\star x_1 +_\star x_2 +_\star x_3), \quad (x_1, x_2, x_3) \in \mathbb{R}_\star^3,$$
$$p = (2, 5, 6),$$
$$v = (1, 2, 1).$$

Find

$$(\partial_\star f /_\star \partial_\star v)(p).$$

**Problem 8.2:** Let

$$V_1 = (t_1^{2\star} +_\star t_1 \cdot_\star t_2) \cdot_\star \partial_\star /_\star \partial_\star t_1 +_\star (t_1 -_\star e^2 \cdot_\star t_2) \cdot_\star \partial_\star /_\star \partial_\star t_2,$$
$$V_2 = (t_1 +_\star t_2^{3\star}) \cdot_\star \partial_\star /_\star \partial_\star t_1 +_\star e^2 \cdot_\star t_2 \cdot_\star \partial_\star /_\star \partial_\star t_2.$$

Find

1. $e^2 \cdot_\star V_1 +_\star V_2.$
2. $V_1 -_\star e^3 \cdot_\star V_2.$
3. $(t_1^{2\star} +_\star t_2) \cdot_\star V_1.$
4. $t_1 \cdot_\star V_2.$

**Problem 8.3:** Let

$$V_1(t_1, t_2) = e^{t_1 + 2t_2^4} \cdot_\star \partial_\star /_\star \partial_\star t_1 +_\star e^{t_1^3 + t_2^2} \cdot_\star \partial_\star /_\star \partial_\star t_2,$$
$$V_2(t_1, t_2) = e^{t_1^2 + t_2^2 + t_1 t_2} \cdot_\star \partial_\star /_\star \partial_\star t_1 +_\star e^{t_1 t_2 + t_1^2 + t_2^3} \cdot_\star \partial_\star /_\star \partial_\star t_2,$$
$$V_3(t_1, t_2) = e^{t_1^4 + t_2 + t_1^2} \cdot_\star \partial_\star /_\star \partial_\star t_1 +_\star e^{t_1^2 + t_2^4} \cdot_\star \partial_\star /_\star \partial_\star t_2,$$
$$V_4(t_1, t_2) = e^{t_1^2 + t_1 t_2} \cdot_\star \partial_\star /_\star \partial_\star t_1 +_\star e^{t_1^2 + t_2^2} \cdot_\star \partial_\star /_\star \partial_\star t_2,$$
$$f(t_1, t_2) = e^{t_1^2 + t_2^2},$$

$(t_1, t_2) \in \mathbb{R}^2_\star$. Find

1. $V_1[V_2]_\star(f)$.
2. $V_1[V_3]_\star(f)$.
3. $V_1[V_4]_\star(f)$.
4. $V_2[V_3]_\star(f)$.
5. $V_2[V_4]_\star(f)$.
6. $V_3[V_4]_\star(f)$.

**Problem 8.4:** Let

$$V_1(t_1, t_2) = t_1^{4\star} +_\star t_1^{3\star} +_\star t_2 \cdot_\star \partial_\star/_\star \partial_\star t_1 +_\star e^{3} \cdot_\star t_1 -_\star e^{4} \cdot_\star t_2^{2\star} \cdot_\star \partial_\star/_\star \partial_\star t_2,$$

$$V_2(t_1, t_2) = (t_1 -_\star t_2^{7\star}) \cdot_\star \partial_\star/_\star \partial_\star t_1 +_\star e^{5} \cdot_\star t_1 \cdot_\star \partial_\star/_\star \partial_\star t_2,$$

$$f(t_1, t_2) = t_1 +_\star t_2^{2\star}, \quad (t_1, t_2) \in \mathbb{R}^2_\star.$$

Find

$$[V_1, V_2]_\star(f).$$

**Problem 8.5:** Let $V_j$, $j \in \{1, 2, 3\}$ be multiplicative vector fields and $a, b \in \mathbb{R}_\star$. Prove that

$$[a \cdot_\star V_1 +_\star b \cdot_\star V_2, V_3]_\star = a \cdot_\star [V_1, V_3]_\star +_\star b \cdot_\star [V_2, V_3]_\star.$$

**Problem 8.6:** Let $V_j$, $j \in \{1, 2, 3\}$, be multiplicative vector fields. Prove that

$$[V_1[V_2, V_3]_\star]_\star +_\star [V_2, [V_3, V_1]_\star]_\star +_\star [V_3, [V_1, V_2]_\star]_\star = 0_\star.$$

# 9

---

## *Multiplicative Riemannian Manifolds*

---

### 9.1 The Notion of a Multiplicative Manifold

**Definition 9.1:** A $k$-dimensional multiplicative differentiable manifold is a set $U$ together with a family $\{U_j\}_{j \in I}$ of subsets so that

1. $U = \cup_{j \in I} U_j$.
2. For any $j \in I$, there is an injective map $\phi_j \colon U_j \to \mathbb{R}_\star^k$ so that $\phi_j(U_j)$ is multiplicative open in $\mathbb{R}_\star^k$.
3. For $U_j \cap U_l \neq \varnothing$, $\phi_j(U_j \cap U_l)$ is multiplicative open in $\mathbb{R}_\star^k$ and

$$\phi_l \circ \phi_j^{-1} \colon \phi_j(U_j \cap U_l) \to \phi_l(U_j \cap U_l).$$

is multiplicative differentiable for any $j, l \in I$.

Each $\phi_j$, $j \in I$, is called a multiplicative chart, $\phi_j^{-1}$, $j \in I$, is reffered as a multiplicative parameterization, $\phi_j(U_j)$, $j \in I$, is said to be multiplicative parameter domain, $\{(U_j, \phi_j)\}_{j \in I}$ is said to be a multiplicative atlas. The maps

$$\phi_l \circ \phi_j^{-1} \colon \phi_j(U_j \cap U_l) \to \phi_l(U_j \cap U_l), \ j, l \in I<$$

are said to be multipolicative coordinate transformations or multiplicative transition functions.

**Example 9.1:** Every multiplicative open subset of $\mathbb{R}_\star^k$ is a $k$-dimensional multiplicative manifold.

**Example 9.2:** Let

$$U = \{x \in \mathbb{R}_\star \colon F(x_1, \ldots, x_n) = 0_\star\},$$

DOI: 10.1201/9781003299844-9

where $F: \mathbb{R}^n_\star \to \mathbb{R}^k_\star$, $F \in \mathscr{C}^1_\star(\mathbb{R}^n_\star)$,

$$\mathrm{Rank}_\star D_\star F = n - k.$$

Then, by the multiplicative implicit function theorem (see the appendix of this book), there are $x_{k+1}, \ldots, x_n$ so that

$$x_{k+1} = x_{k+1}(x_1, \ldots, x_k)$$
$$\vdots$$
$$x_n = x_n(x_1, \ldots, x_k).$$

The map

$$(x_1, \ldots, x_k) \to (x_1, \ldots, x_k, x_{k+1}, \ldots, x_n)$$

is a multiplicative parameterization and the map

$$(x_1, \ldots, x_n) \to (x_1, \ldots, x_k)$$

is a multiplicative chart.

**Definition 9.2:** If $\phi_l \circ \phi_j^{-1} \in \mathscr{C}_\star(\phi_j(U_j \cap U_l))$, for any $j, l \in I$, then $U$ is said to be multiplicative topological manifold.

**Definition 9.3:** If $\phi_l \circ \phi_j^{-1} \in \mathscr{C}^r_\star(\phi_j(U_j \cap U_l))$, $j, l \in I$, $r \in \mathbb{N} \cup \{\infty\}$, then $U$ is said to be $\mathscr{C}^k_\star$-multiplicative manifold.

**Definition 9.4:** A subset $O \subseteq U$ is said to be multiplicative open if $\phi_j(O)$, $j \in I$, is multiplicative open in $\mathbb{R}^k_\star$. This defines a multiplicative topology on $U$ as the set of all multiplicative open sets.

---

## 9.2 Multiplicative Differentiable Maps

Suppose that $U$ is a $m$-dimensional multiplicative manifold, $V$ is a $n$-dimensional multiplicative manifold. Let also, $f: U \to V$ be a given map.

**Definition 9.5:** The map $f$ is said to be multiplicative differentiable if for all multiplicative charts $\phi: U \to \mathbb{R}^m_\star$, $\psi: V \to \mathbb{R}^n_\star$ with $f(U) \subset V$, the map

$$\psi \circ f \circ \phi^{-1}: \mathbb{R}^m_\star \to \mathbb{R}^n_\star$$

is also multiplicative differentiable.

The definition does not depend on the choice of $\phi$ and $\psi$.

**Definition 9.6:** A multiplicative diffeomorphism $f: U \to V$ is defined to be a bijective map which is multiplicative differentiable in both directions. Then the two multiplicative manifolds $U$ and $V$ are said to be multiplicative diffeomorphic.

Two multiplicative diffeomorphic manifolds necessarily have the same dimensions because there is not multiplicative diffeomorphism between $\mathbb{R}_{\star}^m$ and $\mathbb{R}_{\star}^n$, $m \neq n$.

For a multiplicative chart $\phi$, we denote by $(u_1, \ldots, u_k)$ the coordinates of $\mathbb{R}_{\star}^k$ and by $(x_1, \ldots, x_k)$ the corresponding coordinates in $U$. Thus, $x_j(p)$ is the function given by the $j$th coordinate of $\phi(p)$,

$$x_j(p) = u_j(\phi(p)).$$

Then, for a function $f: U \to \mathbb{R}$, we set

$$\partial_\star f /_\star \partial_\star x_j \Big|_p = \partial_\star (f \circ \phi^{-1}) /_\star \partial_\star u_j \Big|_{\phi(p)}.$$

---

## 9.3 Multiplicative Tangent Spaces

Let $U$ be a $n$-dimensional multiplicative manifold and $p \in U$.

**Definition 9.7: (Geometric Definition).** A multiplicative vector at $p$ is any multiplicative tangent vector of any multiplicative curve on $U$ at $p$.

**Definition 9.8: (Algebraic Definition).** Define

$$\mathcal{F}_p(U) = \{f: U \to \mathbb{R}_\star : f \text{ is multiplicative differentiable}\}.$$

A multiplicative tangent vector is any map $X: \mathcal{F}_p(U) \to \mathbb{R}_\star$ with the following two properties

1.

$$X(\alpha \cdot_\star f +_\star \beta \cdot_\star g) = \alpha \cdot_\star X(f) +_\star \beta \cdot_\star X(g)$$

for any $\alpha, \beta \in \mathbb{R}_\star$ and for any $f, g \in \mathcal{F}_p(U)$.

2.

$$X(f \cdot_\star g) = f \cdot_\star X(g) +_\star X(f) \cdot_\star g$$

for any $f, g \in \mathcal{F}_p(U)$.

The set of all multiplicative tangent vectors of $U$ at $p$ will be denoted by $T_{\star p} U$.

**Lemma 9.1:** *Let $X$ be a multiplicative tangent vector and $f$ be a multiplicative constant map. Then*

$$X(f) = 0_\star.$$

*Proof.* We have

$$
\begin{aligned}
X(1_\star) &= X(1_\star \cdot_\star 1_\star) \\
&= 1_\star \cdot_\star X(1_\star) +_\star X(1_\star) \cdot_\star 1_\star \\
&= e^2 \cdot_\star X(1_\star).
\end{aligned}
$$

Hence,

$$X(1_\star) = 0_\star.$$

Take $c \in \mathbb{R}_\star$. Then, using the multiplicative linearity of $X$, we find

$$
\begin{aligned}
X(c) &= X(c \cdot_\star 1_\star) \\
&= c \cdot_\star X(1_\star) \\
&= 0_\star.
\end{aligned}
$$

This completes the proof.

**Theorem 9.1:** *For any multiplicative tangent vector $X$, we have the following representation*

$$X = \sum_{\star j=1}^{n} X(x_j) \cdot_\star \partial_\star /_\star \partial_\star x_j \bigg|_p.$$

*Proof.* Consider a multiplicative chart $\phi\colon U \to V$, where without any loss of generality, we assume that $V$ is a multiplicative open $\varepsilon$-multiplicative ball with $\phi(p) = 0_\star$. Hence,

$$x_1(p) = \cdots = x_n(p) = 0_\star.$$

Let $h\colon V \to \mathbb{R}_\star$ be a multiplicative differentiable function and

$$f = h \circ \phi.$$

Set

$$h_j(y) = \int_{\star 0_\star}^{1_\star} (\partial_\star h /_\star \partial_\star u_j)(t \cdot_\star y) \cdot_\star d_\star t.$$

Observe that

$$
\begin{aligned}
(\partial_\star h /_\star \partial_\star t)(t \cdot_\star y) &= \Sigma_{\star j=1}^n (\partial_\star h /_\star \partial_\star (u_j))(t \cdot_\star y) \\
&\quad \star (d_\star (t \cdot_\star u_j)) /_\star d_\star t \\
&= \Sigma_{\star j=1}^n (\partial_\star h /_\star \partial_\star (u_j))(t \cdot_\star y) \cdot_\star u_j.
\end{aligned}
$$

Hence,

$$
\begin{aligned}
\Sigma_{\star j=1}^n h_j(y) \cdot_\star u_j &= \int_{\star 0_\star}^{1_\star} \Sigma_{\star j=1}^n (\partial_\star h /_\star \partial_\star u_j)(t \cdot_\star y) \cdot_\star u_j \cdot_\star d_\star t \\
&= \int_{\star 0_\star}^{1_\star} (\partial_\star h /_\star \partial_\star t)(t \cdot_\star y) \cdot_\star d_\star t \\
&= h(y) -_\star h(0_\star).
\end{aligned}
$$

Since

$$
\begin{aligned}
f &= h \circ \phi, \\
f_j &= h_j \circ \phi, \\
x_j &= u_j \circ \phi, \quad j \in \{1,\ldots,n\},
\end{aligned}
$$

we find

$$f(q) -_\star f(p) = \sum_{\star j=1}^n f_j(q) \cdot_\star x_j(q).$$

Consequently

$$\left. (\partial_\star f /_\star \partial_\star x_j) \right|_q = f_j(q).$$

Now, using the properties of a multiplicative tangent vector, we find

$$X(f) = X\left( f(p) +_\star \sum_{\star j=1}^n f_j x_j \right)$$

$$= X(f(p)) +_\star X\left( \sum_{\star j=1}^n f_j x_j \right)$$

$$= 0_\star +_\star \sum_{\star j=1}^n X(f_j x_j)$$

$$= \sum_{\star j=1}^n X(f_j(p)) \cdot_\star x_j(p) +_\star \sum_{\star j=1}^n f_j(p) \cdot_\star X(x_j)$$

$$= \sum_{\star j=1}^n f_j(p) \cdot_\star X(x_j)$$

$$= \sum_{\star j=1}^n \left. \left( \partial_\star f /_\star \partial_\star x_j \right) \right|_p \cdot_\star X(x_j)$$

$$= \sum_{\star j=1}^n X(x_j) \cdot_\star \left. \left( \partial_\star /_\star \partial_\star x_j \right) \right|_p (f).$$

Note that

$$(\partial_\star /_\star \partial_\star x_j)(x_l) = \begin{cases} 1_\star & \text{if } j \neq l \\ 0_\star & \text{if } j = l. \end{cases}$$

Thus,

$$\partial_\star /_\star \partial_\star x_j, \quad j \in \{1, \ldots, n\},$$

are multiplicative linearly independent. This completes the proof.

**Theorem 9.2:** *Let $F: U \to V$ be a multiplicative differentiable map, $p \in U, q \in V$ be such that $F(p) = q$. The multiplicative differential of $F$ at $p$ is defined to be the map*

$$D_\star F|_p : T_{\star p} U \to T_{\star q} V,$$

*where*

$$\left( D_\star F|_p (X) \right)(f) = X(f \circ F)$$

*for any $f \in \mathcal{F}_q(V)$. For this multiplicative differential, we have*

$$D_\star (G \circ F) = D_\star G|_{F(p)} \circ D_\star F|_p ,$$

*where $F: U \to V$, $G: V \to W$ or briefly*

$$D_\star (G \circ F) = D_\star G \circ D_\star F.$$

*Proof.* We have

$$
\begin{aligned}
D_\star (G \circ F) |_p (X)(f) &= X(f \circ G \circ F) \\
&= (D_\star F|_p (X))(f \circ G) \\
&= (D_\star G|_q (D_\star F|_p (X)))(f).
\end{aligned}
$$

This completes the proof.

**Definition 9.9:** A multiplicative vector field $X$ on a multiplicative differentiable manifold is an association $p \to X_p \in T_{\star p} U$, $p \in U$, such that in any multiplicative chart $\phi: U \to V$ with coordinates $x_1, \dots, x_n$, the coefficients $\xi_j: U \to \mathbb{R}_\star$ in the representation

$$X_p = \sum_{\star j=1}^{n} \xi_j(p) \cdot_\star (\partial_\star /_\star \partial_\star x_j) \Big|_p$$

are multiplicative differentiable functions.

## 9.4 Multiplicative Riemannian Metrics

Let $U$ be an $n$-dimensional multiplicative manifold and $p \in U$. With $L(T_{\star p} U)$ we will denote the dual space of the space $T_{\star p} U$.

**Definition 9.10:** Multiplicative basis $\{d_\star x_j\}_{j=1}^n$ in $L(T_{\star p}U)$, we define as follows:

$$d_\star x_j\bigg|_p\left((\partial_\star/_\star\partial_\star x_l)\,|_p\right) = \delta_{\star jl}$$

$$= \begin{cases} 1_\star & \text{if } j = l \\ 0_\star & \text{if } j \neq l. \end{cases}$$

**Definition 9.11:** Define

$$L^2(T_{\star p}U) = \{\alpha: T_{\star p}U \times T_{\star p}U \to \mathbb{R}_\star, \alpha \text{ is multiplicative differentiable}\}$$

and in it we define a multiplicative basis

$$d_\star x_j\bigg|_p \otimes_\star d_\star x_l|_p, \quad j, l \in \{1,\dots,n\},$$

as follows:

$$d_\star x_j\bigg|_p \otimes_\star d_\star x_l|_p\left(\left((\partial_\star/_\star\partial_\star x_k)\,|_p\right),\left((\partial_\star/_\star\partial_\star x_r)\,|_p\right)\right) = \delta_{\star ij}\cdot_\star\,\delta_{\star lr}.$$

For the coefficients of the representation

$$\alpha = \sum_{\star i,j} \alpha_{ij}\cdot_\star d_\star x_i \otimes_\star d_\star x_j,$$

we have the expression

$$\alpha_{ij} = \alpha(\partial_\star/_\star\partial_\star x_i, \partial_\star/_\star\partial_\star x_j).$$

**Definition 9.12: (Multiplicative Riemannian Metrics).** A multiplicative Riemann metric $g$ on $U$ is an association $p \to g_p \in L^2(T_{\star p}U)$, $p \in U$, that satisfies the following conditions

1. $g_p(X, Y) = g_p(Y, X)$ for any $X, Y \in T_{\star p}U$.
2. $g_p(X, X) > 0_\star$ for any $X \in T_{\star p}U$, $X \neq 0_\star$.

3. The coefficients $g_{jl}$ in the representation

$$g_p = \sum_{\star j, l} g_{jl} \cdot_\star d_\star x_j |_p \otimes_\star d_\star x_l |_p$$

are multiplicative differential functions.

Then the pair $(U, g)$ is said to be a multiplicative Riemannian manifold. The multiplicative Riemannian metric is also referred as a multiplicative metric tensor.

**Example 9.3:** The pair $(U, g) = (\mathbb{R}^n_\star, g_0)$ is a multiplicative Riemannian manifold, where

$$(g_0)_{ij} = \begin{pmatrix} 1_\star & 0_\star & \cdots & 0_\star \\ 0_\star & 1_\star & \cdots & 0_\star \\ \cdots & \cdots & \cdots & \cdots \\ 0_\star & 0_\star & \cdots & 1_\star \end{pmatrix}$$

and $g_0(\cdot, \cdot)$ is the standard multiplicative inner product. This space is also referred as a multiplicative Euclidean space and it is denoted by $E^n_\star$.

**Example 9.4:** A multiplicative Riemannian metric is given as follows:

$$(g_{ij}) = \begin{pmatrix} 1_\star +_\star x_1^{2_\star} & 0_\star & & \cdots & 0_\star \\ 0_\star & 1_\star +_\star x_2^{2_\star} & & \cdots & 0_\star \\ \cdots & \cdots & & \cdots & \cdots \\ 0_\star & 0_\star & & \cdots & 1_\star +_\star x_n^{2_\star} \end{pmatrix}.$$

A multiplicative Riemannian metric $g$ defines at every point $p$ a multiplicative inner product $g_p$ in the multiplicative tangent space $T_{\star p} U$ and therefore the notation $\langle X, Y \rangle_\star$ instead of $g_p(X, Y)$ is also used.

---

## 9.5 The Multiplicative Riemannian Connection

Suppose that $(U, g)$ is a multiplicative Riemannian manifold.

**Definition 9.13: (Multiplicative Lie Bracket).** Let $X$ and $Y$ be two multiplicative differentiable vector fields on $U$ and $f: U \to \mathbb{R}_\star$ be a multiplicative differentiable function. The multiplicative Lie bracket of $X$ and $Y$ we define as follows:

$$[X, Y]_\star(f) = X(Y(f)) \to_\star Y(X(f)).$$

It is also called the multiplicative Lie derivative $\mathcal{L}_{\star X} Y$ of $Y$ in the multiplicative direction $X$. At a point $p \in U$, we have

$$[X, Y]_{\star p}(f) = X_p(Y(f)) \to_\star Y_p(X(f)).$$

Suppose that $X$, $Y$, $Z$ are multiplicative differentiable vector fields on $U$, $f, h, \phi: U \to \mathbb{R}_\star$ are multiplicative differentiable functions, $\alpha, \beta \in \mathbb{R}_\star$. Below, we will deduct some of the properties of the multiplicative Lie brackets.

1. $[\alpha \cdot_\star X +_\star \beta \cdot_\star Y, Z]_\star = \alpha \cdot_\star [X, Z]_\star +_\star \beta \cdot_\star [Y, Z]_\star.$

*Proof.* We have

$$\begin{aligned}
[\alpha \cdot_\star X +_\star \beta \cdot_\star Y, Z]_\star(f) &= (\alpha \cdot_\star X +_\star \beta \cdot_\star Y)(Z(f)) \\
&\quad \to_\star Z((\alpha \cdot_\star X +_\star \beta \cdot_\star Y)(f)) \\
&= \alpha \cdot_\star X(Z(f)) +_\star \beta \cdot_\star Y(Z(f)) \\
&\quad \to_\star Z(\alpha \cdot_\star X(f) +_\star \beta \cdot_\star Y(f)) \\
&= \alpha \cdot_\star X(Z(f)) +_\star \beta \cdot_\star Y(Z(f)) \\
&\quad \to_\star \alpha \cdot_\star Z(X(f)) \to_\star \beta \cdot_\star Z(Y(f)) \\
&= \alpha \cdot_\star (X(Z(f)) \to_\star Z(X(f))) \\
&\quad +_\star \beta \cdot_\star (Y(Z(f)) - Z(Y(f))) \\
&= \alpha \cdot_\star [X, Z]_\star(f) +_\star \beta \cdot_\star [Y, Z]_\star(f).
\end{aligned}$$

This completes the proof.

2. $[X, Y]_\star = \to_\star [Y, X]_\star.$

*Proof.* We have

$$\begin{aligned}
[X, Y]_\star(f) &= X(Y(f)) \to_\star Y(X(f)) \\
&= \to_\star (Y(X(f)) \to_\star X(Y(f))) \\
&= \to_\star [Y, X]_\star(f).
\end{aligned}$$

This completes the proof.

3.

$$\begin{aligned}
[fX, hY]_\star &= f \cdot_\star h \cdot_\star [X, Y]_\star +_\star f \cdot_\star X(h) \cdot_\star Y \\
&\quad \to_\star h \cdot_\star Y(f) \cdot_\star X.
\end{aligned}$$

*Proof.* We have

$$
\begin{aligned}
[fX, hY]_\star(\phi) &= (fX)(hY(\phi)) -_\star (hY)(fX(\phi)) \\
&= f\cdot_\star X(hY(\phi)) -_\star h\cdot_\star Y(fX(\phi)) \\
&= f\cdot_\star h\cdot_\star X(Y(\phi)) +_\star f\cdot_\star Y(\phi)\cdot_\star X(h) \\
&\quad -_\star h\cdot_\star fY(X(\phi)) -_\star h\cdot_\star Y(f)\cdot_\star X(\phi) \\
&= f\cdot_\star h\cdot_\star (X(Y(\phi)) -_\star Y(X(\phi))) \\
&\quad +_\star f\cdot_\star X(h)\cdot_\star Y(\phi) -_\star h\cdot_\star Y(f)\cdot_\star X(\phi) \\
&= f\cdot_\star h\cdot_\star [X, Y]_\star(\phi) \\
&\quad +_\star f\cdot_\star X(h)\cdot_\star Y(\phi) -_\star h\cdot_\star Y(f)\cdot_\star X(\phi).
\end{aligned}
$$

This completes the proof.

4.

$$
[X, [Y, Z]_\star]_\star +_\star [Y, [Z, X]_\star]_\star +_\star [Z, [X, Y]_\star]_\star = 0_\star.
$$

*Proof.* We have

$$
\begin{aligned}
&[X, [Y, Z]_\star]_\star(\phi) +_\star [Y, [Z, X]_\star]_\star(\phi) +_\star [Z, [X, Y]_\star]_\star(\phi) \\
&= X([Y, Z]_\star(\phi)) -_\star ([Y, Z]_\star)(X(\phi)) +_\star Y([Z, X]_\star(\phi)) \\
&\quad -_\star ([Z, X]_\star)(Y(\phi)) +_\star Z([X, Y]_\star(\phi)) -_\star [X, Y]_\star(Z(\phi)) \\
&= X(Y(Z(\phi))) -_\star X(Z(Y(\phi))) -_\star Y(Z(X(\phi))) +_\star Z(Y(X(\phi))) \\
&\quad +_\star Y(Z(X(\phi))) -_\star Y(X(Z(\phi))) -_\star Z(X(Y(\phi))) +_\star X(Z(Y(\phi))) \\
&\quad +_\star Z(X(Y(\phi))) -_\star Z(Y(X(\phi))) -_\star X(Y(Z(\phi))) +_\star Y(X(Z(\phi))) \\
&= 0_\star(\phi).
\end{aligned}
$$

This completes the proof.

5. $[\partial_\star/_\star\partial_\star x_j, \partial_\star/_\star\partial_\star x_l]_\star = 0_\star, \quad j, l \in \{1,...,n\}.$

*Proof.* We have

$$
\begin{aligned}
[\partial_\star/_\star\partial_\star x_j, \partial_\star/_\star\partial_\star x_l]_\star(\phi) &= (\partial_\star/_\star\partial_\star x_j)(\partial_\star/_\star\partial_\star x_l(\phi)) \\
&\quad -_\star (\partial_\star/_\star\partial_\star x_l)(\partial_\star/_\star\partial_\star x_j(\phi)) \\
&= (\partial_\star/_\star\partial_\star x_j)(\partial_\star/_\star\partial_\star x_l(\phi)) \\
&\quad -_\star (\partial_\star/_\star\partial_\star x_j)(\partial_\star/_\star\partial_\star x_l(\phi)) \\
&= 0_\star(\phi).
\end{aligned}
$$

This completes the proof.

6.

$$\left[\sum_{\star j} \xi_j{}^{\cdot}{}_\star \partial_\star /_\star \partial_\star x_j, \sum_{\star l} \eta_l{}^{\cdot}{}_\star \partial_\star /_\star \partial_\star x_l\right]_\star$$
$$= \sum_{\star j,l} (\xi_j{}^{\cdot}{}_\star (\partial_\star \eta_l /_\star \partial_\star x_j) -_\star \eta_j{}^{\cdot}{}_\star (\partial_\star \xi_l /_\star \partial_\star x_j))$$
$$\star (\partial_\star /_\star \partial_\star x_l).$$

*Proof.* We have

$$\left[\sum_{\star j} \xi_j{}^{\cdot}{}_\star \partial_\star /_\star \partial_\star x_j, \sum_{\star l} \eta_l{}^{\cdot}{}_\star \partial_\star /_\star \partial_\star x_l\right]_\star (\phi)$$

$$= \sum_{\star j} \xi_j{}^{\cdot}{}_\star \left(\partial_\star /_\star \partial_\star x_j \left(\left(\sum_{\star l} \eta_l{}^{\cdot}{}_\star \partial_\star /_\star \partial_\star x_l\right)(\phi)\right)\right)$$

$$-_\star \sum_{\star l} \eta_l{}^{\cdot}{}_\star \left(\partial_\star /_\star \partial_\star x_l \left(\left(\sum_{\star j} \xi_j{}^{\cdot}{}_\star \partial_\star /_\star \partial_\star x_j\right)(\phi)\right)\right)$$

$$= \sum_{\star j} \xi_j{}^{\cdot}{}_\star \left(\sum_{\star l} (\partial_\star \eta_l /_\star \partial_\star x_j){}^{\cdot}{}_\star (\partial_\star \phi /_\star \partial_\star x_l) +_\star \sum_{\star l} \eta_l{}^{\cdot}{}_\star \partial_\star^{2}{}_\star \phi /_\star \partial_\star x_j \partial_\star x_l\right)$$

$$-_\star \sum_{\star l} \eta_l{}^{\cdot}{}_\star \left(\sum_{\star j} (\partial_\star \xi_j /_\star \partial_\star x_l){}^{\cdot}{}_\star (\partial_\star \phi /_\star \partial_\star x_j) +_\star \sum_{\star j} \xi_j{}^{\cdot}{}_\star \partial_\star^{2}{}_\star \phi /_\star \partial_\star x_j \partial_\star x_l\right)$$

$$= \sum_{\star j} \xi_j{}^{\cdot}{}_\star \left(\sum_{\star l} (\partial_\star \eta_l /_\star \partial_\star x_j){}^{\cdot}{}_\star (\partial_\star \phi /_\star \partial_\star x_l)\right)$$

$$-_\star \sum_{\star l} \eta_l{}^{\cdot}{}_\star \left(\sum_{\star j} (\partial_\star \xi_j /_\star \partial_\star x_l){}^{\cdot}{}_\star (\partial_\star \phi /_\star \partial_\star x_j)\right)$$

$$= \sum_{\star j,l} (\xi_j{}^{\cdot}{}_\star (\partial_\star \eta_l /_\star \partial_\star x_j) -_\star \eta_j{}^{\cdot}{}_\star (\partial_\star \xi_l /_\star \partial_\star x_j)){}^{\cdot}{}_\star (\partial_\star \phi /_\star \partial_\star x_l).$$

This completes the proof.

**Definition 9.14:** (The Multiplicative Riemannian Connection). A multiplicative Riemannian connection $\nabla_\star$ on a multiplicative Riemannian manifold $(U, g)$ is a map

$$(X, Y) \to \nabla_{\star X} Y$$

that satisfies the following conditions.

1. $\nabla_{\star X_1 +_\star X_2} Y = \nabla_{\star X_1} Y +_\star \nabla_{\star X_2} Y$.
2. $\nabla_{\star f \cdot_\star X} Y = f \cdot_\star \nabla_{\star X} Y$.
3. $\nabla_{\star X} (Y_1 +_\star Y_2) = \nabla_{\star X} Y_1 +_\star \nabla_{\star X} Y_2$.
4. $\nabla_{\star X} (f \cdot_\star Y) = f \cdot_\star \nabla_{\star X} Y +_\star X(f) \cdot_\star Y$.
5. $X \cdot_\star g(Y, Z) = g(\nabla_{\star X} Y, Z) +_\star g(Y, \nabla_{\star X} Z)$.
6. $\nabla_{\star X} Y -_\star \nabla_{\star Y} X -_\star [X, Y]_\star = 0_\star$.

**Lemma 9.2: (The Multiplicative Koszul Formula).** *For any three multiplicative vector fields we have the following equation*

$$e^2 \cdot_\star \langle Z, \nabla_{\star X} Y \rangle_\star = X \cdot_\star \langle Y, Z \rangle_\star +_\star Y \cdot_\star \langle X, Z \rangle_\star -_\star Z \cdot_\star \langle X, Y \rangle_\star$$
$$-_\star \langle Y, [X, Z]_\star \rangle_\star -_\star \langle X, [Y, Z]_\star \rangle_\star \qquad (9.1)$$
$$-_\star \langle Z, [Y, X]_\star \rangle_\star.$$

*Proof.* By 5), we get

$$X \cdot_\star \langle Y, Z \rangle_\star = \langle \nabla_{\star X} Y, Z \rangle_\star +_\star \langle Y, \nabla_{\star X} Z \rangle_\star,$$
$$Y \cdot_\star \langle X, Z \rangle_\star = \langle \nabla_{\star Y} X, Z \rangle_\star +_\star \langle X, \nabla_{\star Y} Z \rangle_\star.$$
$$-_\star Z \cdot_\star \langle X, Y \rangle_\star = -_\star \langle \nabla_{\star Z} X, Y \rangle_\star -_\star \langle X, \nabla_{\star Z} Y \rangle_\star.$$

Hence,

$$X \cdot_\star \langle Y, Z \rangle_\star +_\star Y \cdot_\star \langle X, Z \rangle_\star -_\star Z \cdot_\star \langle X, Y \rangle_\star$$
$$= \langle Y, \nabla_{\star X} Z -_\star \nabla_{\star Z} X \rangle_\star +_\star \langle X, \nabla_{\star Y} Z -_\star \nabla_{\star Z} Y \rangle_\star$$
$$+_\star \langle Z, \nabla_{\star X} Y +_\star \nabla_{\star Y} X \rangle_\star$$
$$= \langle Y, [X, Z]_\star \rangle_\star +_\star \langle X, [Y, Z]_\star \rangle_\star$$
$$+_\star \langle Z, [Y, X]_\star +_\star e^2 \cdot_\star \nabla_{\star X} Y \rangle_\star.$$

By the last equation, we obtain

$$e^2 \cdot_\star \langle Z, \nabla_{\star X} Y \rangle_\star = X \cdot_\star \langle Y, Z \rangle_\star +_\star Y \cdot_\star \langle X, Z \rangle_\star -_\star Z \cdot_\star \langle X, Y \rangle_\star$$
$$-_\star \langle Y, [X, Z]_\star \rangle_\star -_\star \langle X, [Y, Z]_\star \rangle_\star$$
$$-_\star \langle Z, [Y, X]_\star \rangle_\star.$$

This completes the proof.

**Theorem 9.3:** *On any multiplicative Riemannian manifold $(U, g)$ there is a uniquely determined multiplicative Riemannian connection.*

*Proof.*

1. Uniqueness. Note that for a given $Z$, the right-hand side of (9.1) is uniquely determined. Hence, $\nabla_{\star X} Y$ is uniquely determined.

2. Existence. We define $\nabla_\star$ by the equality (9.1). We will check that it satisfies all requirements of Definition 9.14.

   a. We have

$$e^2 \cdot_\star \langle Z, \nabla_{\star X_1} Y \rangle_\star = X_1 \cdot_\star \langle Y, Z \rangle_\star +_\star Y \cdot_\star \langle X_1, Z \rangle_\star -_\star Z \cdot_\star \langle X_1, Y \rangle_\star$$
$$-_\star \langle Y, [X_1, Z]_\star \rangle_\star -_\star \langle X_1, [Y, Z]_\star \rangle_\star$$
$$-_\star \langle Z, [Y, X_1]_\star \rangle_\star,$$

$$e^2 \cdot_\star \langle Z, \nabla_{\star X_2} Y \rangle_\star = X_2 \cdot_\star \langle Y, Z \rangle_\star +_\star Y \cdot_\star \langle X_2, Z \rangle_\star -_\star Z \cdot_\star \langle X_2, Y \rangle_\star$$
$$-_\star \langle Y, [X_2, Z]_\star \rangle_\star -_\star \langle X_2, [Y, Z]_\star \rangle_\star$$
$$-_\star \langle Z, [Y, X_2]_\star \rangle_\star,$$

$$e^2 \cdot_\star \langle Z, \nabla_{\star X_1} Y \rangle_\star +_\star e^2 \cdot_\star \langle Z, \nabla_{\star X_2} Y \rangle_\star = X_1 \cdot_\star \langle Y, Z \rangle_\star$$
$$+_\star Y \cdot_\star \langle X_1, Z \rangle_\star -_\star Z \cdot_\star \langle X_1, Y \rangle_\star$$
$$-_\star \langle Y, [X_1, Z]_\star \rangle_\star$$
$$-_\star \langle X_1, [Y, Z]_\star \rangle_\star$$
$$-_\star \langle Z, [Y, X_1]_\star \rangle_\star$$
$$+_\star X_2 \cdot_\star \langle Y, Z \rangle_\star +_\star Y \cdot_\star \langle X_2, Z \rangle_\star$$
$$-_\star Z \cdot_\star \langle X_2, Y \rangle_\star$$
$$-_\star \langle Y, [X_2, Z]_\star \rangle_\star -_\star \langle X_2, [Y, Z]_\star \rangle_\star$$
$$-_\star \langle Z, [Y, X_2]_\star \rangle_\star$$

   and

$$e^2 \cdot_\star \langle Z, \nabla_{\star X_1 +_\star X_2} Y \rangle_\star = (X_1 +_\star X_2) \cdot_\star \langle Y, Z \rangle_\star +_\star Y \cdot_\star \langle X_1 +_\star X_2, Z \rangle_\star$$
$$-_\star Z \cdot_\star \langle X_1 +_\star X_2, Y \rangle_\star$$
$$-_\star \langle Y, [X_1 +_\star X_2, Z]_\star \rangle_\star -_\star \langle X_1 +_\star X_2, [Y, Z]_\star \rangle_\star$$
$$-_\star \langle Z, [Y, X_1 +_\star X_2]_\star \rangle_\star$$
$$= X_1 \cdot_\star \langle Y, Z \rangle_\star +_\star X_2 \cdot_\star \langle Y, Z \rangle_\star$$
$$+_\star Y \cdot_\star \langle X_1, Z \rangle_\star +_\star Y \cdot_\star \langle X_2, Z \rangle_\star$$
$$-_\star Z \cdot_\star \langle X_1, Y \rangle_\star -_\star Z \cdot_\star \langle X_2, Y \rangle_\star$$
$$-_\star \langle Y, [X_1, Z]_\star \rangle_\star -_\star \langle Y, [X_2, Z]_\star \rangle_\star$$
$$-_\star \langle X_1, [Y, Z]_\star \rangle_\star -_\star \langle X_2, [Y, Z]_\star \rangle_\star$$
$$-_\star \langle Z, [Y, X_1]_\star \rangle_\star -_\star \langle Z, [Y, X_2]_\star \rangle_\star$$

Consequently

$$e^2 \cdot_\star \left\langle Z, \nabla_{\star X_1 +_\star X_2} Y \right\rangle_\star = e^2 \cdot_\star \left\langle Z, \nabla_{\star X_1} Y \right\rangle_\star +_\star e^2 \cdot_\star \left\langle Z, \nabla_{\star X_2} Y \right\rangle_\star$$

and from here,

$$\nabla_{\star X_1 +_\star X_2} Y = \nabla_{\star X_1} Y +_\star \nabla_{\star X_2} Y.$$

b. We have

$$\nabla_{\star f \cdot_\star X_1} Y = f \cdot_\star \nabla_{\star X_1} Y,$$
$$\nabla_{\star f \cdot_\star X_2} Y = f \cdot_\star \nabla_{\star X_2} Y$$

and

$$\begin{aligned}
\nabla_{\star f \cdot_\star (X_1 +_\star X_2)} Y &= f \cdot_\star \nabla_{\star X_1 +_\star X_2} Y \\
&= f \cdot_\star \left( \nabla_{\star X_1} Y +_\star \nabla_{\star X_2} Y \right) \\
&= f \cdot_\star \nabla_{\star X_1} Y +_\star f \cdot_\star \nabla_{\star X_2} Y.
\end{aligned}$$

c. We have

$$\nabla_{\star X_1} (Y_1 +_\star Y_2) = \nabla_{\star X_1} Y_1 +_\star \nabla_{\star X_1} Y_2,$$
$$\nabla_{\star X_2} (Y_1 +_\star Y_2) = \nabla_{\star X_2} Y_1 +_\star \nabla_{\star X_2} Y_2,$$
$$\nabla_{\star X_1 +_\star X_2} (Y_1 +_\star Y_2) = \nabla_{\star X_1 +_\star X_2} Y_1 +_\star \nabla_{\star X_1 +_\star X_2} Y_2$$

and

$$\begin{aligned}
\nabla_{\star X_1} (Y_1 +_\star Y_2) +_\star \nabla_{\star X_2} (Y_1 +_\star Y_2) &= \nabla_{\star X_1} Y_1 +_\star \nabla_{\star X_1} Y_2 \\
&\quad +_\star \nabla_{\star X_2} Y_1 +_\star \nabla_{\star X_2} Y_2 \\
&= \left( \nabla_{\star X_1} Y_1 +_\star \nabla_{\star X_2} Y_1 \right) \\
&\quad +_\star \left( \nabla_{\star X_1} Y_2 +_\star \nabla_{\star X_2} Y_2 \right) \\
&= \nabla_{\star X_1 +_\star X_2} Y_1 +_\star \nabla_{\star X_1 +_\star X_2} Y_2.
\end{aligned}$$

d. We have

$$\nabla_{\star X_1} (f \cdot_\star Y) = f \cdot_\star \nabla_{\star X_1} Y +_\star X_1(f) \cdot_\star Y,$$
$$\nabla_{\star X_2} (f \cdot_\star Y) = f \cdot_\star \nabla_{\star X_2} Y +_\star X_2(f) \cdot_\star Y$$

and

$$\nabla_{\star X_1 +_\star X_2}(f \cdot_\star Y) = f \cdot_\star \nabla_{\star X_1 +_\star X_2} Y +_\star (X_1 +_\star X_2)(f) \cdot_\star Y$$
$$= f \cdot_\star (\nabla_{\star X_1} Y +_\star \nabla_{\star X_2} Y)$$
$$+_\star X_1(f) \cdot_\star Y +_\star X_2(f) \cdot_\star Y$$
$$= (f \cdot_\star \nabla_{\star X_1} Y +_\star X_1(f) \cdot_\star Y)$$
$$+_\star (f \cdot_\star \nabla_{\star X_2} Y +_\star X_2(f) \cdot_\star Y)$$
$$= \nabla_{\star X_1}(f \cdot Y) +_\star \nabla_{\star X_2}(f \cdot_\star Y).$$

e.  We have

$$X_1 \cdot_\star g(Y, Z) = g(\nabla_{\star X_1} Y, Z) +_\star g(Y, \nabla_{\star X_1} Z),$$
$$X_2 \cdot_\star g(Y, Z) = g(\nabla_{\star X_2} Y, Z) +_\star g(Y, \nabla_{\star X_2} Z)$$

and

$$(X_1 +_\star X_2) \cdot_\star g(Y, Z) = g(\nabla_{\star X_1 +_\star X_2} Y, Z) +_\star g(Y, \nabla_{\star X_1 +_\star X_2} Z)$$
$$= g(\nabla_{\star X_1} Y +_\star \nabla_{\star X_2} Y, Z)$$
$$+_\star g(Y, \nabla_{\star X_1} Z +_\star \nabla_{\star X_2} Z)$$
$$= g(\nabla_{\star X_1} Y, Z) +_\star g(Y, \nabla_{\star X_1} Z)$$
$$+_\star g(\nabla_{\star X_2} Y, Z) +_\star g(Y, \nabla_{\star X_2} Z)$$
$$= X_1 \cdot_\star g(Y, Z) +_\star X_2 \cdot_\star g(Y, Z).$$

f.  We have

$$0_\star = \nabla_{\star X_1} Y -_\star \nabla_{\star Y} X_1 -_\star [X_1, Y]_\star,$$
$$0_\star = \nabla_{\star X_2} Y -_\star \nabla_{\star Y} X_2 -_\star [X_2, Y]_\star$$

and

$$\nabla_{\star X_1 +_\star X_2} Y -_\star \nabla_{\star Y}(X_1 +_\star X_2) -_\star [X_1 +_\star X_2, Y]_\star$$
$$= \nabla_{\star X_1} Y -_\star \nabla_{\star Y} X_1 -_\star [X_1, Y]_\star$$
$$+_\star \nabla_{\star X_2} Y -_\star \nabla_{\star Y} X_2 -_\star [X_2, Y]_\star$$
$$= 0_\star.$$

This completes the proof.

## 9.6 The Multiplicative Christoffel Coefficients

In this section, we will find a representation of

$$\nabla_{\star \partial_\star /_\star \partial_\star x_j} \partial_\star /_\star \partial_\star x_l$$

in the form

$$\nabla_{\star \partial_\star /_\star \partial_\star x_j} \partial_\star /_\star \partial_\star x_l = \sum_{\star k} \Gamma_{ij}^k \partial_\star /_\star \partial_\star x_k.$$

By the Koszul formula, we get

$$
\begin{aligned}
e^2 \cdot_\star & \langle \nabla_{\star \partial_\star /_\star \partial_\star x_j} \partial_\star /_\star \partial_\star x_l, \ \partial_\star /_\star \partial_\star x_k \rangle_\star \\
&= (\partial_\star /_\star \partial_\star x_j) \cdot_\star \langle \partial_\star /_\star \partial_\star x_l, \ \partial_\star /_\star \partial_\star x_k \rangle_\star \\
&+_\star (\partial_\star /_\star \partial_\star x_l) \cdot_\star \langle \partial_\star /_\star \partial_\star x_j, \ \partial_\star /_\star \partial_\star x_k \rangle_\star \\
&-_\star \partial_\star /_\star \partial_\star x_k \cdot_\star \langle \partial_\star /_\star \partial_\star x_j, \ \partial_\star /_\star \partial_\star x_l \rangle_\star \\
&= (\partial_\star g_{lk}) /_\star \partial_\star x_j +_\star (\partial_\star g_{jk}) /_\star \partial_\star x_l \\
&-_\star (\partial_\star g_{jl}) /_\star \partial_\star x_k.
\end{aligned}
$$

Therefore

$$
\begin{aligned}
& \langle \nabla_{\star \partial_\star /_\star \partial_\star x_j} \partial_\star /_\star \partial_\star x_l, \ \partial_\star /_\star \partial_\star x_k \rangle_\star \\
&= (1_\star /_\star e^2) \cdot_\star \left( -_\star (\partial_\star g_{jl}) /_\star \partial_\star x_k +_\star (\partial_\star g_{lk}) /_\star \partial_\star x_j +_\star (\partial_\star g_{jk}) /_\star \partial_\star x_l \right).
\end{aligned}
$$

Let

$$
\Gamma_{jl,k} = (1_\star /_\star e^2) \cdot_\star \left( -_\star (\partial_\star g_{jl}) /_\star \partial_\star x_k +_\star (\partial_\star g_{lk}) /_\star \partial_\star x_j +_\star (\partial_\star g_{jk}) /_\star \partial_\star x_l \right).
$$

Thus, we have

$$
\begin{aligned}
\Gamma_{jl,k} &= \langle \nabla_{\star \partial_\star /_\star \partial_\star x_j} \partial_\star /_\star \partial_\star x_l, \ \partial_\star /_\star \partial_\star x_k \rangle_\star \\
&= \langle \Sigma_{\star m} \Gamma_{jl}^m \cdot_\star \partial_\star /_\star \partial_\star x_m, \ \partial_\star /_\star \partial_\star x_k \rangle_\star \\
&= \Sigma_{\star m} \Gamma_{jl}^m \cdot_\star \langle \partial_\star /_\star \partial_\star x_m, \ \partial_\star /_\star \partial_\star x_k \rangle_\star \\
&= \Sigma_{\star m} \Gamma_{jl}^m \cdot_\star g_{mk}.
\end{aligned}
$$

Let

$$g^{km} = (g_{rs})^{-1\star}.$$

Then

$$\Gamma_{jl}^{m} = \sum_{\star k} \Gamma_{jl,k}\cdot_\star g^{km}.$$

**Definition 9.15:** The symbols $\Gamma_{ij,l}$ and $\Gamma_{ij}^{k}$ will be called the multiplicative Christoffel coefficients.

Let

$$X = \sum_{\star i} \xi_i\cdot_\star \partial_\star/_\star \partial_\star x_i,$$

$$Y = \sum_{\star j} \eta_j\cdot_\star \partial_\star/_\star \partial_\star x_j.$$

$$Y = \sum_{\star j} \eta_j\cdot_\star \partial_\star/_\star \partial_\star x_j.$$

Then

$$\nabla_{\star X} Y = \sum_{\star i} \xi_i\cdot_\star \nabla_{\star \partial_\star/_\star \partial_\star x_i}\left(\sum_{\star j} \eta_j\cdot_\star \partial_\star/_\star \partial_\star x_j\right)$$

$$= \sum_{\star i}\sum_{\star j} \xi_i\cdot_\star \nabla_{\star \partial_\star/_\star \partial_\star x_i}\left(\eta_j\cdot_\star \partial_\star/_\star \partial_\star x_j\right)$$

$$= \sum_{\star i}\sum_{\star j} \xi_i\cdot_\star \eta_j\cdot_\star \nabla_{\star \partial_\star/_\star \partial_\star x_i}\left(\partial_\star/_\star \partial_\star x_j\right)$$

$$+_\star \sum_{\star i}\sum_{\star j} \xi_i\cdot_\star\left(\partial_\star\eta_j/_\star\partial_\star x_i\right)\cdot_\star\left(\partial_\star/_\star\partial_\star x_j\right)$$

$$= \sum_{\star i}\sum_{\star j}\sum_{\star l} \xi_i\cdot_\star\eta_j\cdot_\star\Gamma_{ij}^{l}\cdot_\star(\partial_\star/_\star\partial_\star x_l) +_\star \sum_{\star i}\sum_{\star j}\xi_i\cdot_\star(\partial_\star\eta_j/_\star\partial_\star x_i)$$

$$\cdot_\star(\partial_\star/_\star\partial_\star x_j) = \sum_{\star i}\sum_{\star l}\sum_{\star j}\xi_i\cdot_\star\eta_l\cdot_\star\Gamma_{il}^{j}\cdot_\star(\partial_\star/_\star\partial_\star x_j)$$

$$+_\star \sum_{\star i}\sum_{\star j}\xi_i\cdot_\star(\partial_\star\eta_j/_\star\partial_\star x_i)\cdot_\star(\partial_\star/_\star\partial_\star x_j)$$

$$= \sum_{\star j}\left(\sum_{\star i}\xi_i\cdot_\star(\partial_\star\eta_j/_\star\partial_\star x_i)+_\star\sum_{\star i,l}\xi_i\cdot_\star\eta_l\cdot_\star\Gamma_{il}^{j}\right)\cdot_\star(\partial_\star/_\star\partial_\star x_j).$$

**Definition 9.16:** A multiplicative vector field is said to be multiplicative parallel if

$$\nabla_{\star X} Y = 0_{\star}$$

for any $X$.

**Definition 9.17:** A multiplicative vector field $Y = \sum_{\star i} \xi_i \cdot_{\star} \partial_{\star} /_{\star} \partial_{\star} x_i$, along a multiplicative regular curve $c$ is said to be multiplicative parallel along $c$ if

$$\nabla_{\star c} Y = 0_{\star}.$$

**Definition 9.18:** A multiplicative regular curve $c$ is said to be multiplicative geodesic if

$$\nabla_{\star c^{\star}} c^{\star} = \lambda \cdot_{\star} c^{\star}$$

for some multiplicative scalar function $\lambda$.

# 10

---

## *The Multiplicative Curvature Tensor*

---

### 10.1 Multiplicative Tensors

Suppose that $U$ is a multiplicative differentiable manifold.

**Definition 10.1:** A multiplicative covariant tensor of degree $s$, or briefly $(0, s)$-multiplicative tensor, at a point $p$ on $U$ is the map

$$A_{\star p} \colon \underbrace{T_{\star p} U \times \ldots \times T_{\star p} U}_{s} \to \mathbb{R}_{\star}.$$

An $(1, s)$-multiplicative tensor is the map

$$A_{\star p} \colon \underbrace{T_{\star p} U \times \ldots \times T_{\star p} U}_{s} \to T_{\star p} U.$$

A multiplicative basis in the space of all $(0, s)$-multiplicative tensors is given by

$$\left( d_{\star} x_{j_1} \big|_p \otimes \ldots \otimes d_{\star} x_{j_s} \big|_p \right)_{j_1, \ldots, j_s = 1, \ldots, n},$$

where

$$\left( d_{\star} x_{j_1} \big|_p \otimes \ldots \otimes d_{\star} x_{j_s} \big|_p \right) \left( \partial_{\star} /_{\star} \partial_{\star} x_{l_1}, \ldots, \partial_{\star} /_{\star} \partial_{\star} x_{l_j} \right)$$
$$= \delta_{\star j_1 l_1} \star \cdots \star \delta_{\star j_s l_s}.$$

For the expression

$$A_{\star p} = \sum_{\star} A_{\star j_1 \ldots j_s} \star d_{\star} x_{j_1} \otimes \ldots \otimes d_{\star} x_{j_s},$$

we have the representation

$$A_{\star j_1 \ldots j_s} = A_{\star p} \left( \partial_{\star} /_{\star} \partial_{\star} x_{l_1}, \ldots, \partial_{\star} /_{\star} \partial_{\star} x_{l_j} \right).$$

DOI: 10.1201/9781003299844-10

In the latter case, we have

$$\sum_{\star i} A^i_{\star j_1 \ldots j_s} \partial_\star /_\star \partial_\star x_i = A_{\star p}\left(\partial_\star /_\star \partial_\star x_{l_1}, \ldots, \partial_\star /_\star \partial_\star x_{l_j}\right).$$

A multiplicative differentiable $(0, s)$- or $(1, s)$-multiplicative tensor field $A$ is an association

$$p \to A_{\star p}$$

such that the coefficients $A_{\star j_1 \ldots j_s}$ and $A^i_{\star j_1 \ldots j_s}$ in the above representations are multiplicative differentiable.

Below, we will suppose that all multiplicative tensors are multiplicative differentiable.

**Definition 10.2:** An $s$-multiplicative covariant and $r$-multiplicative contravariant tensor, or briefly $(r, s)$-multiplicative tensor, at a point $p$, is the map

$$A_{\star p}: \underbrace{(T_{\star p}U)^\star \times \ldots \times (T_{\star p}U)^\star}_{r} \times \underbrace{T_{\star p}U \times \ldots \times T_{\star p}U}_{s} \to \mathbb{R}_\star,$$

where $(T_{\star p}U)^\star$ is the dual space of $T_{\star p}U$. A basis in the space of all $(r, s)$-multiplicative tensors is given by

$$(\partial_\star /_\star \partial_\star x_{i_1}\big|_p \otimes \ldots \otimes \partial_\star /_\star \partial_\star x_{i_r}\big|_p$$
$$\otimes d_\star x_{j_1}\big|_p \otimes \ldots \otimes d_\star x_{j_s}\big|_p )_{i_1, \ldots, i_r, j_1, \ldots, j_s = 1, \ldots, n},$$

where

$$\left(\partial_\star /_\star \partial_\star x_{i_1}\big|_p \otimes \ldots \otimes d_\star x_{j_s}\big|_p \right)(d_\star x_{k_1}, \ldots d_\star x_{k_r},$$
$$\partial_\star /_\star \partial_\star x_{l_1}, \ldots, \partial_\star /_\star \partial_\star x_{l_s})$$
$$= \delta_{\star i_1 k_1} \cdot \ldots \cdot_\star \delta_{\star i_r k_r} \cdot_\star \delta_{\star j_1 l_1} \cdot_\star \cdots \cdot_\star \delta_{\star j_s l_s}.$$

For the coefficients of the multiplicative tensor

$$A_{\star p} = \sum_\star A^{i_1, \ldots, i_r}_{\star j_1, \ldots, j_s} \cdot_\star \partial_\star /_\star \partial_\star x_{i_1} \otimes \ldots \otimes \partial_\star /_\star \partial_\star x_{i_r}$$
$$\otimes d_\star x_{j_1} \otimes \ldots \otimes d_\star x_{j_s},$$

we have the representation

$$A^{i_1, \ldots, i_r}_{\star j_1, \ldots, j_s} = A_p\left(d_\star x_{i_1}, \ldots, d_\star x_{i_r}, \partial_\star /_\star \partial_\star x_{j_1}, \ldots, \partial_\star /_\star \partial_\star x_{j_s}\right).$$

**Definition 10.3:** A multiplicative curvature tensor on $U$ is defined to be the map

$$X, Y, Z \to R_\star(X, Y)Z = \nabla_{\star X}\nabla_{\star Y} Z -_\star \nabla_{\star Y}\nabla_{\star X} Z$$
$$-_\star \nabla_{\star[X,Y]_\star} Z.$$

**Example 10.1:** A multiplicative scalar function is a $(0, 0)$-multiplicative tensor.

**Example 10.2:** A multiplicative vector field $X$ is a $(1, 0)$-multiplicative tensor. We have

$$X = \sum_{\star i} \xi_i \cdot_\star \partial_\star /_\star \partial_\star x_i.$$

**Example 10.3:** A one-multiplicative differential form is a $(0, 1)$-multiplica-multiplicative tensor, which is referred as a multiplicative covector field. For the multiplicative differential of $f$, we have

$$d_\star f = \sum_{\star i} (\partial_\star f /_\star \partial_\star x_i) \cdot_\star d_\star x_i.$$

**Example 10.4:** A multiplicative Riemannian metric is a $(0, 2)$-multiplicative tensor.

**Example 10.5:** For a fixed multiplicative vector field, the multiplicative covariant derivative is a $(1, 1)$-multiplicative tensor.

**Example 10.6:** The multiplicative curvature tensor is a $(1, 3)$-multiplicative tensor.

---

## 10.2 Multiplicative Derivatives of Multiplicative Tensor Fields

**Definition 10.4:** Let $A_\star$ be a $(0, s)$- or $(1, s)$-multiplicative tensor field and let $X$ be a fixed multiplicative vector field. We define the multiplicative covariant derivative of $A_\star$ in the multiplicative direction $X$ by the following formula

$$(\nabla_{\star X} A_\star)(Y_1, ..., Y_s) = \nabla_{\star X}(A_\star(Y_1, ..., Y_s))$$
$$-_\star \sum_{\star i=1}^{s} A_\star(Y_1, ..., Y_{i-1}, \nabla_{\star X} Y_i, Y_{i+1}, ..., Y_s). \quad (10.1)$$

Now, we will show that (10.1) is well defined. For this aim, we will prove that

$$(\nabla_{\star X} A_\star)(Y_1, ..., f \cdot_\star Y_j, ..., Y_s) = f \cdot_\star (\nabla_{\star X} A)(Y_1, ..., Y_s).$$

We have

$$\nabla_{\star X} A_\star (Y_1, \ldots, f \cdot_\star Y_j, \ldots, Y_s) = \nabla_{\star X} (A_\star (Y_1, \ldots, f \cdot_\star Y_j, \ldots, Y_s))$$

$$-_\star \sum_{\star i=1}^{s} A_\star (Y_1, \ldots, f \cdot_\star Y_j, \ldots, Y_{i-1}, \nabla_{\star X} Y_i, Y_{i+1}, \ldots, Y_s)$$

$$= f \cdot_\star \nabla_{\star X} (A_\star (Y_1, \ldots, Y_j, \ldots, Y_s))$$

$$+_\star \sum_{\star i=1}^{s} X (f) A_\star (Y_1, \ldots, Y_j, \ldots, Y_s)$$

$$-_\star \sum_{\star i=1}^{s} X (f) \cdot_\star A_\star (Y_1, \ldots, Y_j, \ldots, Y_s)$$

$$-_\star f \cdot_\star \sum_{\star i=1}^{s} A_\star (Y_1, \ldots, Y_{i-1}, \nabla_{\star X} Y_i, Y_{i+1}, \ldots, Y_s)$$

$$= f \cdot_\star \left( \nabla_{\star X} (A_\star (Y_1, \ldots, Y_j, \ldots, Y_s)) \right.$$

$$\left. -_\star \sum_{\star i=1}^{s} A_\star (Y_1, \ldots, Y_{i-1}, \nabla_{\star X} Y_i, Y_{i+1}, \ldots, Y_s)) \right.$$

$$= f \cdot_\star (\nabla_{\star X} A_\star)(Y_1, \ldots, Y_s).$$

**Example 10.7:** Let $f$ be a multiplicative scalar function. Then its multiplicative covariant derivative is

$$d_\star f = D_\star f$$
$$= \nabla_\star f$$

and

$$\nabla_\star f (X) = \nabla_{\star X} f$$
$$= X (f).$$

The multiplicative gradient of $f$ with respect to the multiplicative metric $g$ will be defined by

$$\text{grad}_{\star g} f = \nabla_\star f (X).$$

The second multiplicative covariant derivative of $f$ is

$$\nabla_\star^{2\star} f = \nabla_\star \nabla_\star f.$$

Also, we have

$$\left(\nabla_*^2 f\right)(X, Y) = \left(\nabla_{*X} \nabla_* f\right)(Y)$$
$$= \nabla_{*X}\left(\nabla_* f(Y)\right) -_* \nabla_* f\left(\nabla_{*X} Y\right)$$
$$= \nabla_{*X}\left(\nabla_* f(Y)\right) -_* \nabla_{*X} Y(f).$$

Note that $\nabla_*^2$ is also referred as a multiplicative Hessian of $f$.

**Example 10.8:** Let $A$ be a $(0, 2)$-multiplicative tensor. Then

$$\nabla_{*X} A(Y, Z) = \nabla_{*X}(A(Y, Z)) -_* A(\nabla_{*X} Y, Z)$$
$$-_* A(Y, \nabla_{*X} Z).$$

**Example 10.9:** For the multiplicative curvature tensor

$$R_*(X, Y)Z = \nabla_{*X} \nabla_{*Y} Z -_* \nabla_{*Y} \nabla_{*X} Z$$
$$-_* \nabla_{*[X,Y]_*} Z,$$

the multiplicative covariant derivative is

$$\left(\nabla_{*X} R_*\right)(Y, Z)V = \nabla_{*X}(R_*(Y, Z)V) -_* R_*(\nabla_{*X} Y, Z)V$$
$$-_* R_*(Y, \nabla_{*X} Z)V -_* R_*(Y, Z)\nabla_{*X} V.$$

## 10.3 Properties of the Multiplicative Curvature Tensor

In this section, we will deduct some of the properties of the multiplicative curvature tensor. Suppose that $X, Y, Z$ and $V$ are multiplicative vector fields.

1. $R_*(X, Y)Z = -_* R_*(Y, X)Z.$

*Proof.* We have

$$R_*(X, Y)Z = \nabla_{*X} \nabla_{*Y} Z -_* \nabla_{*Y} \nabla_{*X} Z$$
$$-_* \nabla_{*[X,Y]_*} Z$$
$$= -_*(\nabla_{*Y} \nabla_{*X} Z -_* \nabla_{*X} \nabla_{*Y} Z$$
$$+_* \nabla_{*[X,Y]_*} Z)$$
$$= -_*(\nabla_{*Y} \nabla_{*X} Z -_* \nabla_{*X} \nabla_{*Y} Z$$
$$-_* \nabla_{*[Y,X]_*} Z)$$
$$= -_* R(Y, X)Z.$$

This completes the proof.

2. $R_\star(X, Y)Z +_\star R_\star(Y, Z)X +_\star R_\star(Z, X)Y = 0_\star$.

*Proof.* We have

$$
\begin{aligned}
R_\star(X, Y)Z &+_\star R_\star(Y, Z)X +_\star R_\star(Z, X)Y \\
&= \nabla_{\star X}\nabla_{\star Y}Z -_\star \nabla_{\star Y}\nabla_{\star X}Z -_\star \nabla_{\star[X,Y]_\star}Z \\
&\quad +_\star \nabla_{\star Y}\nabla_{\star Z}X -_\star \nabla_{\star Z}\nabla_{\star Y}X -_\star \nabla_{\star[Y,Z]_\star}X \\
&\quad +_\star \nabla_{\star Z}\nabla_{\star X}Y -_\star \nabla_{\star X}\nabla_{\star Z}Y -_\star \nabla_{\star[Z,X]_\star}Y \\
&= \nabla_{\star X}\nabla_{\star Y}Z -_\star \nabla_{\star Y}\nabla_{\star X}Z -_\star \nabla_{\star[X,Y]_\star}Z \\
&\quad +_\star \nabla_{\star Y}(\nabla_{\star X}Z +_\star [Z, X]_\star) -_\star \nabla_{\star Z}\nabla_{\star Y}X -_\star \nabla_{\star[Y,Z]_\star}X \\
&\quad +_\star \nabla_{\star Z}(\nabla_{\star Y}X +_\star [X, Y]_\star) -_\star \nabla_{\star X}(\nabla_{\star Y}Z +_\star [Z, Y]_\star) \\
&\quad -_\star \nabla_{\star[Z,X]_\star}Y \\
&= \nabla_{\star Y}[Z, X]_\star +_\star \nabla_{\star Z}[X, Y]_\star -_\star \nabla_{\star X}[Z, Y]_\star \\
&\quad -_\star \nabla_{\star[X,Y]_\star}Z -_\star \nabla_{\star[Y,Z]_\star}X -_\star \nabla_{\star[Z,X]_\star}Y \\
&= \nabla_{\star Y}[Z, X]_\star +_\star \nabla_{\star Z}[X, Y]_\star +_\star \nabla_{\star X}[Y, Z]_\star \\
&\quad -_\star \nabla_{\star[X,Y]_\star}Z -_\star \nabla_{\star[Y,Z]_\star}X -_\star \nabla_{\star[Z,X]_\star}Y \\
&= [Y, [Z, X]_\star]_\star +_\star [Z, [X, Y]_\star]_\star +_\star [X, [Y, Z]_\star]_\star \\
&= 0_\star.
\end{aligned}
$$

This completes the proof.

3. $(\nabla_{\star X}R_\star)(Y, Z)V +_\star (\nabla_{\star Y}R_\star)(Z, X)V +_\star (\nabla_{\star Z}R_\star)(X, Y)(V) = 0_\star$.

*Proof.* We have

$$
\begin{aligned}
(\nabla_{\star X}R_\star)(Y, Z)V &= \nabla_{\star X}(R_\star(Y, Z)V) -_\star R_\star(\nabla_{\star X}Y, Z)V \\
&\quad -_\star R_\star(Y, \nabla_{\star X}Z)V -_\star R_\star(Y, Z)\nabla_{\star X}V \\
&= \nabla_{\star X}\big(\nabla_{\star Y}\nabla_{\star Z}V -_\star \nabla_{\star Z}\nabla_{\star Y}V -_\star \nabla_{\star[Y,Z]_\star}V\big) \\
&\quad -_\star R_\star(\nabla_{\star X}Y, Z)V -_\star R_\star(Y, \nabla_{\star X}Z)V \\
&\quad -_\star \big(\nabla_{\star Y}\nabla_{\star Z}\nabla_{\star X}V -_\star \nabla_{\star Z}\nabla_{\star Y}\nabla_{\star X}V -_\star \nabla_{\star[Y,Z]_\star}\nabla_{\star X}V\big) \\
&= \nabla_{\star X}(\nabla_{\star Y}\nabla_{\star Z}V -_\star \nabla_{\star Z}\nabla_{\star Y}V) \\
&\quad -_\star R_\star(\nabla_{\star X}Y, Z)V -_\star R_\star(Y, \nabla_{\star X}Z)V \\
&\quad -_\star \nabla_{\star Y}\nabla_{\star Z}\nabla_{\star X}V +_\star \nabla_{\star Z}\nabla_{\star Y}\nabla_{\star X}V \\
&\quad +_\star \nabla_{\star[Y,Z]_\star}\nabla_{\star X}V -_\star \nabla_{\star X}\nabla_{\star[Y,Z]_\star}V.
\end{aligned}
$$

As in above, we get

$$(\nabla_{\star Y} R)(Z, X)V = \nabla_{\star Y}(\nabla_{\star Z}\nabla_{\star X}V -_\star \nabla_{\star X}\nabla_{\star Z}V)$$
$$-_\star R_\star(\nabla_{\star Y} Z, X)V -_\star R_\star(Z, \nabla_{\star Y} X)V$$
$$-_\star \nabla_{\star Z}\nabla_{\star X}\nabla_{\star Y} V +_\star \nabla_{\star X}\nabla_{\star Z}\nabla_{\star Y} V$$
$$+_\star \nabla_{\star [Z,X]_\star}\nabla_{\star Y} V -_\star \nabla_{\star Y}\nabla_{\star [Z,X]_\star} V$$

and

$$(\nabla_{\star Z} R)(X, Y)V = \nabla_{\star Z}(\nabla_{\star X}\nabla_{\star Y}V -_\star \nabla_{\star Y}\nabla_{\star X}V)$$
$$-_\star R_\star(\nabla_{\star Z} X, Y)V -_\star R_\star(X, \nabla_{\star Z} Y)V$$
$$-_\star \nabla_{\star X}\nabla_{\star Y}\nabla_{\star Z} V +_\star \nabla_{\star Y}\nabla_{\star X}\nabla_{\star Z} V$$
$$+_\star \nabla_{\star [X,Y]_\star}\nabla_{\star Z} V -_\star \nabla_{\star Z}\nabla_{\star [X,Y]_\star} V.$$

Consequently

$$(\nabla_{\star X} R)(Y, Z)V +_\star (\nabla_{\star Y} R)(Z, X)V +_\star (\nabla_{\star Z} R)(X, Y)(V)$$
$$= \nabla_{\star X}(\nabla_{\star Y}\nabla_{\star Z}V -_\star \nabla_{\star Z}\nabla_{\star Y}V)$$
$$-_\star R_\star(\nabla_{\star X} Y, Z)V -_\star R_\star(Y, \nabla_{\star X} Z)V$$
$$-_\star \nabla_{\star Y}\nabla_{\star Z}\nabla_{\star X} V +_\star \nabla_{\star Z}\nabla_{\star Y}\nabla_{\star X} V$$
$$+_\star \nabla_{\star [Y,Z]_\star}\nabla_{\star X} V -_\star \nabla_{\star X}\nabla_{\star [Y,Z]_\star} V$$
$$+_\star \nabla_{\star Y}(\nabla_{\star Z}\nabla_{\star X}V -_\star \nabla_{\star X}\nabla_{\star Z}V)$$
$$-_\star R_\star(\nabla_{\star Y} Z, X)V -_\star R_\star(Z, \nabla_{\star Y} X)V$$
$$-_\star \nabla_{\star Z}\nabla_{\star X}\nabla_{\star Y} V +_\star \nabla_{\star X}\nabla_{\star Z}\nabla_{\star Y} V$$
$$+_\star \nabla_{\star [Z,X]_\star}\nabla_{\star Y} V -_\star \nabla_{\star Y}\nabla_{\star [Z,X]_\star} V$$
$$+_\star \nabla_{\star Z}(\nabla_{\star X}\nabla_{\star Y}V -_\star \nabla_{\star Y}\nabla_{\star X}V)$$
$$-_\star R_\star(\nabla_{\star Z} X, Y)V -_\star R_\star(X, \nabla_{\star Z} Y)V$$
$$-_\star \nabla_{\star X}\nabla_{\star Y}\nabla_{\star Z} V +_\star \nabla_{\star Y}\nabla_{\star X}\nabla_{\star Z} V$$
$$+_\star \nabla_{\star [X,Y]_\star}\nabla_{\star Z} V -_\star \nabla_{\star Z}\nabla_{\star [X,Y]_\star} V$$
$$= -_\star R_\star(\nabla_{\star X} Y, Z)V -_\star R_\star(Y, \nabla_{\star X} Z)V$$
$$-_\star R_\star(\nabla_{\star Y} Z, X)V -_\star R_\star(Z, \nabla_{\star Y} X)V$$
$$-_\star R_\star(\nabla_{\star Z} X, Y)V -_\star R_\star(X, \nabla_{\star Z} Y)V$$
$$+_\star \nabla_{\star [Y,Z]_\star}\nabla_{\star X} V -_\star \nabla_{\star X}\nabla_{\star [Y,Z]_\star} V$$
$$+_\star \nabla_{\star [Z,X]_\star}\nabla_{\star Y} V -_\star \nabla_{\star Y}\nabla_{\star [Z,X]_\star} V$$
$$+_\star \nabla_{\star [X,Y]_\star}\nabla_{\star Z} V -_\star \nabla_{\star Z}\nabla_{\star [X,Y]_\star} V.$$

Note that

$$-_\star R_\star(Z, \nabla_{\star Y} X)V = -_\star R_\star(Z, \nabla_{\star X} Y +_\star [Y, X]_\star)V$$
$$= R_\star(\nabla_{\star X} Y, Z)V +_\star R_\star([Y, X]_\star, Z)V$$
$$= R_\star(\nabla_{\star X} Y, Z)V +_\star \nabla_{\star [Y,X]_\star}\nabla_{\star Z} V$$
$$-_\star \nabla_{\star Z}\nabla_{\star [Y,X]_\star} V -_\star \nabla_{\star [[Y,X]_\star,Z]_\star} V$$
$$= R_\star(\nabla_{\star X} Y, Z)V -_\star \nabla_{\star [X,Y]_\star}\nabla_{\star Z} V$$
$$+_\star \nabla_{\star Z}\nabla_{\star [X,Y]_\star} V -_\star \nabla_{\star [[Y,X]_\star,Z]_\star} V.$$

Then

$$-_\star R_\star(\nabla_{\star X} Y, Z)V -_\star R_\star(Z, \nabla_{\star Y} X)V$$
$$+_\star \nabla_{\star [X,Y]_\star}\nabla_{\star Z} V -_\star \nabla_{\star Z}\nabla_{\star [X,Y]_\star} V$$
$$= -_\star R_\star(\nabla_{\star X} Y, Z)V +_\star R_\star(\nabla_{\star X} Y, Z)V -_\star \nabla_{\star [X,Y]_\star}\nabla_{\star Z} V$$
$$+_\star \nabla_{\star Z}\nabla_{\star [X,Y]_\star} V -_\star \nabla_{\star [[Y,X]_\star,Z]_\star}$$
$$+_\star \nabla_{\star [X,Y]_\star}\nabla_{\star Z} V -_\star \nabla_{\star Z}\nabla_{\star [X,Z]_\star} V$$
$$= -_\star \nabla_{\star [[Y,X]_\star,Z]_\star} V$$
$$= \nabla_{\star [[X,Y]_\star,Z]_\star} V.$$

As above,

$$-_\star R_\star(Y, \nabla_{\star X} Z)V -_\star R_\star(\nabla_{\star Z} X, Y)V$$
$$+_\star \nabla_{\star [Z,X]_\star}\nabla_{\star Y} V -_\star \nabla_{\star Y}\nabla_{\star [Z,X]_\star} V$$

$$= \nabla_{\star [[Z,X]_\star,Y]_\star} V$$

and

$$-_\star R_\star(\nabla_{\star Y} Z, X)V -_\star R_\star(X, \nabla_{\star Z} Y)V$$
$$+_\star \nabla_{\star [Y,Z]_\star}\nabla_{\star X} V -_\star \nabla_{\star X}\nabla_{\star [Y,Z]_\star} V$$
$$= \nabla_{\star [[Y,Z]_\star X]_\star} V.$$

Therefore

$$(\nabla_{\star X} R)(Y, Z)V +_\star (\nabla_{\star Y} R)(Z, X)V +_\star (\nabla_{\star Z} R)(X, Y)(V)$$
$$= \nabla_{\star [[X,Y]_\star,Z]_\star} V +_\star \nabla_{\star [[Z,X]_\star,Y]_\star} V +_\star \nabla_{\star [[Y,Z]_\star,X]_\star} V$$
$$= \nabla_{\star [[X,Y]_\star,Z]_\star +_\star [[Z,X]_\star,Y]_\star +_\star [[Y,Z]_\star,X]_\star} V$$
$$= 0_\star.$$

This completes the proof.

4. $\langle R_\star(X, Y)Z, V \rangle_\star = \neg_\star \langle R_\star(X, Y)V, Z \rangle_\star$.

*Proof.* Firstly, we will note that

$$Y \langle Z, Z \rangle_\star = e^2 \cdot_\star \langle \nabla_{\star Y} Z, Z \rangle_\star$$

and then

$$
\begin{aligned}
X(Y \langle Z, Z \rangle_\star) &= e^2 \cdot_\star X(\langle \nabla_{\star Y} Z, Z \rangle_\star) \\
&= e^2 \cdot_\star \langle \nabla_{\star X} \nabla_{\star Y} Z, Z \rangle_\star \\
&\quad +_\star e^2 \cdot_\star \langle \nabla_{\star Y} Z, \nabla_{\star X} Z \rangle_\star.
\end{aligned}
$$

Then

$$
\begin{aligned}
e^2 \cdot_\star \langle R_\star(X, Y)Z, Z \rangle_\star &= e^2 \cdot_\star \langle \nabla_{\star X} \nabla_{\star Y} Z \neg_\star \nabla_{\star Y} \nabla_{\star X} Z \neg_\star \nabla_{\star [X,Y]_\star} Z, Z \rangle_\star \\
&= e^2 \cdot_\star \langle \nabla_{\star X} \nabla_{\star Y} Z, Z \rangle_\star \\
&\quad \neg_\star e^2 \cdot_\star \langle \nabla_{\star Y} \nabla_{\star X} Z, Z \rangle_\star \\
&\quad \neg_\star e^2 \cdot_\star \langle \nabla_{\star [X,Y]_\star} Z, Z \rangle_\star \\
&= XY \langle Z, Z \rangle_\star \neg_\star e^2 \cdot_\star \langle \nabla_{\star Y} Z, \nabla_{\star X} Z \rangle_\star \\
&\quad \neg_\star YX \langle Z, Z \rangle_\star +_\star e^2 \cdot_\star \langle \nabla_{\star Y} Z, \nabla_{\star X} Z \rangle_\star \\
&\quad \neg_\star e^2 \cdot_\star \langle \nabla_{\star [X,Y]_\star} Z, Z \rangle_\star \\
&= XY \langle Z, Z \rangle_\star \neg_\star YX \langle Z, Z \rangle_\star \\
&\quad \neg_\star e^2 \cdot_\star \langle \nabla_{\star [X,Y]_\star} Z, Z \rangle_\star \\
&= e^2 \cdot_\star \langle XY(Z), Z \rangle_\star \neg_\star e^2 \cdot_\star \langle YX(Z), Z \rangle_\star \\
&\quad \neg_\star e^2 \cdot_\star \langle \nabla_{\star [X,Y]_\star} Z, Z \rangle_\star \\
&= 0_\star,
\end{aligned}
$$

i.e.,

$$\langle R_\star(X, Y)Z, Z \rangle_\star = 0_\star.$$

Hence,

$$
\begin{aligned}
0_\star &= \langle R_\star(X, Y)(Z +_\star V), Z +_\star V \rangle_\star \\
&= \langle R_\star(X, Y)Z, Z \rangle_\star +_\star \langle R_\star(X, Y)Z, V \rangle_\star \\
&\quad +_\star \langle R_\star(X, Y)V, Z \rangle_\star +_\star \langle R_\star(X, Y)V, V \rangle_\star \\
&= \langle R_\star(X, Y)Z, V \rangle_\star +_\star \langle R_\star(X, Y)V, Z \rangle_\star.
\end{aligned}
$$

Therefore

$$\langle R_\star(X, Y)Z, V\rangle_\star = -_\star \langle R_\star(X, Y)V, Z\rangle_\star.$$

This completes the proof.

5. $\langle R_\star(X, Y)Z, V\rangle_\star = \langle R_\star(Z, V)X, Y\rangle_\star$.

*Proof.* We have

$$\begin{aligned}
\langle R_\star(X, Y)Z, V\rangle_\star &= -_\star \langle R_\star(Y, X)Z, V\rangle_\star \\
&= \langle R_\star(X, Z)Y, V\rangle_\star +_\star \langle R_\star(Z, Y)X, V\rangle_\star
\end{aligned}$$

and

$$\begin{aligned}
\langle R_\star(X, Y)Z, V\rangle_\star &= -_\star \langle R_\star(X, Y)V, Z\rangle_\star \\
&= \langle R_\star(Y, V)X, Z\rangle_\star +_\star \langle R_\star(V, X)Y, Z\rangle_\star.
\end{aligned}$$

Hence,

$$\begin{aligned}
e^2 \cdot_\star \langle R_\star(X, Y)Z, V\rangle_\star &= \langle R_\star(X, Z)Y, V\rangle_\star +_\star \langle R_\star(Z, Y)X, V\rangle_\star \\
&\quad +_\star \langle R_\star(Y, V)X, Z\rangle_\star +_\star \langle R_\star(V, X)Y, XZ\rangle_\star.
\end{aligned}$$

Now, we switch $X$ and $Z$, $Y$ and $V$, and we find

$$\begin{aligned}
e^2 \cdot_\star \langle R_\star(Z, V)X, Y\rangle_\star &= \langle R_\star(Z, X)V, Y\rangle_\star +_\star \langle R_\star(X, V)Z, Y\rangle_\star \\
&\quad +_\star \langle R_\star(V, Y)Z, X\rangle_\star +_\star \langle R_\star(Y, Z)V, X\rangle_\star \\
&= \langle R_\star(X, Z)Y, V\rangle_\star +_\star \langle R_\star(Z, Y)X, V\rangle_\star \\
&\quad +_\star \langle R_\star(Y, V)X, Z\rangle_\star +_\star \langle R_\star(V, X)Y, Z\rangle_\star.
\end{aligned}$$

Consequently

$$\langle R_\star(X, Y)Z, V\rangle_\star = \langle R_\star(Z, V)X, Y\rangle_\star.$$

This completes the proof.

## 10.4 The Multiplicative Sectional Curvature

Suppose that $U$ is a multiplicative Riemannian manifold with a multiplicative Riemannian metric $\langle \cdot, \cdot \rangle_\star$.

**Definition 10.5:** The multiplicative standard curvature $R_{\star 1}$ is defined as follows:

$$R_{\star 1}(X, Y)Z = \langle Y, Z \rangle_\star \cdot_\star X -_\star \langle X, Z \rangle_\star \cdot_\star Y.$$

We set

$$k_1(X, Y) = \langle R_{\star 1}(X, Y)Y, X \rangle_\star,$$
$$k(X, Y) = \langle R_\star(X, Y)Y, X \rangle_\star.$$

We have

$$R_{\star 1}(X, Y)Y = \langle Y, Y \rangle_\star \cdot_\star X -_\star \langle X, Y \rangle_\star \cdot_\star Y$$

and then

$$
\begin{aligned}
k_1(X, Y) &= \langle R_\star(X, Y)Y, X \rangle_\star \\
&= \langle \langle Y, Y \rangle_\star \cdot_\star X -_\star \langle X, Y \rangle_\star \cdot_\star Y, X \rangle_\star \\
&= \langle Y, Y \rangle_\star \cdot_\star \langle X, X \rangle_\star -_\star \langle \langle X, Y \rangle_\star \cdot_\star Y, X \rangle_\star \\
&= \langle Y, Y \rangle_\star \cdot_\star \langle X, X \rangle_\star -_\star \langle X, Y \rangle_\star^{2\star}.
\end{aligned}
$$

**Definition 10.6:** Let $\sigma \subset T_{\star p} U$ be a two-dimensional subspace spanned by $X$ and $Y$. The quantity

$$K_\sigma = k(X, Y) /_\star k_1(X, Y)$$

will be called multiplicative sectional curvature of the multiplicative Riemannian manifold with respect to the multiplicative plane $\sigma$.

**Theorem 10.1:** *The multiplicative curvature tensor $R_\star$ can be reconstructed by the multiplicative sectional curvature.*

*Proof.* Note that

$$R_\star(X, Y +_\star Z)(Y +_\star Z) = R_\star(X, Y)Y +_\star R_\star(X, Y)Z +_\star R_\star(X, Z)Y +_\star R_\star(X, Z)Z,$$
$$-_\star R_\star(Y, X +_\star Z)(X +_\star Z) = -_\star R_\star(Y, X)X +_\star R_\star(X, Y)Z +_\star R_\star(Z, Y)X -_\star R_\star(Y, Z)Z,$$
$$0_\star = R_\star(X, Y)Z +_\star R_\star(Y, X)Z.$$

We add these three equations and we find

$$e^3 \cdot_\star R_\star(X, Y)Z = R_\star(X, Y +_\star Z)(Y +_\star Z) -_\star R_\star(Y, X +_\star Z)(X +_\star Z)$$
$$-_\star R_\star(X, Y)Y -_\star R_\star(X, Z)Z +_\star R_\star(Y, X)X$$
$$+_\star R_\star(Y, Z)Z.$$

Next,

$$\langle R_\star(X, Y)Y, Z\rangle_\star = \langle R_\star(Y, Z)X, Y\rangle_\star$$
$$= \langle R_\star(Z, Y)Y, X\rangle_\star.$$

Hence, for any fixed $Y$, we have that

$$\langle R_\star(\cdot, Y)Y, \cdot\rangle_\star$$

is a symmetric bilinear form. Therefore

$$k(X +_\star Z, Y) -_\star k_\star(X, Y) -_\star k(Z, Y) = \langle R_\star(X +_\star Z, Y)Y, X +_\star Z\rangle_\star$$
$$-_\star \langle R_\star(X, Y)Y, X\rangle_\star$$
$$-_\star \langle R_\star(Z, Y)Y, Z\rangle_\star$$
$$= \langle R_\star(Z, Y)Y, X\rangle_\star$$
$$+_\star \langle R_\star(X, Y)Y, Z\rangle_\star$$
$$= e^2 \cdot_\star \langle R_\star(X, Y)Y, Z\rangle_\star.$$

Therefore

$$e^6 \cdot_\star \langle R_\star(X, Y)Z, V\rangle_\star = e^2 \cdot_\star e^3 \cdot_\star \langle R_\star(X, Y)Z, V\rangle_\star$$
$$= e^2 \cdot_\star (\langle R_\star(X, Y +_\star Z)(Y +_\star Z), V\rangle_\star$$
$$-_\star \langle R_\star(Y, X +_\star Z)(X +_\star Z), V\rangle_\star -_\star \langle R_\star(X, Y)Y, V\rangle_\star$$
$$-_\star \langle R_\star(X, Z)Z, V\rangle_\star +_\star \langle R_\star(Y, X)X, V\rangle_\star +_\star \langle R_\star(Y, Z)Z, V\rangle_\star)$$
$$= k(X +_\star V, Y +_\star Z) -_\star k(X, Y +_\star Z)$$
$$-_\star k(V, Y +_\star Z) -_\star k(Y +_\star V, X +_\star Z)$$
$$+_\star k(Y, X +_\star Z) +_\star k(V, X +_\star Z) -_\star k(X +_\star V, Y) +_\star k(X, Y)$$
$$+_\star k(V, Y) -_\star k(X +_\star V, Z) -_\star k(X, Z) +_\star k(V, Z) +_\star k(Y$$
$$+_\star V, Y) -_\star k(Y, X) -_\star k(V, X) +_\star k(Y +_\star V, Z) -_\star k(Y, Z)$$
$$-_\star k(V, Z).$$

The last formula gives an explicit representation of $R_\star$. This completes the proof.

**Corollary 10.1:** *Suppose that the multiplicative sectional curvature $K_\sigma$ does not depend on the multiplicative plane $\sigma$, but only on the choice of $p$, meaning that it is a multiplicative scalar function $K: U \to \mathbb{R}_\star$. Then*

$$R_\star = K \cdot_\star R_{\star 1}.$$

*Proof.* By the definition of the multiplicative sectional curvature, it follows that

$$k(X, Y) = K \cdot_\star k_1(X, Y).$$

Since the explicit formula for $R_\star$ contains only terms $k$ and

$$k_1(X, Y) = \langle R_{\star 1}(X, Y)Y, X \rangle_\star,$$

we obtain that

$$R_\star = K \cdot_\star R_{\star 1}.$$

This completes the proof.

**Definition 10.7:** If on a multiplicative Riemannian manifold $K_\sigma$ is a constant, or equivalently, $R_\star = K \cdot_\star R_{\star 1}$, $K \in \mathbb{R}_\star$, the multiplicative manifold is said to be a space of multiplicative constant curvature.

## 10.5 The Multiplicative Ricci Tensor

Let $U$ be an $n$-dimensional Riemannian manifold.

**Definition 10.8:**

1. Let $A_\star$ be a $(1, 1)$-multiplicative tensor, $A_{\star p}: T_{\star p}U \to T_{\star p}U$. We define the multiplicative contraction or multiplicative trace of $A_\star$, as follows:

$$CA_\star|_p = Tr_\star(A_{\star p})$$
$$= \sum_{\star j} \langle A_{\star p} E_j, E_j \rangle_\star,$$

where $E_1, \cdots, E_n$ is a multiplicative basis on $T_{\star p}U$. If $b_1, \cdots, b_n$ be a multiplicative basis and

$$A_\star b_j = \sum_{\star j} A^k_{\star j} b_k,$$

then

$$\mathrm{Tr}_\star A_\star = \sum_{\star k} A^k_{\star k}.$$

2. Let $A_\star$ be a $(1, s)$-multiplicative tensor. Then, for any fixed $k \in \{1, \ldots, s\}$ and any fixed multiplicative vector fields $X_j, j \ne k$,

$$A_\star(X_1, \ldots, X_{k-1}, \cdot, X_{k+1}, \ldots, X_n)$$

is a $(1, 1)$-multiplicative tensor whose multiplicative contractions are denoted by $C_k A_\star$ and they are given by

$$C_k A_\star(X_1, \ldots, X_{k-1}, X_{k+1}, \ldots, X_n)$$
$$= \sum_{\star j} \langle A_\star(X_1, \ldots, X_{k-1}, E_j, X_{k+1}, \ldots, X_n), E_j \rangle_\star.$$

Therefore $C_k A_\star$ are $(0, s-1)$-multiplicative tensor.

3. The multiplicative divergence of a multiplicative vector field $Y$ is defined to be the multiplicative trace $\nabla_\star Y$, i.e.,

$$\mathrm{div}_\star Y = C \nabla_\star Y$$
$$= \sum_{\star k} \left\langle \nabla_{\star E_k} Y, E_k \right\rangle_\star.$$

4. The multiplicative divergence of a symmetric $(0, 2)$-multiplicative tensor $A_\star$ is defined by

$$(\mathrm{div}_\star A_\star)(X) = \sum_{\star k} \left( \nabla_{\star E_k} A_\star \right)(X, E_k).$$

**Definition 10.9:** The first multiplicative curvature of the multiplicative curvature tensor $R_\star(X, Y)$ is given by

$$(C_1 R_\star)(Y, Z) = \sum_{\star j} \langle R_\star (E_j, Y)Z, E_j \rangle_\star$$

and it is called the multiplicative Ricci tensor. It is denoted by $\mathrm{Ric}_\star(Y, Z)$

The multiplicative trace of the multiplicative Ricci tensor is called by the multiplicative scalar curvature $S$. We have

$$S = \sum_{\star i,j} \langle R_\star (E_j, E_i)E_i, E_j \rangle_\star.$$

**Lemma 10.1:** *For any* $(1, 1)$-*multiplicative tensor* $A_\star$, *one has*

$$C(\nabla_{\star X} A_\star) = \nabla_{\star X}(CA_\star).$$

*Proof.* We have

$$CA_\star = \sum_{\star j} \langle A_\star E_j, E_j \rangle_\star.$$

Hence,

$$\nabla_{\star X}(CA_\star) = \sum_{\star j} \langle \nabla_{\star X}(A_\star E_j), E_j \rangle_\star +_\star \sum_{\star j} \langle A_\star E_j, \nabla_{\star X} E_j \rangle_\star$$

and

$$C\nabla_{\star X} A_\star = \sum_{\star j} \langle (\nabla_{\star X} A_\star)E_j, E_j \rangle_\star$$
$$= \sum_{\star j} (\langle \nabla_{\star X}(A_\star E_j), E_j \rangle_\star -_\star \langle A_\star(\nabla_{\star X} E_j), E_j \rangle_\star).$$

Since

$$\nabla_{\star X} E_j = \sum_{\star k} w_{jk}(X) \cdot_\star E_k$$

and

$$w_{jk} +_\star w_{kj} = 0_\star,$$

we have

$$\sum_{\star j} \left( \langle A_\star E_j, \nabla_{\star X} E_j \rangle_\star +_\star \langle A_\star (\nabla_{\star X} E_j), E_j \rangle_\star \right)$$

$$= \sum_{\star j} \left( \left\langle A_\star E_j, \sum_{\star k} w_{jk}(X) \cdot_\star E_k \right\rangle_\star +_\star \left\langle A_\star \left[ \sum_{\star k} w_{jk}(X) \cdot_\star E_k \right], E_j \right\rangle_\star \right)$$

$$= \sum_{\star j,k} \left( w_{jk}(X) \cdot_\star \langle A_\star E_j, E_k \rangle_\star +_\star w_{kj}(X) \cdot_\star \langle A_\star E_j, E_k \rangle_\star \right)$$

$$= 0_\star.$$

Consequently

$$\nabla_{\star X}(CA_\star) = C\nabla_{\star X} A_\star.$$

This completes the proof.

**Remark 10.1:** As above, one has

$$C_k(\nabla_{\star X} A_\star) = \nabla_{\star X}(C_k A_\star)$$

for any $(1, s)$-multiplicative tensor $A_\star$.

---

## 10.6 The Multiplicative Einstein Tensor

**Definition 10.10:** A multiplicative Riemannian manifold $(U, g)$ is said to be a multiplicative Einstein space if

$$\text{Ric}_\star(X, Y) = \lambda \cdot_\star g(X, Y)$$

for any $X, Y$, with $\lambda : U \to \mathbb{R}_\star$.
  If $(U, g)$ is a multiplicative Einstein space, then

$$S = e^{\lambda n}.$$

**Definition 10.11:** The expression

$$\text{ric}_\star(X) = \text{Ric}_\star(X, X) /_\star g(X, X)$$

is also called the multiplicative Ricci curvature in the direction $X$.
  Before to deduct some of the properties of the multiplicative Einstein spaces, we will prove the following important relations.

**Lemma 10.2:** *We have*

$$(\nabla_{\star X} R_\star)(Y, Z)V = -_\star (\nabla_{\star X} R_\star)(Z, Y)V,$$

$$\langle (\nabla_{\star X} R_\star)(Y, Z)V, U \rangle_\star = -_\star \langle (\nabla_{\star X} R_\star)(Y, Z)U, V \rangle_\star,$$

$$\langle (\nabla_{\star X} R_\star)(Y, Z)V, U \rangle_\star = \langle (\nabla_\star R_\star)(V, U)Y, Z \rangle_\star$$

*and*

$$\mathrm{Tr}_\star(\nabla_{\star X} Ric_\star) = e^2 \cdot_\star div_\star(Ric_\star X).$$

*Proof.* We have

$$
\begin{aligned}
(\nabla_{\star X} R_\star)(Y, Z)V &= \nabla_{\star X}(R_\star(Y, Z)V) -_\star R_\star(\nabla_{\star X} Y, Z)V \\
&\quad -_\star R_\star(Y, \nabla_{\star X} Z)V -_\star R_\star(Y, Z)\nabla_{\star X} V \\
&= -_\star \nabla_{\star X}(R_\star(Z, Y)V) +_\star R_\star(V, \nabla_{\star X} Y)V \\
&\quad +_\star R_\star(\nabla_{\star X} Z, Y)V +_\star R_\star(Z, Y)\nabla_{\star X} V \\
&= -_\star (\nabla_{\star X} R_\star)(Z, Y)V
\end{aligned}
$$

and

$$
\begin{aligned}
\langle (\nabla_{\star X} R_\star)(Y, Z)V, U \rangle_\star &= \langle \nabla_{\star X}(R_\star(Y, Z)V) -_\star R_\star(\nabla_{\star X} Y, Z)V \\
&\quad -_\star R_\star(Y, \nabla_{\star X} Z)V -_\star R_\star(Y, Z)\nabla_{\star X} V, U \rangle_\star \\
&= \langle \nabla_{\star X}(R_\star(Y, Z)V), U \rangle_\star -_\star \langle R_\star(\nabla_{\star X} Y, Z)V, U \rangle_\star \\
&\quad -_\star \langle R_\star(Y, \nabla_{\star X} Z)V, U \rangle_\star -_\star \langle R_\star(Y, Z)\nabla_{\star X} V, U \rangle_\star \\
&= \langle \nabla_{\star X}(R_\star(Y, Z)V), U \rangle_\star +_\star \langle R_\star(\nabla_{\star X} Y, Z)U, V \rangle_\star \\
&\quad +_\star \langle R_\star(Y, \nabla_{\star X} Z)U, V \rangle_\star +_\star \langle R_\star(Y, Z)U, \nabla_{\star X} V \rangle_\star \\
&= \langle \nabla_{\star X}(R_\star(Y, Z))V, U \rangle_\star +_\star \langle R_\star(Y, Z)\nabla_{\star X} V, U \rangle_\star \\
&\quad +_\star \langle R_\star(\nabla_{\star X} Y, Z)U, V \rangle_\star +_\star \langle R_\star(Y, \nabla_{\star X} Z)U, V \rangle_\star \\
&\quad -_\star \langle R_\star(Y, Z)\nabla_{\star X} V, U \rangle_\star \\
&= -_\star \langle \nabla_{\star X}(R_\star(Y, Z))U, V \rangle_\star +_\star \langle R_\star(\nabla_{\star X} Y, Z)U, V \rangle_\star \\
&\quad +_\star \langle R_\star(Y, \nabla_{\star X} Z)U, V \rangle_\star \\
&= -_\star \langle \nabla_{\star X}(R_\star(Y, Z)U), V \rangle_\star +_\star \langle R_\star(Y, Z)\nabla_{\star X} U, V \rangle_\star \\
&\quad +_\star \langle R_\star(\nabla_{\star X} Y, Z)U, V \rangle_\star +_\star \langle R_\star(Y, \nabla_{\star X} Z)U, V \rangle_\star \\
&= -_\star \langle (\nabla_{\star X} R_\star)(Y, Z)U, V \rangle_\star,
\end{aligned}
$$

and

$$\langle (\nabla_{\star X} R_\star)(Y, Z)V, U\rangle_\star = \langle \nabla_{\star X}(R_\star(Y, Z)V) -_\star R_\star(\nabla_{\star X}Y, Z)V$$
$$-_\star R_\star(Y, \nabla_{\star X}Z)V -_\star R_\star(Y, Z)\nabla_{\star X}V, U\rangle_\star$$
$$= \langle \nabla_{\star X}(R_\star(Y, Z)V), U\rangle_\star -_\star \langle R_\star(\nabla_{\star X}Y, Z)V, U\rangle_\star$$
$$-_\star \langle R_\star(Y, \nabla_{\star X}Z)V, U\rangle_\star -_\star \langle R_\star(Y, Z)\nabla_{\star X}V, U\rangle_\star$$
$$= \langle \nabla_{\star X}(R_\star(Y, Z)V), U\rangle_\star +_\star \langle R_\star(\nabla_{\star X}Y, Z)U, V\rangle_\star$$
$$+_\star \langle R_\star(Y, \nabla_{\star X}Z)U, V\rangle_\star +_\star \langle R_\star(Y, Z)U, \nabla_{\star X}V\rangle_\star$$
$$= \nabla_\star(\langle R_\star(Y, Z)V, U\rangle_\star) -_\star \langle R_\star(Y, Z)V, \nabla_{\star X}U\rangle_\star$$
$$-_\star \nabla_\star(\langle R_\star(V, U)Y, Z\rangle_\star) +_\star \langle \nabla_{\star X}(R_\star(V, U)Y), Z\rangle_\star$$
$$-_\star \langle R_\star(\nabla_{\star X}V, U)Y, Z\rangle_\star -_\star \langle R_\star(V, U)\nabla_{\star X}Y, Z\rangle_\star$$
$$= \langle \nabla_{\star X}(R_\star(V, U)Y), Z\rangle_\star -_\star \langle R_\star(V, \nabla_{\star X}U)Y, Z\rangle_\star$$
$$-_\star \langle R_\star(\nabla_{\star X}V, U)Y, Z\rangle_\star -_\star \langle R_\star(V, U)\nabla_\star Y, Z\rangle_\star$$
$$= \langle (\nabla_\star R_\star)(V, U)Y, Z\rangle_\star.$$

Hence, using that

$$C_1(\nabla_{\star X} R_\star) = \nabla_{\star X}(C_1 R_\star),$$

we get

$$\mathrm{Tr}_\star(\nabla_{\star X}\mathrm{Ric}_\star) = \mathrm{Tr}_\star(C_1(\nabla_{\star X}R_\star))$$
$$= \sum_{\star i,j} \langle (\nabla_{\star X}R_\star)(E_i, E_j)E_j, E_i\rangle_\star$$
$$= -_\star \sum_{\star i,j} (\langle (\nabla_{\star E_i}R_\star)(E_j, X)E_j, E_i\rangle_\star$$
$$+_\star \langle (\nabla_{\star E_j}R_\star)(X, E_i)E_j, E_i\rangle_\star)$$
$$= \sum_{\star i,j} (\langle (\nabla_{\star E_i}R_\star)(E_j, X)E_i, E_j\rangle_\star$$
$$+_\star \langle (\nabla_{\star E_j}R_\star)(E_i, X)E_j, E_i\rangle_\star)$$
$$= e^2\cdot_\star \sum_{\star i,j} \langle (\nabla_{\star E_i}R_\star)(E_j, X)E_i, E_j\rangle_\star$$
$$= e^2\cdot_\star \sum_{\star i} C_1(\nabla_{\star E_i}R_\star)(X, E_i)$$
$$= e^2\cdot_\star \sum_{\star i} \nabla_{\star E_i}(C_1 R_\star)(X, E_i)$$
$$= e^2\cdot_\star \sum_{\star i} \nabla_{\star E_i}\mathrm{Ric}_\star(X, E_i)$$
$$= e^2\cdot_\star\mathrm{div}_\star(\mathrm{Ric}_\star)(X).$$

This completes the proof.

**Definition 10.12:** The multiplicative Einstein tensor $G_\star$ is defined as follows:

$$G_\star = \text{Ric}_\star -_\star (S/_\star e^2) \cdot_\star g.$$

**Theorem 10.2:** *On an arbitrary multiplicative Riemannian manifold, we have*

$$div_\star (G_\star) = 0_\star$$

*and*

$$div_\star (\text{Ric}_\star) = div_\star ((S/_\star e^2) \cdot_\star g).$$

*Proof.* We have

$$
\begin{aligned}
e^2 \cdot_\star div_\star (\text{Ric}_\star X) &= \text{Tr}_\star (\nabla_{\star X} \text{Ric}_\star X) \\
&= \nabla_{\star X} S \\
&= \sum_{\star j} \left( \nabla_{\star E_j} S \right) g(X, E_j) \\
&= \sum_{\star j} \left( \nabla_{\star E_j} (Sg)(X, E_j) -_\star S \cdot_\star \nabla_{\star E_j} (g)(X, E_j) \right) \\
&= \sum_{\star j} \nabla_{\star E_j} (Sg)(X, E_j) \\
&= div_\star (Sg).
\end{aligned}
$$

This completes the proof.

# Appendix A

## The Multiplicative Lipschitz Condition

Let $U \subseteq \mathbb{R}_\star$.

**Definition A.1:** We say that a function $f \colon U \to \mathbb{R}_\star$ satisfies the multiplicative Lipschitz condition on $U$ if there exists a constant $L > 0$ so that

$$\left| f(x) \rightarrow_\star f(y) \right|_\star \leq L_\star \left| x \rightarrow_\star y \right|_\star \tag{A.1}$$

for $x, y \in U$. The constant $L$ will be called the multiplicative Lipschitz constant.

In fact, condition (A.1) can be rewritten in the form

$$\left| \frac{f(x)}{f(y)} \right|_\star \leq L_\star \left| \frac{x}{y} \right|_\star, \quad x, y \in U,$$

or

$$\left| \frac{f(x)}{f(y)} \right|_\star \leq e^{\log L \log \left| \frac{x}{y} \right|_\star}, \quad x, y \in U. \tag{A.2}$$

We have the following cases. Let $x, y \in U$.

1. If $f(x) \geq f(y)$ and $x \geq y$, then condition (A.2) is equivalent to the condition

$$\frac{f(x)}{f(y)} \leq e^{\log L \log \frac{x}{y}},$$

or

$$f(x) \leq f(y) e^{\log L \log \frac{x}{y}}.$$

2. If $f(x) \geq f(y)$ and $x \leq y$, then condition (A.2) is equivalent to the condition

$$f(y) \leq f()x)e^{\log L \log \frac{y}{x}}.$$

3. If $f(x) \leq f(y)$ and $x \geq y$, then condition (A.2) is equivalent to the condition

$$f(x) \leq f(y)e^{\log L \log \frac{x}{y}}.$$

4. If $f(x) \leq f(y)$ and $x \leq y$, then condition (A.2) is equivalent to the condition

$$f(y) \leq f(x)e^{\log L \log \frac{y}{x}}.$$

**Example A.1:** Let

$$f(x) = e^{-x}, \quad x \in \mathbb{R}_\star.$$

We will prove that the function $f$ satisfies the multiplicative Lipschitz condition on $\mathbb{R}_\star$ with a constant $e$. Let $x, y \in \mathbb{R}_\star$ and $x \geq y$. Then condition (A.1) is equivalent to the condition

$$\left| e^{-x} \rightarrow_\star e^{-y} \right|_\star \leq e \cdot_\star \left| x \rightarrow_\star y \right|_\star,$$

or

$$\left| e^{-x+y} \right|_\star \leq e \cdot_\star \left( \frac{x}{y} \right),$$

or

$$e^{y-x} \leq e^{\log \frac{x}{y}},$$

or

$$e^{y-x} \leq e^{\log x - \log y},$$

or

$$y - x \le \log x - \log y,$$

or

$$y + \log y \le x + \log x.$$

Let

$$g(t) = t + \log t, \quad t \in \mathbb{R}_\star.$$

Then

$$g'(t) = 1 + \frac{1}{t}$$
$$> 0, \quad t \in \mathbb{R}_\star.$$

Thus, $g$ is monotone increasing on $\mathbb{R}_\star$ and then

$$g(x) \ge g(y).$$

Therefore the function $f$ satisfies the multiplicative Lipschitz condition with a multiplicative Lipschiotz constant $e$.

**Example A.2:** Let

$$f(x) = \frac{1}{1+x}, \quad x \in \mathbb{R}_\star.$$

We will show that the function $f$ satisfies the multiplicative Lipschitz condition with a multiplicative Lipschitz constant $e$. Let $x, y \in \mathbb{R}_\star$ and $x \ge y$. Condition (A.1) is equivalent to the condition

$$\left| \frac{1}{1+x} -_\star \frac{1}{1+y} \right|_\star \le e \cdot_\star |x -_\star y|_\star$$

or

$$\left| \frac{1+y}{1+x} \right|_\star \le e \cdot_\star \left| \frac{x}{y} \right|_\star,$$

or

$$\frac{1+x}{1+y} \leq e^{\log x - \log y},$$

or

$$\frac{1+x}{1+y} \leq \frac{x}{y},$$

or

$$\frac{1+x}{x} \leq \frac{1+y}{y}.$$

Let

$$g(t) = \frac{1+t}{t}, \quad t \in \mathbb{R}_\star.$$

Then

$$g'(t) = \frac{t - (1+t)}{t^2}$$

$$= -\frac{1}{t^2}$$

$$< 0, \quad t \in \mathbb{R}_\star.$$

Thus, $g$ is monotone decreasing on $\mathbb{R}_\star$ and then

$$\frac{1+x}{x} \leq \frac{1+y}{y}.$$

Therefore $f$ satisfies the multiplicative Lipschitz condition with a multiplicative Lipschitz constant $e$.

**Theorem A.1:** *Let* $f \in \mathscr{C}^1_\star(U)$. *If*

$$|f^\star(x)|_\star \leq L, \quad x \in U,$$

*then $f$ satisfies the multiplicative Lipschitz condition with a multiplicative Lipschitz constant $L$.*

*Proof.* Take $x, y \in \mathbb{R}_\star$ arbitrarily. We apply the multiplicative mean value theorem and we get that there is a $c$ between $x$ and $y$ so that

$$|f(x) -_\star f(y)|_\star \leq |f^\star(c)|_\star \cdot_\star |x -_\star y|_\star$$
$$\leq L \cdot_\star |x -_\star y|_\star.$$

This completes the proof.

# Appendix B

## *The Multiplicative Implicit Function Theorem*

Let $a, b \in \mathbb{R}_\star$, $a < b$, and $(x_0, y_0) \in [a, b] \times \mathbb{R}_\star$. Take $c \in \mathbb{R}_\star$, $c > 0_\star$, and

$$D = \{(x, y) \in \mathbb{R}_\star \times \mathbb{R}_\star : a \leq x \leq b, \quad y_0 \rightarrow_\star c \leq y \leq y_0 +_\star c\}.$$

We have that $(x_0, y_0) \in D$.

**Theorem B.1:** *Let* $f: D \to \mathbb{R}_\star$ *be a continuous function that satisfies the multiplicative Lipschitzcondition with a constant* $L \in (0_\star, 1_\star)$ *with respect to its second argument. Let also*

$$f(x_0, y_0) = 0_\star. \tag{B.1}$$

*Then there exists a* $\delta > 0_\star$ *for which there exists a unique function* $\phi \in \mathscr{C}([x_0 \rightarrow_\star \delta, x_0 +_\star \delta] \cap [a, b])$ *such that*

$$\phi(x) = y_0 +_\star f(x, \phi(x)), \quad x \in [x_0 \rightarrow_\star \delta, x_0 +_\star \delta] \cap [a, b], \quad \text{and} \quad y_0 = \phi(x_0).$$

*Proof.* Let $q > 0_\star$ be chosen so that

$$q < (1_\star \rightarrow_\star L) \cdot_\star c.$$

Since $f(\cdot, y_0)$ is a continuous function, by (B.1), it follows that there is a $\delta > 0_\star$ so that

$$|f(x, y_0) \rightarrow_\star f(x_0, y_0)|_\star < q, \quad x \in [x_0 \rightarrow_\star \delta, x_0 +_\star \delta] \cap [a, b].$$

Define

$$X = \{g \in \mathscr{C}([x_0 \rightarrow_\star \delta, x_0 +_\star \delta] \cap [a, b]): |g(x) \rightarrow_\star y_0|_\star \leq c,$$
$$x \in [x_0 \rightarrow_\star \delta, x_0 +_\star \delta] \cap [a, b]\}$$

and in $X$ define the metric

$$d_\star(y, z) = \max_{x \in [x_0 -_\star \delta, x_0 +_\star \delta] \cap [a,b]} |y(x) -_\star z(x)|, \quad y, z \in X.$$

Note that $X$ is a complete metric space with respect to the metric $d$. For $y \in X$, define the operator

$$Fy(x) = y_0 +_\star f(x, y(x)), \quad x \in [x_0 -_\star \delta, x_0 +_\star \delta] \cap [a, b]. \quad (B.2)$$

We have that $Fy \in \mathscr{C}([x_0 -_\star \delta, x_0 +_\star \delta] \cap [a, b])$ for any $y \in X$ and

$$
\begin{aligned}
|Fy(x) -_\star y_0|_\star &= |f(x, y(x))|_\star \\
&\le |f(x, y(x)) -_\star f(x, y_0)|_\star +_\star |f(x, y_0)|_\star \\
&\le L \cdot_\star |y(x) -_\star y_0|_\star +_\star (1_\star -_\star L) \cdot_\star c \\
&\le L \cdot_\star c +_\star (1_\star -_\star L) \cdot_\star c \\
&= c, \quad x \in [x_0 -_\star \delta, x_0 +_\star \delta] \cap [a, b],
\end{aligned}
$$

for any $y \in X$. Therefore $F: X \to X$. Next, take $y_1, y_2 \in X$. Then

$$
\begin{aligned}
Fy_1(x) &= y_0 +_\star f(x, y_1(x)), \\
Fy_2(x) &= y_0 +_\star f(x, y_2(x)), \quad x \in [x_0 -_\star \delta, x_0 +_\star \delta] \cap [a, b],
\end{aligned}
$$

and

$$
\begin{aligned}
|Fy_1(x) -_\star Fy_2(x)|_\star &= |y_0 +_\star f(x, y_1(x)) -_\star y_0 -_\star f(x, y_2(x))|_\star \\
&= |f(x, y_1(x)) -_\star f(x, y_2(x))|_\star \\
&\le L \cdot_\star |y_1(x) -_\star y_2(x)|_\star \\
&\le L \cdot_\star d_\star(y_1, y_2), \quad x \in [x_0 -_\star \delta, x_0 +_\star \delta] \cap [a, b].
\end{aligned}
$$

Therefore

$$d_\star(Fy_1, Fy_2) \le L \cdot_\star d_\star(y_1, y_2)$$

and $F: X \to X$ is a multiplicative contraction. Therefore there exists a unique $\phi \in X$ so that

$$\phi(x) = y_0 +_\star f(x, \phi(x)), \quad x \in [x_0 -_\star \delta, x_0 +_\star \delta] \cap [a, b].$$

Now, construct the sequence

$$y_0(x) = y_0,$$
$$y_n(x) = y_0 +_\star f(x, y_{n-1}(x)), \quad x \in [x_0 -_\star \delta, x_0 +_\star \delta] \cap [a, b], \quad n \in \mathbb{N}.$$

We have

$$
\begin{aligned}
|y_1(x) -_\star y_0|_\star &= |f(x, y_0)|_\star \\
&\leq |f(x, y_0) -_\star f(x_0, y_0)|_\star +_\star |f(x_0, y_0)|_\star \\
&\leq (1_\star -_\star L) \cdot_\star c \\
&< c, \\
|y_2(x) -_\star y_1(x)|_\star &= |f(x, y_1(x)) -_\star f(x, y_0)|_\star \\
&\leq L \cdot_\star |y_1(x) -_\star y_0|_\star \\
&\leq L \cdot_\star c, \\
|y_3(x) -_\star y_2(x)|_\star &= |f(x, y_2(x)) -_\star f(x, y_1(x))|_\star \\
&\leq L \cdot_\star |y_2(x) -_\star y_1(x)|_\star \\
&\leq L^2 \cdot_\star c, \quad x \in [x_0 -_\star \delta, x_0 +_\star \delta] \cap [a, b].
\end{aligned}
$$

Assume that

$$|y_n(x) -_\star y_{n-1}(x)|_\star \leq L^{n-1} \cdot_\star c, \quad x \in [x_0 -_\star \delta, x_0 +_\star \delta] \cap [a, b],$$

for some $n \in \mathbb{N}$. Then

$$
\begin{aligned}
|y_{n+1}(x) -_\star y_n(x)|_\star &= |f(x, y_n(x)) -_\star f(x, y_{n-1}(x))|_\star \\
&\leq L \cdot_\star |y_n(x) -_\star y_{n-1}(x)|_\star \\
&\leq L^n \cdot_\star c, \quad x \in [x_0 -_\star \delta, x_0 +_\star \delta] \cap [a, b].
\end{aligned}
$$

Therefore

$$\sum_{\star n=0}^{\infty} |y_{n+1}(x) -_\star y_n(x)|_\star \leq c \cdot_\star \sum_{\star n=0}^{\infty} L^n_\star$$

$$= c /_\star (1_\star -_\star L), \quad x \in [x_0 -_\star \delta, x_0 +_\star \delta] \cap [a, b].$$

Consequently

$$y_n \to \phi, \quad \text{as} \quad n \to \infty,$$

multiplicative uniformly on $[x_0 -_\star \delta, x_0 +_\star \delta] \cap [a, b]$. Note that

$$y_1(x_0) = y_0 +_\star f(x_0, y_0)$$
$$= y_0,$$
$$y_2(x_0) = y_0 +_\star f(x_0, y_1(x_0))$$
$$= y_0 +_\star f(x_0, y_0)$$
$$= y_0.$$

Assume that

$$y_n(x_0) = y_0$$

for some $n \in \mathbb{N}$. Then

$$y_{n+1}(x_0) = y_0 +_\star f(x_0, y_n(x_0))$$
$$= y_0 +_\star f(x_0, y_0)$$
$$= y_0.$$

Therefore

$$y_0 = \lim_{n \to \infty} y_n(x_0)$$
$$= \phi(x_0).$$

This completes the proof.

**Theorem B.2:** *Let* $f: D \to \mathbb{R}_\star$ *be a continuous function and the classical derivative* $f'_y$ *exists and it is continuous on D. Let also,*

$$f(x_0, y_0) = 0_\star,$$
$$f'_y(x_0, y_0) \ne 0.$$

*Then there exists a* $\delta > 0_\star$ *for which there exists a unique function* $\phi \in \mathscr{C}([x_0 -_\star \delta, x_0 +_\star \delta] \cap [a, b])$ *such that*

$$f(x, \phi(x)) = 0_\star, \quad x \in [x_0 -_\star \delta, x_0 +_\star \delta] \cap [a, b], \quad and \quad y_0 = \phi(x_0).$$

*Proof.* Let

$$\lambda = -_\star\left(1/_\star f_y^\star\,(x_0,\,y_0)\right)$$

and

$$g\,(x,\,y) = y -_\star\ y_0 +_\star\ \lambda\cdot_\star f\,(x,\,y), \quad (x,\,y) \in D.$$

Then

$$g_y^\star\,(x,\,y) = 1_\star +_\star\ \lambda\cdot_\star f_y^\star\,(x,\,y), \quad (x,\,y) \in D,$$

and

$$
\begin{aligned}
g_y^\star\,(x_0,\,y_0) &= 1_\star +_\star\ \lambda\cdot_\star f_y^\star\,(x_0,\,y_0)\\
&= 1_\star -_\star\ \left(f_y^\star\,(x_0,\,y_0)/_\star f_y^\star\,(x_0,\,y_0)\right)\\
&= 0_\star.
\end{aligned}
$$

Take $k \in (0_\star,\,1_\star)$. Then, using that $f_y^\star$ is continuous on $D$, there exists a $\delta_1 > 0_\star$ so that

$$
\begin{aligned}
D_1 = \{(x,\,y) &\in [a,\,b] \times \mathbb{R}_\star\colon x_0 -_\star\ \delta_1 \le x \le x_0 +_\star\ \delta_1, \quad y_0 -_\star\ \delta_1 \le y\\
&\le y_0 +_\star\ \delta_1\} \subset D
\end{aligned}
$$

and

$$|g_y^\star\,(x,\,y)\,|_\star\ \le k, \quad (x,\,y) \in D_1.$$

Let

$$
\begin{aligned}
D_2 = \{(x,\,y) &\in D_1\colon x_0 -_\star\ (\delta_1/_\star e^2) \le x \le x_0 +_\star\ (\delta_1/_\star e^2), \quad y_0 -_\star\ (\delta_1/_\star e^2)\\
&\le y \le y_0 +_\star\ (\delta_1/_\star e^2)\}.
\end{aligned}
$$

Applying the multiplicative mean value theorem, we get

$$
\begin{aligned}
|g\,(x,\,y_1) -_\star\ g\,(x,\,y_2)\,|_\star\ &= |g_y^\star\,(x,\,\xi)\,|_\star \cdot_\star |y_1 -_\star\ y_2|_\star\\
&\le k\cdot_\star |y_1 -_\star\ y_2|_\star, \quad (x,\,y_1),\,(x,\,y_2) \in D_2,
\end{aligned}
$$

where $\xi$ is between $y_1$ and $y_2$. Note that

$$g(x_0, y_0) = y_0 \rightarrow_\star y_0 +_\star \lambda \cdot_\star f(x_0, y_0)$$
$$= 0_\star.$$

Therefore $g$ satisfies all conditions of Theorem B.1. Then there exists a $\delta > 0_\star$ for which there exists a unique continuous function $\phi \in \mathscr{C}$ $([x_0 \rightarrow_\star \delta, x_0 +_\star \delta] \cap [a, b])$ and

$$\phi(x) = y_0 +_\star g(x, \phi(x)), \quad x \in [x_0 \rightarrow_\star \delta, x_0 +_\star \delta] \cap [a, b], \qquad \text{(B.3)}$$

and

$$y_0 = \phi(x_0).$$

By (B.3), we get

$$\phi(x) = y_0 +_\star \phi(x) \rightarrow_\star y_0 +_\star \lambda \cdot_\star f(x, \phi(x))$$
$$= \phi(x) +_\star \lambda \cdot_\star f(x, \phi(x)), \quad x \in [x_0 \rightarrow_\star \delta, x_0 +_\star \delta] \cap [a, b],$$

whereupon

$$f(x, \phi(x)) = 0_\star, \quad x \in [x_0 \rightarrow_\star \delta, x_0 +_\star \delta] \cap [a, b].$$

This completes the proof.

# Bibliography

D. Aniszewska, Multiplicative Runge-Kutta Method, *Nonlinear Dynamics* 50 (1–2) (2007) 265–272.

A. Bashirov, E. Kurpinar, A. Özyapici, Multiplicative Calculus and its Applications, *Journal of Mathematical Analysis and its Applications* 337 (1) (2008) 36–48.

F. Córdova-Lepe, The Multiplicative Derivative as a Measure of Elasticity in Economics, *TEMAT-Theaeteto Atheniensi Mathematica* 2(3) (2006), online.

S. Georgiev. *Focus on Calculus*, Nova Science Publisher, 2020.

B. Gompertz. On the Nature of the Function Expressive of the Law of Human Mortality, and on a New Mode of Determining the Value of Life Contingencies, *Philosophical Transactions of the Royal Society of London* 115 (1825) 513–585.

M. Grossman, R. Katz, *Non-Newtonian Calculus*, Pigeon Cove, Lee Press, Massachusetts, 1972.

M. Grossman, *Bigeometric Calculus: A System with a Scale-Free Derivative*, Archimedes Foundation, Rockport, Massachusetts, 1983.

W. Kasprzak, B. Lysik, M. Rybaczuk, *Dimensions, Invariants Models and Fractals*, Ukrainian Society on Fracture Mechanics, SPOLOM, Wroclaw-Lviv, 2004.

R.R. Meginniss, Non-Newtonian Calculus Applied to Probability, Utility, and Bayesian Analysis, Manuscript of the report for delivery at the 20th KDKR-KSF Seminar on Bayesian Inference in Econometrics, Purdue University, West Lafayette, Indiana, 23 May 1980.

M. Riza, A. Özyapici, E. Misirli, Multiplicative Finite Difference Methods, *Quarterly of Applied Mathematics* 67 (4) 745–754, 2009.

M. Rybaczuk, A. Kedzia, W. Zielinski, The Concepts of Physical and Fractional Dimensions II. The Differential Calculus in Dimensional Spaces, *Chaos Solutions Fractals* 12 (2001), 2537–2552.

D. Stanley, A Multiplicative Calculus, *Primus IX* (4) (1999) 310–326.

# Index

Printed in the United States
by Baker & Taylor Publisher Services